TECHNOLOGY OF CARBON AND GRAPHITE FIBER COMPOSITES

TECHNOLOGY OF CARBON AND GRAPHITE FIBER COMPOSITES

JOHN DELMONTE

VNR VAN NOSTRAND REINHOLD COMPANY
NEW YORK CINCINNATI ATLANTA DALLAS SAN FRANCISCO
LONDON TORONTO MELBOURNE

Van Nostrand Reinhold Company Regional Offices:
New York Cincinnati Atlanta Dallas San Francisco

Van Nostrand Reinhold Company International Offices:
London Toronto Melbourne

Library of Congress Catalog Card Number: 80–17980
ISBN: 0-442-22072-3

Manufactured in the United States of America

Published by Van Nostrand Reinhold Company
135 West 50th Street, New York, N.Y. 10020

Published simultaneously in Canada by Van Nostrand Reinhold Ltd.

15 14 13 12 11 10 9 8 7 6 5 4 3 2 1

Library of Congress Cataloging in Publication Data

Delmonte, John.
Technology of carbon and graphite fiber composites.

Includes bibliographical references and index.
1. Fibrous composites. 2. Carbon fibers.
3. Graphite fibers. I. Title.
TA418.9.C6D43 620.1'97 80-17980
ISBN 0-442-22072-3

To
My Darling Jill

Preface

Carbon and graphite fibers for reinforcing plastics and other materials have been available since the early 1960s. These reinforcing fibers and their composites have been the subject of numerous studies and reports, funded principally by U.S. government agencies that have recognized that the intrinsic qualities of the materials were valuable to aerospace applications. Carbon and graphite fibers soon joined glass and boron fibers, and in turn were joined by synthetic aramid fibers to form the basis of advanced composites, which excelled in structural applications. The availability and relative low cost of glass fibers have propelled these materials to the forefront of industrial, recreational, and military applications. The compelling reasons for continued interest in carbon and graphite fibers lay in their stiffness qualities (high elastic moduli), which surpassed the best performers of other fibers. In spite of their relative high cost, carbon and graphite fibers are moving to the forefront of advanced structural materials.

Advanced composite materials basically consist of two fundamental components: the reinforcing fiber and the matrix resin binder, which bonds them into structural members. In the following chapters, the more successful resin binders, as well as the reinforcing fibers are examined in detail. Attention is also given to non-resin binders for carbon and graphite fibers, such as metals and glasses. Likewise, considerable attention is directed to hybrid structures involving multiple fibers and resins, looked upon by some as the direction for future growth in the field of advanced composites. As the reader will observe from the chapter headings, developments of advanced composites are categorized by their temperature performance, and we have, for example, chapters covering resin matrices for service to 200°C, service to 300°C, service above 300°C, and thermoplastic binders, the last reference concerned primarily with molding materials.

Heterogeneous products such as advanced composites necessitate complex analyses, and a number of excellent publications are referenced in the bibliographies at the end of each chapter. In fact, the plethora of fine papers and reports have made it difficult to choose the more relevant papers to cross-reference. Because the technology of carbon and graphite fibers has improved materially within the last ten years, some of the observations prior to the 1970s have been considered obsolete and hence more current material is introduced. Furthermore, published material has not always adequately categorized the resin binders or the candidate fibers. For these reasons, I have attempted to present a balanced viewpoint in examining the technology of carbon and graphite fibers and have drawn extensively upon those elements of published reports which will update this growing subject. At the Delsen Testing Laboratory where I am associated, we have the opportunity to evaluate and to test many types of advanced composites, and hence have a "feel" for the important qualities of these products.

Acknowledgments are many, and I applaud the fine papers published by those capable individuals mentioned in the references of the bibliographies. Collectively their efforts have provided important stepping stones in the growing technology of carbon and graphite fibers. Particular attention is directed to technology of carbon and graphite fibers. Attention is directed to professional societies such as SAMPE (Society for Advancement of Materials and Process Engineering); American Chemical Society (ACS): Society of Plastics Industry (SPI); Society of Plastic Engineers (SPE); American Society of Testing Materials (ASTM); Society of Automotive Engineers (SAE); and other groups such as National Aeronautics and Space Administration; Air Force Material Laboratory; Technology Conferences; and American Carbon Committee, all of whom have extended approval to me for use of certain illustrative material appearing in this book.

In recognizing my own limitations in covering such a diverse group of disciplines required for the technology of carbon and graphite fibers, I solicited the following individuals for reviewing specific chapters which I had written: Dr. H. Volk, Union Carbide Corp. (Chapter 2 on C-fiber), Dr. F. Hirosawa, Furane Plastics Division of M & T Chemicals (Chapter 4 on resin matrices), Mr. L. McCrory, Delsen Testing Lab (Chapter 7 on mechanical properties), Mr. J. Hertz, General Dynamics Corp. (Chapter 9 on environmental influences), Mr. L. McCrory, Delsen Testing Laboratory and Dr. Paul McMahon, Celanese Corp. (Chapter 10 on test methods), Mr. E. Crossland, Hercules Corp. (Chapter 11 on aerospace applications), Mr. D. Bauer, Super-Temp Division of B.F. Goodrich, (Chapter 13 on CVD carbon-composites), and Dr. J. Leslie, consultant, (Chapter 14 on manufacturing methods). To these individuals I express my gratitude for their advice and many helpful suggestions.

JOHN DELMONTE

Contents

1 Origins of Carbon and Graphite Fibers

The origins of high strength carbon and graphite fiber composites may be found during the 1960's, when it became apparent that carbon and graphite fibers which contributed significantly to the strength and stiffness of structural products could be developed.

Carbon fiber was first employed by Edison as a filamentary material in early development of the electric light. However, it gave way to more efficient and durable materials within a few years. The role of fibrous products became identified with structural developments in the aerospace and automotive fields, as glass fibers reached a pre-eminent role in many thousands of successful applications in transportation; building construction; marine, electrical, and thermal insulation; and many industrial products. The urgency of structural requirements for the aircraft and aerospace industry led to improved high strength and high modulus fibers and their composites. Developments in boron fibers and carbon fibers, in combination with new, improved binders, formed the basis of advanced composites. These materials, though in limited supply as fibers, met some of the early physical objectives. The continuing interest, principally of the U.S. Air Force, the Navy, and space agencies, have encouraged and financed further efforts to look for other materials, and the appearance of carbon/graphite fibers with enhanced physical properties has been gratifying. While the available glass fibers have possessed high tensile strength qualities, there has not been success in developing high modulus qualities which would achieve stiffness superior to the structural metals. This goal has been attained by advanced composites of boron and of graphite composites. The superior performance of composites is summarized in Fig. 1-1, which cites compelling reasons for active consideration of composites in structure—even to replacement of metal.

The role of glass fibers in composites technology has been adequately documented in the technical press. This book will emphasize the growing position of carbon and graphite fibers, with a few comparative references to the other com-

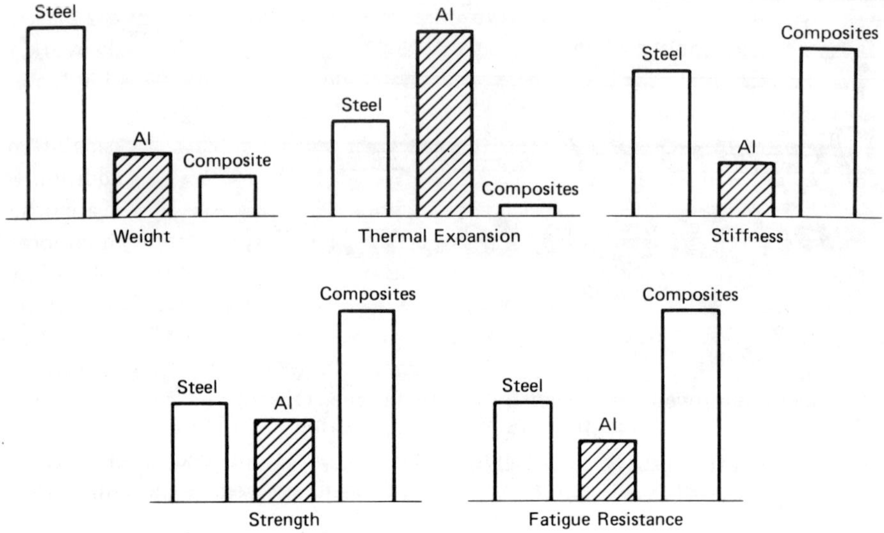

Fig. 1-1. Comparison of composites with metals (Deutsch, G.S.-(NASA)-Sampe 23d Symposium May 1978).

petitive fibers. This is important because of the growing position of hybrid composites. Though, at the time of this publication, the price of carbon fibers is high—due to the complex technology required for developing superior properties—the fact remains that carbon is one of our more abundant elements, with the likelihood of substantial reductions in price in the not-too-distant future. Even from the early 1970's to the late 1970's, we have seen the price of the fibers fall from \$100/lb to under \$20/lb.

There are many industrial examples of early laminated plastics with fiber reinforcement in the aircraft and automotive field, dating back some 40 or 50 years, though their role was generally identified as a secondary structural product, as shields or covers, and seldom as a major unit for carrying stress. One of the first documents to focus on structural plastics for aircraft was by Johnson of the Army Air Corps.[1] At that time (1941) the desired minimum values for mechanical properties at +160°, 75°, and -65°F were stated as follows:

Specific tensile strength—22,000 psi
Specific flexural strength—13,000 psi
Specific modulus of elasticity—3.6×10^6 psi

Considering the time at which these targets were set, this was a realistic goal.

With the advent of the war years, and then the tremendous growth of the plastics industry, new roles and new objectives have been evolved. The early work is cited because it identifies the early role that armed forces have played in fostering advanced composites.

Several semi-technical documents have been instrumental in focusing attention on the emerging role of composites. We are reminded that, in addition to the obvious strength-weight advantages of improved composites, there are other considerations, particularly thermal properties and performance at high temperatures. For example, there is a limit to the operating temperatures of turbine engines because of the oxidation of some metallic components. Some of the non-oxidizable metals are too weak at the higher temperatures. Reinforced with strong, stiff carbon/graphite fibers, these chemically resistant metals have shown significant improvements in stiffness and fatigue. One of the chapters of this book deals with the subject of metal matrices for graphite fibers. The growing requirements for space vehicles, capable of re-entry, point to the need for structural materials that possess high modulus, high strength, high temperature resistance, and low density. With advanced composites employing carbon/graphite fiber reinforcements and high temperature resistant matrices, the materials appear at hand.

A composite is formed when two or more materials are judiciously combined, usually with the intent of achieving better results than can be obtained by using the individual materials by themselves. This simple concept has wide implications, as it introduces new ways of planning the development and application of materials. Design problems are now complicated by the fact that we are no longer dealing with a uniform and homogeneous substance, but rather with a mixture of products which may differ in physical, thermal, and electrical properties. Carbon/graphite fiber composites have an organic resin binder, or matrix, to weld them into a solid body. The carbonaceous fiber portion contributes stiffness and strength, a large reduction in thermal expansion coefficient, a large improvement in fatigue strength, and decreased electrical resistivity. The technology of carbon/graphite composites entails examination of how to combine such diverse materials into useful engineering products, using different resin materials, and hybridization with other fibers, notably glass and aramids.

A designer recognizes the potentialities of materials engineering when he begins to think, not necessarily in terms of monolithic materials, but in terms of optimizing performance. A case in point would be application to a helicopter rotor or an aircraft wing spar. Greatest stresses arise at the positions of attachment to the main body or to the hub of the rotor. From a weight saving point of view, there is nothing to be gained by extending the larger structural support at the point of attachment to the opposite end or tip—where stresses are minimal. Cross-sections may be reduced, and, as possible in composite technology, fiber reinforcement may be planned to position the stress absorbing fibers to those regions where they will do the most good. Several materials may be used in order to distribute effectively the required materials of construction. For example, the stiffer graphite fibers may find that impact resistance may be enhanced or micro-crack formation of composites may be alleviated by the inclusion of more yieldable fibers.

With carbon/graphite fiber reinforcements, there are numerous classes of composites, such as:

1. Randomly distributed chopped carbon and graphite fibers in a resin matrix used as molding compounds.
2. Continous collimated filaments in a resin matrix, as an unidirectional tape, which may be positioned in accord with engineering design.
3. Woven carbon/graphite fabrics of a multitude of structures, which are laminated together with resin bonding agents. Included in this group are three dimensional fabrics for special purposes, felts, and impregnated rovings for packings, etc.
4. Honeycomb constructions, important to lightweight aerospace structures, which are also manufactured with carbon/graphite fiber face plies.
5. Continuous long pultrusions of carbon/graphite fiber and resin and/or metal.
6. Carbon/carbon composites, wherein carbon and graphite fibers are used to reinforce carbonized resin matrices for operation at very high temperatures.

In the chapters on applications, there will be numerous unique examples of composites that will be referenced. The utilization of carbon/graphite fibers may be motivated by many reasons, such as superior stiffness, better fatigue strength, low shrinkage and low expansion coefficients, high heat resistance, and good chemical resistance. In combination with resin matrices, special processing methods must be followed to achieve high performance composites.

From a practical viewpoint based upon many years of experience with such materials and their applications, emphasis will be placed upon the more promising materials and directions, to the exclusion of certain products. Among the thermosetting resins, for example, interest is currently high on the epoxy and the polyimide resin systems. In this text, the relative advantages of other candidate thermosets and thermoplastics will be reviewed.

As a means of directing attention to early review articles on composites, including carbon/graphite, some publications which have helped to set the stage for the newer materials of construction should be cited.

A study by Hoffman,[2] of the Rand Corp., entailed a review of high strength fibers in tension structural elements. He concluded that there were no major obstacles to utilization of fibers to improve present-day structural materials.

A survey of ceramic and graphite fibers was published in 1965.[3] The report attempts to compile a unified source of authoritative scientific and technological information on whiskers and fibers for use above 1200°F as fillers, flexible structures, and insulations, and for reinforcements in composite materials. Hundreds of references and patents, selected as being representative of new fiber developments, were reviewed. The authors recognized the impossibility at that time of keeping abreast of the many developments in this field. However, they commented that among those materials capable of meeting aerospace requirements were new high strength glasses; carbon filaments; the polycrystalline filaments of metallic oxides, carbides, and boron; and the single crystal whiskers. Approximately eight patents on carbon/graphite were reviewed in the survey. This has been expanded and updated at the end of this chapter.

In a 1963 article on composites, it was stated that ". . . today's composites will be inefficient by tomorrow's standards. Whether theoretical understanding, the development of valid design techniques and the generation of useful data will keep pace with material developments and application requirements remains to be seen. It is the very number and complexity of composites . . . [that holds] their greatest promise."[4]

As Alexander[5] commented in headlining his article in *Fortune* in 1966: "Airplanes could be lighter, turbines, more efficient—if the amazingly strong materials promised by the new fiber technology work out. It's a big 'if,' but getting smaller." At that time, boron fiber was the material being concentrated upon. As Peterson, Head of Advanced Filaments and Composites Division of the Wright-Patterson Air Force Base, said: the Air Force is not committed to boron, but is looking for any high strength, high modulus fiber that will work.

In a *Modern Plastics* article, Grimes[6] compares fibrous and wire reinforcements with metal and organic matrices. At the time of this report (1966), graphite fibers were rated at 160×10^3 psi tensile strength and modulus of 20×10^6 psi. In the ratio of modulus to density, they were behind boron and beryllium fibers. More recent technology led to large increases in the graphite strength and modulus, and the highest specific rating to graphite. With a specific gravity of 1.86, beryllium must always be considered, though the unavailability of adequate sources of raw materials and the toxicity of its salts have discouraged major efforts in this direction.

Improvements in the physical properties of carbon and graphite were soon forthcoming (1968). Simon and Prosen[7] reported tensile strength of fibers at 400,000 psi and moduli of elasticity of 50×10^6 psi. They investigated 17 types of carbonaceous fibers at this time, bonding with epoxy resins into unidirectional composites and testing physical properties. Composite moduli and tensile strengths showed effective translation of fiber properties to composite properties—with the exception of low shear strength. Improvements were made with shear strength by growing silicon carbide whiskers on the surface of the carbon fibers. This research emphasized the necessity for further effort to enhance the adhesion of matrix resins to the fiber surface, through suitable surface treatment.

As noted in a 1970 report, Dwyer[8] comments that "it has only been in the last few years that the state of the art of composites has reached the point where these new metals and plastics could be used effectively—mainly in airframes and aircraft engine components—and present a serious challenge to the dominance of steel, aluminum, and titanium in this field." By this time, production graphite fibers were reported at 40 $\times$ 10^6 psi elastic modulus—appearing in the forefront as the best candidate for high specific modulus (E/density). Illustrated in this paper is an F-5 wing tip, by Hercules, Inc., reported to be the first graphite/epoxy part flown (weight saving over aluminum counterpart, 25%). Kuebler and Jordan showed at the 1970 meeting of the Society of Manufacturing Engineers the dominating position of strong graphite and boron fibers. It is of current interest to note that research is in progress on depositing boron on graphite fibers in place of a tungsten core. Comparable strengths and low densities are obtained.[9]

The following year (1971) heightened the competitive interest between boron and graphite fibers. While the modulus of graphite fibers had continued to improve, this was offset by a fall-off in tensile and shear properties. Boron had better compressive properties, owing to the large filaments diameter—about 100 microns (0.004 inches), compared to 6.6 microns for graphite. On the other hand, the smaller diameter graphite fiber reinforcement makes it possible to utilize the more common glass fiber technology. A comparison of the composite properties of the two filaments in composites was made by Powers,[10] and is reproduced below. These were based upon unidirectionally-oriented fiber reinforced resin laminates bonded with epoxy resin, containing 50% fiber by volume.

BORON-EPOXY COMPARISON (1971)

PROPERTY	BORON FIBER COMPOSITE	GRAPHITE FIBER COMPOSITE
Density (g/ml)	2.08	1.61
Tensile strength (psi)	225,000	190,000
Modulus of elasticity (psi)	32 $\times$ 10^6	27 $\times$ 10^6
Beam interlaminar shear (psi)	17,000	3000–10,000*

*Higher values for special surface treatment.

Preferences were still toward the graphite fibers because of their lower density, higher specific strength to weight ratios, and easier processability. Major carbon and graphite fiber producers who listed[11] anticipated production for 1971 were:

Hercules, Inc.	40,000 lb
Celanese Corp.	20,000 lb
Union Carbide Corp.	10,000 lb
Great Lakes Carbon Corp.	7000 lb
Whittaker-Morgan, Inc.	4600 lb
Hitco	3600 lb

Courtaulds of England and Kureha, Nippon Chemical, and Toray of Japan are offering carbon and graphite fibers at this time. At the end of Chapter 2, there is a listing of current activity in carbon and graphite fibers as of early 1980. Current 1980 production is approaching 3,000,000 lb.

An effective review of advanced composites was made by Lovelace[12] in 1974, who commented that the modern composites as they are known today owe their genesis to glass fiber-polyester composites developed over the 1940's. The intensive research and development programs in the U.S., together with important contributions by the United Kingdom and Japan, have enabled the composites to fulfill their initial forecast of structural performance. While boron fiber was the first to achieve prominence as a high modulus fiber for advanced composites, others, such as graphite (as we have seen) and beryllium oxide and boron carbide fibers, were superior to the best that could be produced from glass (S-glass exceeded the modulus of standard E-glass by approximately 20%). Lovelace highlights the problems attendant upon translating the properties of experimental whiskers into high strength composites. Fabrication by batch processes, whisker orientation, non-uniform aspect ratio, and quality control yielded products which did not live up to expectation—leaving the activity to continuous fibrous filaments in laminated structures. Developments in filament winding equipment and in pultrusion equipment made it possible to realize the full potential in these fibers, and in particular high modulus graphite fibers.

The properties of the lamina depend upon the fiber orientation. In terms of uniaxial properties, the 0° orientation has the highest stiffness and strength. These properties decrease as the angle of orientation moves away from 0°. In fact, physical property determination at 90° to the fiber orientation is a fair compromise of the properties of the resin matrix. Among unusual features of advanced composite laminates is the high elastic modulus of graphite fiber reinforced composites, considerably superior to glass fiber reinforced composites. In addition, the low, controllable thermal expansion coefficient has made possible stable space antennas and optical mounting hardware. These special situations and the contributions of the carbon/graphite fiber composites will be discussed in separate chapters on applications.

Fig. 1–2. Replica of Edison's first electric lamp with a carbon fiber filament. *(Courtesy: Pacific Gas & Electric Co.)*

To demonstrate flight worthiness, one of the first applications of advanced composites was an F–5 wing tip manufactured by Hercules, Inc., showing a 25% weight reduction over its aluminum counterpart. Successful performance led to other adaptations, and in a chapter on aerospace applications, there are cited military and commercial aircraft which have logged many satisfactory flight hours. Weight savings were substantial in most cases. Major air carriers have been involved in flying carbon/graphite fiber composites substituted for metal components, as part of a program for promoting confidence levels, and examining materials for any untoward environmental effects.

With this overview of the trends in the field of composites, we can commence a more detailed study of the evolution of carbon and graphite fibers, and the technical problems which had to be solved in achieving commercial success. A convenient way for accomplishing this, and as a means of ascertaining their strengths and weaknesses, is to review salient U.S. patents bearing upon this field. From an informative and educational standpoint, such a review is very helpful. As a reference and as a guidepoint for future research, the data summarized on the following pages should prove valuable. Should the reader desire more information, the original patent specifications should be procured from the U.S. Patent Office.

Thomas A. Edison was the first to successfully employ carbonized filaments of cellulosic origin for incandescent lighting (see Fig. 1-2). His preferred filaments for use in electric light bulbs were prepared by dissolving clean cotton in a zinc chloride solution, and squirting the mass through a die into a fluid hardener. After drying the filament, it was carbonized out of contact with air. As the art progressed, it was the practice to immerse the filament into a hydrocarbon, such as benzene, and pass an electric current through to heat the filament, decompose the hydrocarbon, and leave the resulting carbon on the filament. The filament base was a slightly impure, dense, hard carbon base surrounded by a shell or coating of carbon.

U.S. Patent 223,398—"Electric Lamp"
Edison, T.A.
Filed November 1, 1879 Granted January 27, 1880

The invention consists of a light giving body of coiled carbon wire or sheets, which offer resistance to the passage of electric current. This body is sealed in a glass container in a nearly perfect vacuum to prevent oxidation and injury to the conductor by the atmosphere. The current is conducted into the vacuum bulb through platinum wires sealed into the glass. Cotton threads and vegetable fibers are carbonized and coiled and clamped to the metal lead-in wires.

U.S. Patent 230,309—"Process of Manufacturing Carbon Conductors"
Maxim, H.S.
Filed March 22, 1880 Granted July 20, 1880

An improved process of making carbon filaments for electric lamps by exposing carbonaceous substances to a hydrocarbon gas (coal gas, for example) and heating in a furnace is described. The substances, such as cardboard, paper, silk, or cotton, are cut or bent to desired shape, and then charred in a retort, while hydrocarbon vapor is flowing through. The carbon conductors were described as more dense and tougher than those made by other processes.

U.S. Patent 2,582,764—"Manufacture of Carbon Electrodes"
Bailey, B., Great Lakes Carbon Corp.
Filed May 28, 1948 Granted January 15, 1952

Methods of preparing carbon for shaping into electrodes are covered in the patent claims. These are measures which would also influence subsequent extrusion processes. A heavy liquefiable hydrocarbon from petroleum oil is coked to a volatile content exclusive of water of about 8 to 20%. This raw uncalcined coke is mixed with about 10 to 20% plasticizing agent (coal tar derivative, coal tar pitch, anthracene oil), and the plasticizing agent is largely converted to carbon under baking conditions so that it is not decomposable below 500°F. The coke particles are partially alloyed with plasticizing agent and molded at 100° to 400°F before becoming non-flowable. The shaped bodies are baked with a gradual upheat to 1000° to 2000°F.

U.S. Patent 2,631,982—"Process for Treating Tar"
Donegan, J.W., to Allied Chemical Corp.
Filed March 9, 1951 Granted March 17, 1953

The process describes the technique of removing finely divided particles from tars and pitches used for industrial processes. By mixing 0.2 to 0.5 volume of a predominantly aromatic solvent (xylene or toluene) into the pitch and by heating at 170° to 450°F, the fine suspended particles are flocculated into coarse aggregates, which may be removed by filtration.

U.S. Patent 2,635,994—"Production of Carbon Resistors"
Tierman, M., to Sprague Electric Co.
Filed April 27, 1950 Granted April 21, 1953

A process is described for producing a stable carbon type of resistor by exposing a heat resistant base to a carbon arc in a carbon arc chamber having an inert atmosphere. A thin layer of pure carbon is deposited upon said base. The temperature of base affected the rate of deposit of the carbon layer.

U.S. Patent 2,765,354—"Polymeric Carbon and Method of Production"
Carpenter, G.B., Blanchard, E.R., and Busselli, A., to Air Reduction Co.
Filed March 23, 1953 Granted October 2, 1956

This process provides polymeric carbon formed by dehydrogenation of polyacetylene. It was discovered that when polyacetylene is heated above 320°C in a vacuum or in an inert gas, it undergoes dehydrogenation and an increase in density. The polymeric carbon may be formed and molded. Volume resistivity of the polymeric carbon is controlled from 10^{13} ohm-cm to about 10^{4} ohm-cm by degree of dehydrogenation.

U.S. Patent 2,796,331—"Process for Making Fibrous Carbon"
Kaufmann, H.F., Griffiths, D., and Mackay, J.S., to Pittsburgh Coke and Chemical Co.
Filed June 9, 1954 Granted June 18, 1957

The process of producing fibrous carbon by slow pyrolysis of a hydrocarbon vapor methane with hydrogen or coke oven gas, by passing a diluted gas through a zone maintained at 1150° and 1450°C is described. A residence time of at least 0.4 seconds deposits fibrous carbon on inert surfaces within the heating chamber. Methane gas, diluted with hydrogen, was used with a small amount of hydrogen sulfide (0.3 to 4.0%) as a catalyst. Carbon fibers in the tube were spread out in the direction of gas flow, and represented approximately 7% of the carbon fed as methane. The filament diameters averaged 4 microns.

U.S. Patent 2,822,321—"New Carbon Product and Method of Manufacture"
Pickard, J.A.
Filed January 31, 1955 Granted February 4, 1958

Carbon products with fibrous formation, suitable for filtration purposes, were produced by this invention. Peat moss free of inorganic matter was disintegrated in a hammer mill. Sufficient moisture was present to avoid brittleness of fibers. The milled material was carbonized at 450° to 950°C to produce carbon particles having an interlocked fibrous branching formation and a reticulated structure.

U.S. Patent 2,913,802—"Thermal Modifications of Acrylonitrile Yarns"
Barnett, I., to Johns-Manville Corp.
Filed July 16, 1953 Granted November 24, 1958

This patent discloses heat treatment of acrylonitrile fibers to improve their thermal stability. Heat treating from 300°F to the depolymerization temperature of the organic fibers (~650°F) in air for several hours resulted in a marked change in the appearance and the thermal characteristics of the fiber. The fabric turned black in color, was no longer thermoplastic, was flame resistant, was dimensionally stable, and achieved insolubility in a common polyacrylonitrile solvent such as dimethylformamide. The heat treated fabric retained the excellent abrasion resistance and strengths of the original filamentary material.

U.S. Patent 2,957,756—"Filamentary Graphite and Method for Producing the Same"
Bacon, R., to Union Carbide Corp.
Filed March 18, 1953 Granted October 25, 1960

The patent describes articles of manufacture consisting of high tensile strength and electrically conductive graphite filaments. Graphite filaments which were short, thick, and brittle had been produced by decomposing various hydrocarbons on heated carbon or graphite. In this invention, a graphite rod is placed over a carbon block in a closed vessel, inert gases introduced, and a direct current arc drawn so that 3500° to 4000°K is reached. Pressure is maintained between 1150 and 1400 psi. This pressure is critical and must be maintained below the triple point of graphite–1470 psi. A fast growing stalagmite-like matrix is formed, in which are embedded fine graphite filaments as the rod is moved away from the block under voltage and gas pressure. The preferred condition for graphite whisker growth using a 1/2 inch diameter graphite rod, is 1300 psi., argon pressure, with an arc operating at 80 volts and 80 amperes. Filaments of tensile strength of 3,000,000 lb/inch^2 and diameters of 0.1 to 10 microns were reported. The fibers are substantially monocrystalline, giving sharp X-ray patterns.

U.S. Patent 3,011,981—"Electrically Conductive Fibrous Carbon"
Soltes, W.T.
Filed April 21, 1958 Granted December 5, 1961

The invention is directed toward the preparation of electrically conductive carbonaceous products from high carbon pure cellulosic fibers, such as cotton, or synthetics, such as rayon. Methods described in the specification subject fiber materials in the absence of oxygen to staged heat treatments in the range of 300° to about 1500°F for a time sufficient to effect carbonization of the contained fibers and to volatilize extraneous matter from the fibers. The end products noted are electrical conductors, which are highly absorptive and very flexible. Air and liquid filters were prepared from the carbonized products.

U.S. Patent 3,027,222—"Fireproof Acrylonitrile Copolymers"
Wilkinson, W.K., To du Pont de Nemours Co.
Filed September 3, 1957 Granted March 27, 1962

Fireproofing of acrylonitrile polymers (85 to 99% polymer) by heating in air at 220° to 275°C to achieve fireproofness in the fibers is described. The fibers had been previously drawn to at least 300% of their original as-spun length. Several copolymers of acrylonitrile were also designated, including methylacrylate and a styrene sulfonate salt.

U.S. Patent 3,053,775—"Method for Carbonizing Fibers"
Abbott, W.F., to Carbon Wool Corp.
Filed November 12, 1959 Granted September 11, 1962

The objective was to provide strong carbon fibers with high fiber density from synthetic fibers, such as regenerated cellulose (viscose rayon) and rayon acetate. Natural fibers, such as cotton, were deemed not suitable. Rayon fibers were processed continuously in a tunnel with inert gas at 300° to 500°F, requiring 8 to 120 minutes to attain the desired carbonization. The fiber produced was subjected to steam at this higher temperature. Tensile strengths were in the range of 5000 to 10,000 psi. About 50% of the fiber precursor was converted to carbon fiber. Some of the hydrocarbon gases liberated during fiber pyrolysis were retained to provide a non-oxidizing atmosphere.

U.S. Patent 3,107,152—"Fibrous Graphite"
Ford, C.E. and Mitchell, C.V., to Union Carbide Corp.
Filed September 12, 1960 Granted October 5, 1963

This invention covers the process for graphitizing woven and non-woven cellulosic fibers to produce electrically conductive and chemically resistant graphite fibers of high strength and flexibility. Carefully controlled heating of viscose rayon was practiced—slowly at first, 10° to 50°C/hour from 100° to 400°C—increasing up to 100°C/hour from 400° to 900°C, and then a rapid rise to 3000°C to complete graphitization, with an inert gaseous atmosphere. When using non-woven filaments or yarns (for example, 1100/720 rayon tire cord), the graphitized fibers were 5 to 7 microns in diameter, with tensile strengths of 108 to 130 ksi, and a specific resistance of 2000 to 5400 micro-ohm-cm. Prewoven rayon products were also graphitized. X-ray diffraction patterns showed a high degree of uniformity and oriented crystallinity. Tensile strengths of several monofilaments were shown.

DIAMETER IN MICRONS	TENSILE STRENGTH (psi)
5–7	108.000–130.000
20–25	48.000– 53.000
146 (lamp filaments)	28.000– 33.000

U.S. Patent 3,107,180—"Process for Deposition of Pyrolytic Graphite"
Diefendorf, R.J., to General Electric Co.
Filed January 26, 1961 Granted October 15, 1963

In carrying out this invention, a carbonaceous material, such as methane, is fed to an enclosure at a temperature of 1000° to 2300°C. As the gas decomposes, it is mixed with 0.1 to 5.0 volume % of nitric oxide to produce a non-sooting carbon vapor. Claims also cover prior mixing of methane gas with nitric oxide. Pyrolytic graphite was deposited on articles placed in this enclosure.

U.S. Patent 3,116,975—"Artificial Graphite Process"
Cross, C., Ecker, D., and Stein, O.L., to Union Carbide Corp.
Filed February 8, 1961 Granted January 7, 1964

The patent describes problems relating to thermal conversion of flexible cellulosic material into flexible graphite. These are related to careful drying prior to carbonization and the control of gases evolved during graphitization of the material. The process developed was applicable to regenerated cellulose, and entailed continuous purging of the furnace with non-oxidizing and non-carbonaceous gases, impinging in large volume against materials being graphitized.

U.S. Patent 3,125,404—"Heat Resistant Fibrous Products and Method of Making Same"
Crawley, W.P., Carborundum Co.
Filed October 24, 1957 Granted March 17, 1964

Crawley discloses the heat treatment of polyacrylonitrile fibers in air, above 160°C, to exhibit superior resistance to degradation on short term exposure to high temperatures. Even in a flame, the non-flammability of the acrylic fiber is manifest. These heat resistant fibers are readily formed into yarns and cloth, including the blending in of aluminum silicate ceramic fibers. Insulating cloths, filter elements, gaskets, and packings of unusual heat resistance are formed. The process entails heating such fabrics in an oxidizing atmosphere at 200° to 310°C.

U.S. Patent 3,174,895—"Graphite Cloth Laminates"
Gibson, D.W., McGhee, K.B., and Stroup, R., to Union Carbide Corp.
Filed September 7, 1960 Granted May 23, 1965

The patent relates to fabrication of carbon or graphite bonded laminates of graphitized cellulosic cloth. The laminate is made up of a plurality of woven or unwoven graphitized cellulosic sheets, bound together by a carbonized binder. Carbonaceous binders mentioned in the claims include furfuraldehyde-ketone and furfuraldehyde-furfuryl alcohol condensation products. The laminated stack is cured under pressure at 165°C and carbonized in a non-oxidizing atmosphere at 900°C, and further graphitized at 2900°C in a non-oxidizing atmosphere.

U.S. Patent 3,179,605—"Manufacture of Carbon Cloth"
Ohsol, E., to Haveg Industries, Inc.
Filed October 12, 1962 Granted April 20, 1965

The objective of this patent was to achieve reproducible results in obtaining high quality, high strength carbon fibers from regenerated cellulose. Claims emphasized washing of the cellulose until the sodium content was below 25 ppm, and then carrying out a staged, long time heating in a non-oxidizing atmosphere to achieve the carbon fibers. In washing the regenerated cellulose, acid rinses (5% acetic acid) were also employed. Test data showed improved strengths in carbon cloth as the initial sodium content was reduced.

U.S. Patent 3,235,323—"Heat Resistant Black Fibers and Fabrics Derived from Rayon"
Peters, E.M., to Minnesota Mining and Manufacturing Co.
Filed January 21, 1965 Granted February 15, 1966

This patent discusses the heat treatment of rayon (regnerated cellulose) fabrics and fibers which have been impregnated with a water soluble nitrogenous salt. Ammonium sulfate, ammonium sulfamate, ammonium chloride, and ethyl amine salts of phosphoric acid were mentioned in the specification. The main objective was to obtain a black converted cloth, by heat treating above 450°F, which had the same pliancy as the original rayon cloth, with much improved chemical resistance, flame resistance, and abrasion resistance. A shorter conversion period is claimed when fibers are pretreated with the above salts. When a small amount of boric acid is included and the fabrics finally heat treated at graphitization temperature, the trace amount of boron present is claimed to yield fibers with a lower order of electrical resistivity and a positive temperature coefficient of resistivity.

U.S. Patent 3,242,000—"Impregnated Acrylic Textile Product and Method for Producing Same"
Lynch, J., to Deering Milliken Research Corp.
Filed August 30, 1963 Granted March 22, 1966

Improved heat resistant textiles are claimed in this patent, which relates to the impregnation of an acrylic fiber textile material with a solution of a refractory metal cation (exs-zirconium acetate, isobutyl titanate, molybdenum chloride, and titanium chloride). The pick-up attained was 25 to 50% by weight of metal cation based on the dry weight of the fabric. The impregnated material was carbonized at a temperature up to 3000°C, though about 550°C was preferred.

U.S. Patent 3,285,696—"Preparation of Flexible Carbon Fiber"
Tsunoda, Y. (Japan)
Filed April 5, 1963 Granted November 15, 1966

Fibrous carbon and graphite prepared by this patent are stronger than those made by direct carbonization. Yarns of polyacrylonitrile (PAN) and copolymers of PAN are pretreated by heating in an oxygen atmosphere from 180° to 550°C for a sufficient time to "blacken" the fibers (usually 5 to 60 minutes). The fibers are then carbonized in an inert atmosphere between 700° and 1200°C for not less than 1 hour. The graphitized fibers are obtained by heating the carbonized fibers at 1200° to 3000°C in an inert atmosphere.

U.S. Patent 3,301,472—"Laminate Comprising Carbon Fibers, Carbonized Resin, and Inorganic Oxide Fibers"
Noland, R.L., Travis, W.C., and Dagneau, M.B., to Haveg Industries, Inc.
Filed June 23, 1961 Granted January 31, 1967

A laminated structure for exposure to the high temperatures and high gaseous velocities of rocket motors, comprising a layer of carbon fibers bound together by a carburized resin selected from thermosetting resins, united to inorganic oxide fibers. Claims cite phenol-formaldehyde binders and silicate fibers from silica fiber cloth. The patent specification describes the mixing of a chopped 1/2 inch square of fibrous graphite in cloth form with a phenolic resin. This is then preformed. The chopped silica cloth and phenolic resin are also preformed. The two preforms are molded together. The resulting article withstood the firings of rocket nozzles for 60 seconds at 6000°F.

U.S. Patent 3,304,148—"Carbon Cloth Annealing Process"
Gallagher, J.B., to Haveg Industries (Subsidiary of Hercules Powder)
Filed June 17, 1963 Granted February 14, 1967

The patent process is directed towards improvement in the quality of carbon fibers from woven regenerated cellulose, which are heated in an inert atmosphere. During the pyrolysis of the carbon fibers, highly cross-linked brittle resinous material sometimes appears on the surface, reducing the strength of the carbon cloth. The relatively brittle resinous layer is burnt off by further heating the cloth in air, at a temperature of 1000° to 2000°F for a short time (5 seconds to 5 minutes). The burst and tear strength of the carbonized cloth are materially increased.

U.S. Patent 3,305,315—"Process for Manufacturing Flexible Carbonaceous Materials"
Bacon, R., Moyer, R.O., and Cranch, G., to Union Carbide Corp.
Filed September 20, 1962 Granted February 21, 1967

This process, developed about fibers of cellulosic origin (cotton, jute, linen, rayon, etc.), starts by immersing materials into a 7% by volume phosphoric acid denatured ethanol solution, followed by air drying to evaporate the ethanol. Heat treating in an oxidizing atmosphere at 250°C achieves rupture of the carbon-oxygen and carbon-hydrogen bonds, accompanied by evolution of water, but falling short of scission of the main cellulose molecule. This is followed by slow heating to 700°C in a nitrogen atmosphere. Graphitization takes place at 2700°C with the material surface swept by nitrogen. The process yields a product capable of forming flexible fabrics with good drape. It was felt that the use of the acid pretreatment permits the heat treating process to remove water, to take place at lower temperatures.

U.S. Patent 3,313,597—"Method for Continuous Graphitization of Carbonaceous Thread"
Cranch, G. and Shinko, J.S., to Union Carbide Corp.
Filed January 24, 1963 Granted April 11, 1967

Described is a continuous process for graphitizing a carbonaceous thread comprising at least 90% carbon, by providing a pair of spaced electrical contacts between which the thread passes. An electric current passing through the contacts and the thread establishes a graphitization temperature of 1800° to 3000°C in an atmosphere free of oxygen. Patent specifications further emphasize the necessity of a carefully controlled heating schedule for cloths, felts, braids, yarns, knits, weaves, and monofilaments—in which the first stage, essentially a batch process, heats material up to 900°C, in which the rate of heating is below 100°C/hour. In the second graphitization stage (1300° to 3000°C), the rate of heating is much more rapid. The process is directed towards rayon (cellulose) and polyacrylonitrile fibers.

U.S. Patent 3,333,926—"Carbonizing Cellulosic Textile Materials"
Moyer, R.O., Ecker, D.R., and Spry, W.J., to Union Carbide Corp.
Filed October 30, 1963 Granted August 1, 1967

This is a process for producing carbon textiles by carbonizing a cellulosic textile material in an oxygen free, non-carbonaceous atmosphere containing 5% chlorine, at temperatures of 25° to 650°C. After purging the gas in the furnace with nitrogen, continuing carbonization to 800°C produces the carbon textile. The material may be graphitized by heating to higher temperatures—preferably in 20% chlorine and 80% nitrogen. It is claimed that higher strength products are obtained.

U.S. Patent 3,367,812—"Process of Producing Carbonized Articles"
Watts, W.H., to Union Carbide Corp.
Filed November 14, 1962 Granted February 6, 1968

In this process for the manufacture of formed carbonaceous articles from the parallel fibers of yarns and tows, derived from cellulosic origin, the articles are treated with carbonizable binders from thermosetting or thermoplastic resins. A specific binder cited in one of the claims comprises 25% furfuryl alcohol, 25% furfuryl-aldehyde, and 50% phenolic formaldehyde resin. The articles are subjected to graphitizing temperatures. Fiber diameters of 5 to 100 microns are mentioned in the specification.

U.S. Patent 3,378,345—"Process for Producing Pyrolytic Graphite Whiskers"
Bourdeau, R. and Papalegis, F.E., to Union Carbide Corp.
Filed March 22, 1965 Granted April 16, 1968

This patent covers an improved process for preparing pyrolytic graphite whiskers through the decomposition of hydrocarbons in the presence of an active gas of carbon dioxide, water, or a combination of both at reduced pressure. The preferred ratios of hydrocarbon gas (methane) to the carbon dioxide or steam is 25 to 1 to about 50 to 1. The temperature in the pyrolysis zone was raised gradually from about 700°C (low at first to promote nucleation) to at least 1400°C. A high concentration of 10 to 45 micron diameter graphite whiskers, 2 to 3 inches long, was produced.

U.S. Patent 3,392,216—"Producing Carbon Structures From Molten Baked Substances"
Otani, S., to Kureha (Tokyo)
Filed October 26, 1964 Granted July 9, 1968

Organic substances from polyvinyl chloride, polyacrylonitrile, petroleum pitch, and coal pitch are melt-extruded into fibers or films after being heated in nitrogen in the range of 300° to 500°C. The materials are then contacted with an oxidizing gas at a temperature under 100°C, ozone being identified as improving the product. Final carbonization in an inert atmosphere took place at temperatures above 500°C. The ozone treatment step was claimed to enhance the strength of the resulting carbon filaments.

U.S. Patent 3,395,970—"Method of Carbonizing Polyacrylonitrile, Impregnated Cellulose, Cyanoethylated Cellulose and Acrylonitrile Graft Copolymerized Cellulose Textiles"
Machell, G., to Deering Milliken Research Corp.
Filed October 30, 1963 Granted August 6, 1968

This patent relates to heat resistant textile materials and, more specifically, to acrylonitrile treated carbonized textile materials. The claims indicate that at least 5% acrylonitrile is picked up, based on the initial weight of the dry cellulose textile material. Carbonizing takes place in an inert (non-oxidizing) atmosphere at a temperature in the range of 300° to 3000°C. When relatively small quantities of acrylonitrile are present in the cellulosic fabric, an inert nitrogen or argon atmosphere is used during carbonization.

U.S. Patent 3,399,252—"Method and Apparatus for Manufacture of High Strength and High Modulus Carbon Filaments"
Hough, R.L. and Schwartz, R.T., to U.S.A.
Filed April 15, 1966 Granted August 27, 1968

Carbonized organic polymers or carbonized cellulosic materials are brought to temperatures of 2500° to 3000°C, suitable for graphitization. While heated, the yarn received tensile stresses which effect elongations up to 100% of its initial length. In the preferred method and apparatus of this invention, the heating and stretching of the yarn are accomplished simultaneously in an electric glow discharge device. The carbon yarn functions as a positively charged electrode with instant attainment of processing temperature. An auxiliary electrode is positioned in close proximity to the carbon strands passing through the tube.

U.S. Patent 3,404,061—"Flexible Graphite Material"
Shane, J.H., Russell, R.J., and Bochman, R.A., to Union Carbide Corp.
Filed April 15, 1963 Granted October 1, 1968

The principal object of this invention is to provide a flexible sheet material consisting of essentially graphite, free of bonding material. The flexible sheet is prepared from graphite particles which have first been appreciably expanded or intumesced in the presence of oxidizing agents. Compressive pressures upon the voluminously expanded graphite particles force them into inter-engagement with each other to form a sheet with a high degree of flexibility. Tensile strengths are in the range of 500 to 1500 psi.

U.S. Patent 3,406,126—"Conductive Synthetic Resin Composition Containing Carbon Filaments"
Litant, I., to AVCO, Inc.
Filed December 7, 1966 Granted October 15, 1968

Described is an electrically conducting composite consisting of a non-conducting moldable material and 0.5 to 5.0% of unwoven carbon filaments, with a dc resistivity of 10^3 ohms-cm or less at 25°C. The carbon yarn is chopped into 1/2 inch lengths and the individual filaments must be separated from one another when they are incorporated into a plastic matrix, in order to achieve good conductivity. The carbonized fibers used were about 10 microns in diameter, and were worked into the compositions by a slow speed, low shearing blade.

U.S. Patent 3,412,062—"Production of Carbon Fibers and Compositions Containing Said Fibers"
Johnson, W., Phillips, L., and Watt, W.
Filed April 19, 1965 Granted November 19, 1968

Carbon fibers having high tensile strength and high Young's modulus are made by conversion of organic polymer fibers, such as polyacrylonitrile, by the combined effect of heating in a non-oxidizing atmosphere to a carbonizing temperature of about 1000°C, and by the application of longitudinal tension at some stage of the conversion. As a preliminary step, the polyacrylonitrile is heated in an oxidizing atmosphere from 200° to 250°C for sufficient time to attain complete permeation of oxygen. The fiber is held under longitudinal tension to limit the shrinkage of fibers during heating to not more than 12% of the length of the fiber. Further heat treatment of the fiber in a non-oxidizing atmosphere (argon) at 2500° to 3000°C attains a Young's modulus of 38×10^6 psi. Initial heat treatment gave a carbonized density of 1.5 g/cc; a carbon filament diameter of 11 to 13 microns; a tensile strength of individual fibers of 110×10^3 psi; a Young's modulus of 16.5×10^6 psi; and a failing strain of 0.7%. Further heat treatment to 2500°C gave these fibers a carbonized density of 1.90 g/cc; a tensile strength of 200×10^3 psi; and a Young's modulus of 35×10^6 psi.

U.S. Patent 3,427,120—"Producing Method of Carbon or Carbonaceous Material"
Shindo, A., Fujii, R., and Souma, I., to Agency of Industrial Science and Technology (Japan)
Filed December 17, 1963 Granted February 11, 1969

Polyvinyl alcohol fibers and fabrics are oxidized in air at 180° to 200°C, further heated at a rate of 10°C, and then more rapidly heated in nitrogen to 1000°C for carbon fibers, followed by heating to 2500°C for graphite fibers. Tensile strengths of 5300 kg/cm^2 are reported for thc fibrous graphite.

U.S. Patent 3,443,899—"Process for Making Graphite Type Fibers"
Accountius, O.E., North American Rockwell
Filed July 22, 1966 Granted May 13, 1969

This invention relates to the extrusion and subsequent pyrolysis of para-polyphenylene into carbonaceous fibers having aligned crystallinity. Para-polyphenylene powder is compacted and extruded under pressures of 10,000 to 100,000 psi at temperatures of 485° to 515°F. After extrusion, the fibers are pyrolyzed in an inert atmosphere at temperatures of 800° to 1400°C. Minimum tensile strengths of 20,000 psi and elastic moduli of 50 $\times$ 10^6 psi were reported for the fibers. Fiber diameters of 3 to 50 mils were indicated.

U.S. Patent 3,449,077—"Direct Production of Graphite Fibers"
Staetz, D.E., to Celanese Corp.
Filed February 13, 1967 Granted June 10, 1969

The continuous preparation of uniform high modulus graphite fibers from pre-oxidized polybenzimidazole fibers is described. A particular useful sub-class of polybenzimidazoles for fiber production consists of

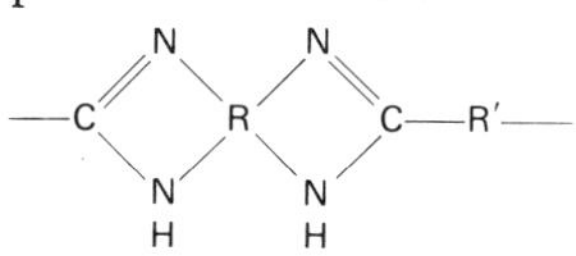

R = Aromatic nucleus (preferred—(3,3′-4, 4′-bisphenylene)
R^1 = Carbocyclic aromatic or heterocyclic ring.

The most important polybenzimidazole commercially is poly-2,2′-*m*-phenylene-5, 5′-bibenzimidazole. In the processing of fibers from this material, preoxidizing in air at 400° to 550°C for 2 to 15 minutes, with an acetylene fuel for heating, followed a resident time in the flame of 6 to 17 seconds. Tensile strengths of 150 to 200 ksi and initial moduli of 20 to 30 $\times$ 10^6 psi are reported.

U.S. Patent 3,454,362—"Process for Producing Fibrous Graphite"
Spry, W.S. to Union Carbide Corp.
Filed August 26, 1965 Granted July 8, 1969

This is an improved process for the production of filaments or yarn from carbonized polymeric fibrous materials by heating the starting material to a temperature to permit it to undergo plastic flow and longitudinal stretching of at least 1%, while heating the elongated yarn to graphitizing temperatures (2800° to 2830°C) and cooling the permanently elongated yarn. Individual monofilaments attain 100,000 psi, while the twisted yarn made from a plurality of the same filaments reach only 15,000 psi tensile strength, and whereas, by this invention, the monofilaments attain 360,000 psi and the yarn reaches 120,000 psi tensile strengths. Graphitizing was performed in an argon atmosphere. Claims cover polyacrylonitrile, polyvinyl alcohol, and regenerated cellulose.

U.S. Patent 3,461,082—"Method for Producing Carbonized Lignin Fiber"
Otani, S., Kiryo-Shi, Fuknoka, Y., Bunjiro, I., and Sasaki, K.
Filed October 4, 1965 Granted August 12, 1969

Lignin, recovered as a by-product in the paper manufacturing industry, has been used for preparing fibers which have been carbonized. These were considered as lower cost precursors than the carbonized fibers prepared from cellulose or polyacrylonitrile. Dry spinning the lignin fiber in an inert gas at 150° to 200°C and removing foreign matter is practiced; or dissolving the lignin in dimethyl sulfoxide may be practiced. The weak lignin fibers were sometimes reinforced by addition of polyvinyl alcohol or polyacrylonitrile. The lignin fiber is pretreated in an oxidizing atmosphere at 50° to 400°C, and then heated at a rate of less than 50°C/minute to a temperature between 600° and 1000°C in an inert gas. Tensile strengths of the carbonized fiber were reported as 3 to 5×10^3 kg/cm^2.

U.S. Patent 3,462,289—"Process for Producing Reinforced Carbon and Graphite Bodies"
Rohl, C.W. and Robinson, J.H., to Carborundum Co.
Filed August 5, 1965 Granted August 19, 1969

The process is described for forming and shaping carbon or graphite reinforcing fibers by:

1. Forming the fibers without any binder matrix.
2. Holding the shape under vacuum.
3. Pressure impregnating the fibers with a carbonizable resin matrix, such as phenolic resins, epoxy resins, coal tar pitch, or, preferably, furfuryl alcohol.
4. Placing the impregnated shape under heat and pressure to cure the matrix.
5. Additional impregnation and baking.
6. Carbonization by slow heating to 800°C in an inert atmosphere.
7. Final graphitization at 2500° to 2800°C.

U.S. Patent 3,462,340—"Fiber Containing Pyrolytic Composite Material"
Hough, R.L., U.S.A.F.
Filed July 28, 1965 Granted August 19, 1969

This invention describes composites of pyrolitic fibers, such as graphite, borides, nitrides, and carbides, which are embedded in a pyrolytic matrix from the group of graphite, borides, nitrides, and carbides. These materials were developed for application to high speed gases of nozzles of jet engines or for use on nose cones of atmospheric re-entry vehicles. Plural layers are used whereby one layer is perpendicular to the surface of the layer and at least one layer is parallel to the surface of the composite. Graphite fibers and matrices, as well as silicon carbide fibers and matrices, were cited in the claims.

U.S. Patent 3,472,730—"Heat Curable Filament-Reinforced Resinous Sheeting and Laminating Process"
Frigstad, R.A., to Minnesota Mining and Manufacturing Co.
Filed October 10, 1966 Granted October 14, 1969

This patent discloses resinous sheets reinforced with layers of collimated carbon fibers, and the attainment of high interlaminar strength. The laminate comprises a thin, flexible layer of high strength reinforcing filaments (glass or carbon); heat curable epoxy/anhydride resin coated on filaments; and a separate exterior film 1/2 to 4 mils in thickness, consisting of a thermosetting phenolic resin modified with carboxylated acrylonitrile-butadiene rubber for toughness.

U.S. Patent 3,476,703—"Treatment of Carbon Fibers and Composite Materials"
Wadsworth, N. and Watt, W., to National Research Development Corp. (England)
Filed February 16, 1968 Granted November 4, 1969

This is a process of treating carbon fibers with an ultimate tensile strength over 100×10^3 psi and a Young's modulus of at least 16×10^6 psi—to improve bond to epoxy matrix—wherein the carbon fibers are heated in air of at least 350°C—to provide a pitted surface and weight loss of 0.05 to 6%. The patent specification showed preference for epoxy resin matrices in preparing composite structural materials.

U.S. Patent 3,503,708—"Graphite Yarn"
Spry, W.J., to Union Carbide Corp.
Filed December 27, 1968 Granted March 31, 1970

The patent describes a permanently elongated, creep resistant, high temperature refractory carbonaceous yarn of cellulosic origin, prepared from a plurality of graphite monofilaments, each with a Young's modulus of over 10×10^6 psi and a breaking strength of 120×10^3 psi. Carbonaceous yarns are also described with a Young's modulus of 55×10^6 psi and break strengths of 370×10^3 psi. The carbonaceous yarns, which comprise a yarn previously carbonized at 1200°C, are subjected to additional tensional forces while heating to graphitizing temperatures (up to 3000°C). Marked improvements in strength and stiffness occur by relieving unequal filament to filament stresses and orienting the crystallites in the graphitic layers in a direct or parallel direction to longitudinal axes of the monofilaments.

U.S. Patent 3,529,934—"Process for the Preparation of Carbon Fibers"
Shindo, A., to Nippon Carbon Co. (Japan)
Filed January 4, 1968 Granted September 22, 1970

Carbonaceous and carbon fibers are produced from each of cellulosic, polyvinyl alcohol, and acrylic fibers by subjecting them to heat treatment in an atmosphere containing gaseous hydrogen chloride (HCl). The process directs attention to treating the fibers in the HCl atmosphere at 250° to 450°C.

Fibers were also held under tension during heat treatment at 2500° to 3000°C to obtain higher moduli and tensile strengths. The patent claims that the prior treatment of fibers in the HCl atmosphere will improve the yield of carbonaceous fibers.

U.S. Patent 3,533,743—"Manufacture of Continuous High Modulus Carbon Yarns and Monofilaments"
Prescott, R. and Orr, I.W., to Great Lakes Carbon Corp.
Filed May 28, 1968 Granted October 13, 1970

Polyacrylonitrile (PAN) yarns or monofilaments are subjected to a two stage process: 1) a continuous stage carried out under tension, and 2) a discontinuous skein stage. Under tension, the PAN is stretched in steam to at least 150% of its original length, and then oxidized in air for one hour at 230° to 250°C. The yarn is formed into skeins and subjected to further oxidation, carbonization, and, if desired, graphitization (the last two steps in a non-oxidizing media). This patent emphasizes the production of long fiber lengths.

U.S. Patent 3,552,923—"Production of Carbon Fibers"
Carpenter, W.G., Johnson, W., Lloyd, T., McMullen, P., and Watt, W. (England)
Filed June 26, 1967 Granted January 5, 1971

Described is an apparatus for the production of continuous lengths of carbon fibers of high tensile strength and high Young's modulus. The polymer fibers are held taut as they are passed through the oxidizing oven, restricting the lengthwise shrinkage to not more than 10%, and prior to oxidizing to stretch the fibers not more than 20% of their original length. Complete penetration is attained by the oxygen throughout the fibers at temperatures of 200° to 250°C. This is followed by a carbonizing treatment in which a non-oxidizing atmosphere is maintained, as the fibers are allowed to shrink as they are raised in temperature to, and held at, 1000°C. The carbonizing treatment may be followed by further processing to temperatures up to 3000°C. Test data reported in the specification for polyacrylonitrile fibers show ultimate tensile strengths of 320,000 psi and Young's modulus of 30 $\times$ 10^6 psi.

U.S. Patent 3,629,379—"Carbon Filaments from Low Priced Pitches"
Otani, S., to Kureha (Tokyo)
Filed November 6, 1969 Granted December 21, 1971

Low priced pitch (from petroleum, coal tar, or natural asphalts), having carbon content from 91 to 96.5% by weight and a mean molecular weight of at least 400 to 2000, is readily melt-spun into a filament. The filaments are formed by heating at 300° to 500°C, and then subjected to an oxidation treatment at temperature below the fusion temperature and spinning temperature of the fibers. The resulting filaments are rapidly carbonized by firing in an inert atmosphere below 2000°C. Graphitization may also be introduced by heating above 2000°C. Considerable attention is given in the patent specification on enhancing the quality of the pitch by treatment with additives such as mercaptans and peroxides to increase molecular weight of the pitch before spinning.

U.S. Patent 3,632,385—"Carbon Composite Structures"
Schmitt, C.R. and Schreyer, J.M., Oak Ridge
Filed March 17, 1970 Granted January 4, 1972

Porous carbon structures such as honeycomb and reticulated carbon foam contain cavities or interstices which are filled with cellular carbon foam to provide high strength, thermal insulating products. Furfuryl alcohol resins achieve a good bond between the carbon foam filler and porous structure. Further high temperatures are applied to carbonize the cured furfuryl alcohol binder and a furfuryl alcohol foam producing admixture. Temperature range for the further carbonization is reported at 900° to 1100°C.

U.S. Patent 3,635,675—"Preparation of Graphite Yarns"
Ezekiel, H.B., to U.S.A.
Filed May 28, 1968 Granted January 18, 1972

The patent describes the process for graphitizing polymeric fibers by first stabilizing the fibers by heating in an oxidizing atmosphere at 200° to 500°C, and thereafter rapidly heating the fibers in an inert atmosphere to a temperature of 1800° to 3200°C in a matter of several minutes. The process was directed principally at acrylonitrile polymers, polybenzimidazoles, polyoxadiazoles, and aromatic polyimides. One of the claims notes the optimum stabilization taking place by heating 3 hours at 265° to 290°C. Remarkably high moduli (up to 70×10^6 psi) and tensile strengths (over 300×10^3 psi) are shown for single filaments in examples outlined in the patent specification.

U.S. Patent 3,660,140—"Treatment of Carbon Fibers"
Scola, D. and Roth, H., to United Aircraft Corp.
Filed June 18, 1970 Granted May 2, 1972

The method is described for treating high modulus, high strength carbon fibers to improve bonding characteristics in a resin matrix by immersing the fiber in concentrated nitric acid (70%) for 4 to 8 hours at refluxing temperature. Several graphite yarns were tested and flexural properties and short beam shear strengths were reported in the patent to demonstrate the advantages gained by the nitric acid pretreatment before bonding with an epoxy resin matrix.

U.S. Patent 3,671,192—"Method of Stabilizing Acrylic Polymer Fibers Prior to Graphitization"
Ezekiel, H.M., to U.S.A.
Filed May 28, 1968 Granted June 20, 1972

The method of stabilizing acrylonitrile polymer fibers prior to carbonization, which comprises the step of oxidizing the polymer at 265° to 290°C in an oxygen containing atmosphere for at least three hours. The polymer fibers are wound on a metal tube to dissipate the heat of reaction generated during oxidation. Fibers are preferably in the form of a yarn and are under tension. Copolymers of acrylonitrile with methyl acrylate and with vinyl acetate are mentioned. When stabilization is followed by graphitization at 1800° to 3200°C, graphite fibers of high tensile strength and high modulus of elasticity are obtained.

U.S. Patent 3,677,804—"Moldable Needle-like Assemblies"
Kalnin, I. and Breckenridge, G.J. Celanese Corp.
Filed September 19, 1969 Granted July 18, 1972

Graphitic fibers (1/4 to 3/4 inch in length) are dispersed into an aqueous slurry comprising a minor amount of epoxy resins ("Epi-rez" 510, "Epon" 828, Der-332) and a curing agent of nadic methyl anhydride and imidazole catalyst, plus glycerol to prevent agglomeration and a water soluble non-ionic surface active agent. This is agitated at an elevated temperature while the needle-like fibers are impregnated. About 10 to 30 parts by weight of fibers are introduced per 100 parts of epoxy resin and curing agent. The emulsion is heated until the epoxy resin advances to a stage B cure and the coated fibers are separated by filtration. Treated fibers are washed further, stored under refrigeration, and subsequently molded. Flexural strengths of 90,600 psi, a short beam shear of 8200 psi, and a density of 1.36/cc are observed on molded parts.

U.S. Patent 3,702,054—"Production of Graphite Fibers"
Araki, T., Asano, K., and Yamada, J., to Kureha
Filed July 27, 1971 Granted November 7, 1972

Graphite fibers having superior strength and modulus of elasticity are obtained by twisting carbon fibers at a specified twisting rate and then graphitizing the twisted yarn. It is claimed that the space or porosity between filaments in the yarn may be reduced as the stress applied to carbon fibers during twisting acts during graphitization at high temperatures to flatten portions of the circular carbon fibers into more favorable polygonal cross-sections. Thus, a linear or planar contact is established along the lengthwise direction of the fiber. For the carbon fibers to exhibit this deformation requires in excess of 2000°C—preferably 2500 to 3500°C in an inert argon or helium atmosphere. It was found that fibers spun at high speed and twisted within the breaking strength of the fibers showed a remarkable improvement in strength and moduli of elasticity.

U.S. Patent 3,705,236—"Method of Producing Carbon Fibers"
Ishikawa, T. and Teranishi, H., to Nippon Carbon Co. (Japan)
Filed October 28, 1970 Granted December 5, 1972

This is a method for producing carbon fibers with tensile strengths over 150 kg/mm^2 and a Young's modulus of more than 32,000 kg/mm^2 by applying a vibrational energy to the fibers while they are under tension (20 to 120 mg/denier). The frequency of the vibrational energy is described in the range of 30 cycles/second (preferred) to 1,000,000 cycles/second for a period of

5 minutes to 10 hours while processing the carbon fiber through its heat treatments up to graphitization at 2500°C. It is claimed that fiber diameter is reduced and that the degree of orientation of fibrous molecules is improved. The heat treatments include treatment of organic fibers at 150° to 300°C non-oxidizing atmosphere; carbonized up to 1000°C in vacuo or in an inert atmosphere: and graphitization in the range of 2500° to 3000°C. Polyacrylonitrile (PAN) fibers are mentioned most frequently, and a table of the patent specification shows for PAN fibers the following properties:

	30 MINUTES VIBRATION AT 50 CYCLES PER SECOND	NO VIBRATION
Tensile strength (kg/mm^2)		
Carbonized	186	164
Graphitized	174	150
Young's modulus (kg/mm^2)		
Carbonized	21,000	13,000
Graphitized	57,000	30,000

U.S. Patent 3,716,607—"Heat Treatment of Molten Carbonaceous Material Prior to Conversion to Carbon Fibers"
Otani, S., to Kureha
Filed July 7, 1970 Granted February 13, 1973

A method is provided for the production of carbon shaped articles in the form of filaments, yarns, ribbons, films, tubes, etc., which consists of:

1. Heating organic substances, such as polyvinyl chloride, polyacrylonitrile, petroleum pitch, or coal tar pitch, in the presence of inert gas at 300° to 500°C to induce the materials to fuse.
2. Forming the substance in an inert atmosphere.
3. Oxidizing the shaped article at temperatures below the shaping temperature.
4. Subjecting the article to carbonization by heating in an inert atmosphere substantially above 500°C.

The invention describes materials suitable for thermal insulation and also the manufacture of electrodes.

U.S. Patent 3,718,574—"Method for Heavying Polycylic Substances"
Araki, T., Asano, K., Takita, H., and Awao, J. to Kureha
Filed December 28, 1970 Granted February 27, 1973

An improved method is described for the production of a petroleum pitch with increased carbon yield. Hydrocarbons composed primarily of condensed polycyclic aromatic compounds are oxidized at temperatures of 0° to 400°C, and subsequently treating the oxidized material with ammonia or amines within the same temperature range. The "heavied" or thickened material prepared in this manner appears to maintain the integrity of the matrix structure, as compared to other processes using Friedel-Crafts reagents ($AlCl_3$). Further heat treatment for carbonization and then graphitization results in improved graphite crystallite orientation. The process for preparing carbon fibers is described as follows:

Spinning of pitch fibers of 10 microns average diameter in a centrifugal spinning machine having a rotary cylinder 150 mm in diameter and a depth of 30 mm, and provided with 120 nozzles of 0.7 mm in diameter, with a spinning rate of 2000 m/MW at a temperature of 300° to 330°C. Reported strengths in the patent specification for the graphite fibers are 220 ksi tensile strength, and tensile modulus of 38×10^6 psi.

U.S. Patent 3,720,536—"Treatment of Carbon Fibers"
Scola, D., to United Aircraft Corp.
Filed June 18, 1970 Granted March 13, 1973

This is a method of improving bonding characteristics of carbon fiber to a resin matrix without causing significant degradation of mechanical properties of the fiber by heating the fiber to 1000° to 1500°C (1200° to 1300°C—preferred) in a nitrogen atmosphere for 0.1 to 60 seconds. The nitrogen atmosphere had an oxygen concentration of 1.8% by volume. Optimum time indicated was 1.7 to 6.7 seconds. Numerous charts are presented to show the effect of processing variables on the ultimate strength (see Fig. 1-3).

U.S. Patent 3,723,157—"Production of Resin Impregnated Fibrous Graphite Ribbons"
Druin, M., to Celanese Corp.
Filed November 7, 1969 Granted March 27, 1973

A process is provided wherein resin impregnated graphite ribbons may be efficiently produced which may be used in the manufacture of fiber reinforced composite structures. A plurality of multifilament bundles of a fibrous material

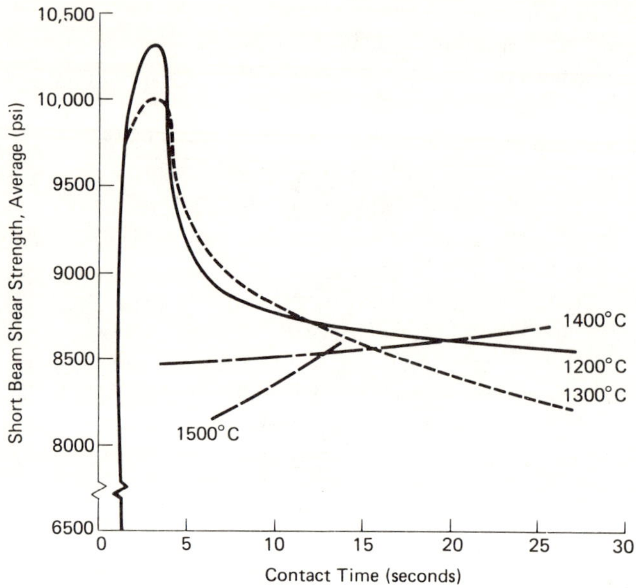

Fig. 1-3. Time-temperature influence on short beam strength.

capable of undergoing graphitization are continuously passed through a graphitization zone from which they are continuously conveyed to and through a coating zone in which they are impregnated with a thermosetting resin. Throughout the continuous process of the invention, the multifilament bundles are provided in an essentially parallel relationship in the form of a ribbon and are continuously passed in the direction of their length. The ribbon product of the invention is particularly suited for use in the formation of strong lightweight composites by filament winding, or other composite forming techniques.

U.S. Patent 3,723,605—"Production of Continuous Length of Graphitic Fibrous Material"
Ram, M.J., to Celanese Corp.
Filed June 10, 1970 Granted March 27, 1973

Volatilization of a catalytic quantity of boron within a heating zone enables graphitization to proceed more efficiently. Graphitic fibrous materials exhibiting improved tensile strength (to 350 ksi) and higher Young's modulus (to 100×10^6 psi) are prepared in this manner in a continuous process. Boron compounds capable of volatilization below 2000°C are introduced to the walls of the heating chamber before the fibrous material is brought into the heating zone. Materials undergoing graphitization are under tension, heated in an inert atmosphere to temperatures of 2400 to 3100°C. The boron compound is volatilized to a concentration of 200 to 2000 ppm by volume in the heating zone.

U.S. Patent 3,764,662—"Process for Making Carbon Fiber"
Roberts, B.W., to General Electric Co.
Filed April 21, 1971 Granted October 9, 1973

This invention discloses a process improvement for obtaining both higher strength and high modulus in graphite fibers. Polyacrylonitrile based fibers held under tension in an oxidizing atmosphere at 200° to 250°C, carbonized under hydrogen atmosphere to 1000°C, and then carbonized in an inert atmosphere for at least 1 hour at 1300° to 1800°C. This invention discusses a post-heat treatment for 30 to 90 seconds in an inert atmosphere at 2950°C to obtain a fiber with a tensile strength of 350 to 400 ksi and a modulus of 45 to 50 $\times$ 10^6 psi. Effect of post-heat treatment is disclosed in the invention.

U.S. Patent 3,767,741—"Making Carbon Fibers from Solvent Extracted Residues of Petroleum"
Toyoguchi, M., Iijima, H., and Morijya, K., to Mitsubishi Oil Co.
Filed December 10, 1970 Granted October 23, 1973

In the production of carbon fibers, a process is outlined for removing low molecular weight saturates fractions from the vacuum distillation residue of petroleum by solvent extraction with propane. The vacuum distillation residue of petroleum comprises low molecular weight (400 to 800) aliphatic saturates (an oily component); aromatics having a higher molecular weight; resins which are higher in molecular weight than the aromatics; and asphaltenes having the highest molecular weight, which are colloidally dispersed in said saturates, aromatics, and resins. The bituminous material is then air blown at a temperature of 260° to 280°C, generally for 10 to 18 hours, to reduce the saturates below 1% by weight. The resulting material, with a softening point range of 130° to 200°C and a penetration of 0 to 2 (ASTM-D5 at 25°C, 100 g, and 5 seconds), is extrusion spun at a temperature of 200° to 300°C to obtain carbon fibers 5 to 20 microns in diameter. The fibers are heated in an oxidizing atmosphere at 200° to 300°C for 2 to 10 hours. They become stiff and do not deform on further heating. Further heat treatment in an inert atmosphere up to 1000°C for 1 to 6 hours yields carbon fibers with a tensile strength of 10 $\times$ 10^3 kg/cm^2. The fibers may be extrusion spun through a nozzle or centrifugally spun using a rotating disc.

U.S. Patent 3,787,541—"Graphitization of Mesophase Pitch Fibers"
Grindstaff, L.L. and Whittaker, M.P.
Filed October 26, 1971 Granted January 22, 1974

A high boiling petroleum hydrocarbon cokable feedstock, obtained as a residue of vacuum distillation or coal tar pitch, has its mesophase contents substantially increased by heating at a temperature of 400° to 450°C for several hours. At least 75% quinoline or pyridine insoluble mesophase pitch is formed. The mesophase pitch is pulverized prior to extrusion into filaments or fibers in a temperature range of 400° to 700°C under high pressure (10,000 psi). Further heat treatments are practiced to obtain the carbon/graphite fibers.

U.S. Patent 3,791,840—"Treatment of Carbon Fibers to Improve Shear Strength in Composites"
Barr, J.B., to Union Carbide Corp.
Filed October 21, 1970 Granted February 12, 1974

Disclosures are made of the improvement of the interlaminar shear strength of carbon fiber-plastic composites by treating the carbon fibers with an aqueous solution of hypochlorous acid having a pH below 6, before the fibers are incorporated into a plastic matrix. Fifty-one claims are granted in this patent. The hypochlorous acid solution was maintained at 20° to 30°C. The solution was prepared from a commercial bleach solution having a pH of 11 and containing 5.25% sodium hypochlorite and 4.8% sodium chloride. This solution was acidified to a pH of 4 to 5 by slowly bubbling chlorine through. Better torsional shear strengths were reported for composites with hypochlorous acid solution, compared with straight sodium hypochlorite solution.

U.S. Patent 3,801,350—"High Modulus Graphite Fibers Having Improved Bonding Properties"
Dauksys, R., to U.S.A.
Filed May 5, 1972 Granted April 2, 1974

Bonds to high modulus graphite fibers are improved by treating them with a solution of sodium iodate (25 to 200 g/l) in aqueous dioxane, washing and drying the treated fibers; and contacting the fibers with tin tetrachloride in 2-butanone. This treatment is to enhance the bond of epoxy matrices to graphite fibers. The tin tetrachloride was about 0.2 to 1.0 volume percent in solvent, and the period of contact was from 15 seconds to 30 minutes. Interlaminar shear strengths of the graphite fibers in epoxy matrices were improved by 50%, according to limited data referenced in the patent specification.

U.S. Patent 3,837,904—"Method of Sizing Carbon Fibers"
Hill, J., to Great Lakes Carbon Corp.
Filed February 29, 1972 Granted September 24, 1974

Methods are described for sizing carbon fibers by coating with a sizing containing 5 to 60 volume percent of a reactive functionally terminated polymer (reactive end groups such as hydroxyl, carboxyl, and thiol). Claims note that discrete particles at least 1000 Angstroms in diameter are formed on the fibers from the sizing solution. Butadiene-acrylonitrile polymer solutions are mentioned most frequently.

U.S. Patent 3,853,610—"Composite Materials Comprising Epoxy Resin Matrix and Carbon Fibers"
Byrne, J.B. and Ledbetter, H.D., to Dow Chemical Co.
Filed March 10, 1972 Granted December 10, 1974

This invention is directed towards the development of a continuous coating of amorphous carbon on the surface of carbon fibers (modulus in excess of 12×10^6 psi) prior to admixture with resin matrix. A thermally carbonizable organic precursor material is deposited on the fibers, followed by pyrolysis at temperatures below 1200°C, in a nitrogen atmosphere containing 900 to 3200 ppm of oxygen. The amorphous carbon coating, derived from polyvinyl chloride, polyacrylonitrile, or polyvinylidene chloride, has good utility as a sizing agent. Carbon coating thickness is estimated at 150 Angstroms or more, with crystallite basal planes at random with respect to the surface. Table VI of the patent specification shows the following results on 60% by volume of graphitized fibers:

SAMPLES	FIBER TENSILE STRENGTH (psi)	INTERLAMINAR SHEAR STRENGTH, EPOXY COMPOSITE (psi)
Untreated fibers	289,000	2950
Amorphous carbon coating–carbonizing temperature		
825°C	355,000	4560
950°C	323,000	12,980
1010°C	261,000	13,300

U.S. Patent 3,894,863—"Graphite Composite"
Lachman, W.L., Penty, R.A., and Jahn, A.F., Fiber Materials, Inc.
Filed October 3, 1974 Granted July 15, 1975

Graphite fibers are coated with titanium boride and titanium carbide by a vapor deposition technique. The coating promotes wetting of the fibers by metals such as aluminum, copper, lead, zinc, and magnesium. The coatings are in the range of 100 to 10,000 Angstroms in thickness. Fibers were exposed to a gas mixture of principally argon, plus $TiCl_4$, BCl_3, and zinc for 30 minutes at 650°C, and transferred to a molten metal bath.

U.S. Patent 3,931,392—"Enhancement of Tensile Strength of Carbon Fibers"
Dietz, V., to U.S.A.
Filed January 10, 1974 Granted January 6, 1976

A process is described for enhancing the tensile strength of carbon fibers by immersion in liquid bromine, followed by its removal with an inert gas. Immersion takes place at ambient temperatures for at least one hour. Data shown for polyacrylonitrile carbon fibers indicate tensile strengths are increased 40%.

U.S. Patent 3,971,669—"Carbon Fiber Composites"
Wrzesien, A., Whitney, I., and Katona, J., to Hyfil, Ltd.
Filed July 17, 1973 Granted July 27, 1976

A method is given for producing a composite shaped article from low density, high bulk felt from continuous carbonizable filaments in an oxidized state. This is accomplished by carbonizing high bulk, low density felt; presizing felt with a very dilute solution of thermosensitive resins, which will, upon solvent removal, lock fibrous elements together by adhesion at cross-over locations; followed by further resin to presized felt to occupy voids between adhered fibrous elements. Further molding and heat treatment produces a high temperature resistant product.

U.S. Patent 3,997,654—"Method for Production of Carbonaceous Articles, Particularly Strands"
Kolling, G. and Romey, I., Bergwerksverband, GmbH
Filed April 22, 1975 Granted December 14, 1976

A process is described for heating a bituminous, hard coal-tar pitch having a Kramer-Sarnow softening point between 70° and 190°C to a temperature of

40° to 100°C above the softening point. A filtering operation is carried out to remove solid and pseudo-solid particles which occur naturally in the pitch. The remainder of molten pitch is further heated in an inert atmosphere to remove lower molecular weight volatiles. The pitch is then extruded into fibers (12 microns in diameter is mentioned) and dusted with finely divided activated carbon, which has been treated with an oxidizing agent (nitro aromatic compounds). Carbon strands produced at 1000°C show tensile strengths of about 50 ksi.

U.S. Patent 4,005,183—"High Modulus, High Strength Carbon Fibers From Mesophase Pitch"
Singer, L.S., to Union Carbide Corp.
Filed March 5, 1973 Granted January 25, 1977

High modulus, high strength carbon fibers having a highly oriented structure, with carbon crystallites aligned parallel to the fiber axis, are produced from carbonaceous pitches which have been partly transformed to a liquid crystal or "mesophase" state. Emphasis is placed on attaining the three dimensional crystalline structure characteristic of polycrystalline graphite. Petroleum pitches containing a mesophase content of 45 to 65% by weight, which exhibit a viscosity of 30 to 60 poises at 340° to 380°C, are used to spin fibers of 10 to 20 micron diameters. These fibers are rendered thermosetting by heating in air at 300° to 390°C. They are carbonized at temperatures up to 1500° to 1700°C for 1 to 5 minutes in an inert atmosphere, after a preliminary stage at 700° to 900°C. Graphitization takes place at temperatures above 2500°C. Fibers prepared by this process have electrical resistivities from 150×10^{-6} to 200×10^{-6} ohm-cm, tensile strengths in excess of 250×10^3 psi, and Young's moduli of elasticity greater than 75×10^6 psi.

U.S. Patent 4,061,520—"Method of Making High Strength to Weight Structure"
Cecka, A.M. and Dano, P.O.L., Fansteel, Inc.
Filed September 3, 1976 Granted December 6, 1977

Structures such as tennis racket frames are manufactured by providing expandable foamed urethane plastic cores. The cores are wrapped around by resin impregnated graphite fibers. During cure, the foam expands, generating pressure which forces the core against the resin impregnated graphite fibers, achieving a tight band within the closed mold. The foamable resin composition has an expansion ratio to 1.5 to 3.0. Angular graphite ply distribution of the graphite fibers is plus or minus 45°. Good performance of tennis rackets is achieved at low weight, with an effective distribution of carbon/graphite fibers within their structure.

U.S. Patent 4,065,340—"Composite Lamination Method"
Dickerson, G. to U.S.A.
Filed April 28, 1977 Granted December 27, 1977

A process is described for preparing minimum void, thick composite laminates from graphite fiber/epoxy prepregs such as Hercules 3501-AS and Narmco's 5201 systems. Preimpregnated unidirectional continuous carbon/epoxy fiber filament prepregs are specified in tape form from .002 inch to .007 inch thicknesses. After cutting to the desired lengths, the tapes are stacked and *partially cured* at an elevated temperature under vacuum. The mass of stacked tapes is cooled, cut into a plurality of strips along the parallel orientation of the fibers, restacked with uncured prepreg tape between each of the plurality of partially cured laminated strips. This assembly is fully cured and the final structure machined to desired size and shape. Thick void, free carbon/graphite laminates are claimed.

U.S. Patent 4,065,549—"High Tensile Strength, High Young's Modulus Carbon Fibers"
Kinoshita, Y., Toray (Japan)
Filed August 12, 1976 Granted December 27, 1977

High tensile strength, high Young's modulus carbon fibers, substantially free from holes and voids (less than 2% by plasma etching test) show values of 310 ksi tensile strength and a Young's modulus of at least 31 msi. Excellent homogeneity of internal structure is obtained by a process of repeated contacting polyacrylonitrile precursor filaments with heated rolls at 260° to 380°C for less than one second in an oxidizing atmosphere. Copolymer of 99 mole percent of acrylonitrile and one mole percent of methyl acrylate are mentioned most frequently in the examples. The oxidized fibers processed in this manner are further carbonized in an inert atmosphere above 800°C, preferably at 1000° to 1600°C. Fibers were examined under a scanning electron microscope after exposure to a special plasma etching technique to reveal holes that had not been observed earlier. The process covered by this invention reduces the number of holes and voids and yields a more consistent and uniform fiber structure. The process of spinning acrylic fibers was continuously connected to the above process of oxidation. The speed of the organic fiber in the oxidizing step is in excess of 20 m/minute.

U.S. Patent 4,070,446—"Process for Production of Carbon Fibers"
Horikiri, S., Iseki, J., and Minobe, M., Sumitomo Chemical Co.
Filed February 1, 1974 Granted January 24, 1978

Described is a process for production of carbon fibers by sulfonating polyethylene fiber with chlorosulfonic acid or sulfuric acid (at a temperature of 60° to 90°C) and carbonizing the resulting precursor fiber by heating at a temperature of 600° to 2000°C in an inert gas. The polyethylene fiber should have a density of 0.94 g/cm^3 or more and a melt index of 0.1 to 10. Further processing at 2000 to 3000°C is also claimed. Sulfonation and carbonization steps under tension of 20 to 200 mg/denier are also claimed, yielding carbon fibers approximately 8 microns in diameter. Tensile strength of 180 ksi and elastic moduli of 12 msi were reported for the fibers.

U.S. Patent 4,072,084—"Graphite Fiber Reinforced Structure for Supporting Machine Tools."
Charles, E.K., Kovach, L., and Hurst, J.S.
Filed December 17, 1976 Granted February 7, 1978

Graphite fiber reinforced reinforced tool structures are described which exhibit minimal deflection, thermal expansion, and vibration characteristics. Graphite fibers or yarn are disposed in a 0°/90° pattern and bonded with an epoxy resin. Claims are directed towards specific distribution of the graphite fibers parallel to the main axes of the tubular body support (90% of volume of the fibers), whereas the balance are circumferentially wound in situ.

U.S. Patent 4,072,516—"Graphite Fiber/Metal Composite"
Pepper, R.T. and Zack, T., to Fiber Materials, Inc.
Filed September 15, 1975 Granted February 7, 1978

Graphite fibers with a coating of a mixture of silicon oxide and silicon carbide are dispersed in a solid matrix of metal. Aluminum, magnesium, titanium, nickel are among metals mentioned. The coating protects the graphite surface from attack by carbide forming matrix metal. The coating is formed by an intermediate temperature vapor deposition technique (at 1550°C. for five minutes) involving the reduction of silicon tetrachloride in presence of hydrogen and carbon dioxide (mixture of 67% vol $SiCl_4$; 32% H_2; and 1% CO_2 was cited). Coatings in the range of 100 to 10,000 Å were mentioned on the graphite fibers. PAN precursor graphite yarn containing 10,000 individual fibers were used. Titanium boride and carbon coatings of 200 Å. thickness were also deposited on top of the silicon carbide/oxide coating, functioning as wetting agents for the molten aluminum alloys.

U.S. Patent 4,079,122—"Preparation of Carbon Fibers"
McLaughlin, V.C., Moreton, R., and Watt, W., to National Research Development Corp. (England)
Filed October 22, 1976 Granted March 14, 1978

Heating organic polymer precursor fiber to 200° to 400°C in an inert atmosphere, while the natural shrinkage of the fiber is at least restrained, followed by further heat treatment in the range of 800° to 3000° in a non-oxidizing atmosphere, is described. The organic polymer is a copolymer of acrylonitrile (100 molar parts); 2 to 20 molar parts of vinyl chloride or vinylidene chloride; 0.5 to 5 molar parts of itaconic acid; and up to 5 molar parts of other comonomers.

U.S. Patent 4,089,190—"Carbon Fiber Drive Shaft"
Worgan, G.P. and Smith, D.R., to Union Carbide Corp.
Filed April 14, 1976 Granted May 16, 1978

Distribution and orientation of carbon fibers (at least 20%) in a resin matrix are discussed in detail in the construction of a tubular drive shaft. Thirty-seven patent claims are granted for the construction. Good noise and vibration dampening characteristics are claimed, and the drive shaft is designed for operation at 6000 rpm, said shaft having a maximum longitudinal modulus of elasticity of 8×10^6 psi. From 10 to 20% of the carbon fibers (having a modulus below 26×10^6 psi) are aligned parallel to the longitudinal axis of the drive shaft. The balance of fibers are circumferentially and helically disposed about the axes—at 85° to 90° angles to the longitudinal axis.

U.S. Patent 4,154,807—"Process for Production of Carbon Fibers"
Kishimoto, S. and Ohana, I., to Japan Exlan Co.
Filed March 7, 1978 Granted May 15, 1979

Spinning an acrylonitrile polymer (at least 90 mole percent) and 0.01 to 1.0 mole percent SO_2H groups, which are partly converted to SO_3X (where X is a monovalent metal carbon or ammonium ion) is described. The spun fiber bundle is heat stretched, followed by carbonizing and graphitizing.

U.S. Patent 4,157,409—"Method of Making Metal Impregnated Graphite Fibers"
Plevitt, A. and Band, H.
Filed August 28, 1978 Granted June 5, 1979

This is a process for immersion of graphite into a liquid room temperature alloy of sodium and potassium maintained under anhydrous conditions and in inert atmosphere. The treatment enhances wettability and infiltration of graphite fibers by metal matrices. Metal matrices for impregnating the treated graphite fibers include aluminum, magnesium, copper, lead, and zinc. On cooling, after metal impregnation and hardening of metal matrix, the graphite reinforced metal is fabricated.

U.S. Patent 4,163,003—"Unsaturated Epoxides as Coupling Agents for Carbon Fibers"
Paul, J.T. and Weldy, W.E., to Hercules Inc.
Filed May 23, 1975 Granted July 31, 1979

Adhesion between carbon fibers and unsaturated matrix resins used in the preparation of composite structures is improved by the use of an unsaturated epoxide coupling agent such as allyl glycidyl ether or glycidyl acrylate. Unsaturated polyester resins and unsaturated epoxide resins are referenced in the patent claims. Various examples show a 20 to 40% improvement in composite short beam strength of graphite fiber/matrix resin systems using 1 to 5 parts by weight of coupling agent per 100 parts of matrix resin. Subject also covered in Brit. Patent 1,527,524, October 4, 1979, to Hercules.

U.S. Patent 4,186,179—"Process for Oxidized Carbon Fibers"
Katsuki, K. and Watanobe, S., to Toray Ind. (Japan)
Filed—Feb. 14, 1979 Granted January 29, 1980

This patent describes process for manufacture of improved strength carbon fibers through intermittent contact with heated rolls at 200 to 400°C. in an oxidizing atmosphere. Product then passes to a non-oxidizing atmosphere at 800°C. Interlaced filament yarns without crimps or loops are specified. Refer to U.S. Patent 4,065,549.

U.S. Patent 4,187,135—"Fiber Reinforced Composite Tubular Shaft"
Yates, D.N. and Presta, J.C., to Celanese Corp.
Granted February 5, 1980

Structural and assembly details are shown on the application of carbon-fiber prepregs used in the fabrication of composite tubular shafts. Metal sleeves with circumferential recesses are employed in conjunction with carbon/graphite fiber products to fabricate torsion transmitting connections.

REFERENCES

1. Johnson, J.B., "Structural Plastics in the Aviation Industry," *Society of Plastics Industry*, October 1941.
2. Hoffman, G.A. (Rand Corp.), "Fibered Materials for Flight Structures," Report RM-1868, 63 pages, February 1970.

3. McCreight, L.R., Rauch, H.W., and Sutton, W.H., "Survey of State of Art of Ceramic and Graphite Fibers," Technical Report, AFML-TR-65-105, 379 pages, May 1965.
4. Editors, "The Promise of Composites," *Materials in Design Engineering,* page 126, September 1963.
5. Alexander, T., *Fortune,* page 152, April 1966.
6. Grimes, D.L., "Composite Structures–They Start Where Metals Leave Off," *Modern Plastics,* Vol. 43, page 221, May 1966.
7. Simon, R.A. and Prosen, S.P., "Carbonaceous Fiber Composites," Report NOL-TR-68-132, October 1968.
8. Dwyer, J.J., "Composites," *American Machinist,* Special Report No. 643, July 13, 1970.
9. Economy, J., "High Strength Fibers," *SAMPE Journal,* Vol. 12, page 6, December 1976.
10. Powers, W., *Twenty-sixth Annual Technical Conference,* SPI, RP/C Institute, Section 8G, Washington, D.C., February 1971.
11. Anon., *Industry Week,* March 25, 1970.
12. Lovelace, A.M., "Advanced Composites," *Eleventh Von Karman Lecture Journal of Aircraft,* Vol. 11, No. 9, page 501, September 1974.

2
Preparation and Properties of Carbon and Graphite Fibers

Carbon is an abundant element on this planet and its utilization by industry has been and continues to be the subject of many studies. Elemental carbon is sixth in the periodic table, with an atomic weight of 12.011 and consisting of 98.9%C_{12} and 1.1%C_{13}. Carbon and its graphite derivatives have emerged prominently in twentieth century technology—more recently as moderators in nuclear power facilities and as high strength fibers, the latter being the subject of this book. There are other high strength fibers, such as glass fibers, but the developments in high modulus carbon-based fibers have established a unique category. In developing the background of carbon fiber technology, there will be references to the preparation and adaptation of finely divided carbon products to chemical processing industries, where there are experiences applicable to the carbon fiber products. Finely divided graphite products, valuable for lubrication purposes, were the subject of Acheson's patents, which appeared early in the twentieth century.[1]

Commercial carbon and graphite fibers are derived either from fibrous organic precursors or by the extrusion of pitch. Patents reviewed in the first chapter described processes of heat treatment to remove oxygen, nitrogen, and hydrogen from organic fibers. Earlier patents were understandably highly empirical, but, as knowledge was enriched by new contributions, more refined techniques evolved. In fact, at the time of publication of this book, there is a growing distinction between results as obtained upon earlier developed carbon/graphite (C/G) fibers and those of more recent origin. The consistency and quality of the C/G fibers have been markedly improved, leading to wider acceptance. Some of the earlier manufacturers of carbon fibers have withdrawn from the marketplace, primarily because the hoped-for growth has been slow in being achieved. Other major suppliers of C/G fibers in the United States are beginning to rely more on fibers produced in this country, rather than depending upon overseas sources (which have been Japan or England). The high performance requirements of C/G composites for aerospace dictate the ultimate source from whichever country or manufacturer is best prepared. The stringency of the quality control measures have of necessity raised manufacturing costs, and C/G fibers are not inexpensive. While their selling price has come down from about \$500/lb to nearer \$25/lb, they are comparable with cost effectiveness of lower priced materials for aerospace applications. It has been apparent that the processing and equipment costs for the production of C/G fibers dwarf the fiber raw material cost.

Nomenclature as used in this text will reflect the nature of the C/G fibers. Graphite fibers consist essentially of carbon atoms, which by high temperature heat treatment, have obtained a crystalline state with a high degree of three-dimensional order as determined by X-ray diffraction patterns. Carbon fibers show a lower degree of three-dimensional order or may be almost amorphous. Graphite fibers have higher modulus and strength, and are better electrical and thermal conductors. The structural composites on which considerable time and discussion will be placed will refer to the carbon/graphite fibers (C/G fibers) used for reinforcement. This measure is in recognition of the carbon origin of the graphite as well as the likelihood of less than perfect graphite crystalline structure, with its probable regions of some amorphous carbon. There will be further industrial applications–in the use of pure carbon fibers (for example, in portland cement reinforcement), where the expense of achieving graphitic structures will serve no useful purposes. The reader will also find references to "hybrids" of fibers, where there will not only be mixtures of two or more totally different fibers, but also of C/G fibers of different heat treatments and processing.

PROPERTIES OF BULK GRAPHITE

Some of the more promising materials and processes for aerospace appeared in a 1965 report by the technical staff of Air Force Materials Laboratory.[2] Significant attention was given to graphite and graphite composites though emphasis and data concentrated upon blocks and cylinders of high heat resistant graphite with superior physical properties. The bulk graphites were prepared by hot working processes, which applied mechanical pressure to graphite heated to temperatures within the plastic range. Bulk densities up to 2.15 g/cm^3 were achieved. The combination of excellent erosion resistance and high with-grain strength made this graphite a leading candidate for rocket nozzle applications. A summary from this report of the high temperature properties of bulk graphite (ZTA) measuring 14 inches in diameter, is given in Table 2-1.

Table 2-1. High Temperature Properties of Bulk Graphite ZTA.[2]

	TEMPERATURE (°C)	WITH GRAIN (AVERAGE)	ACROSS GRAIN (AVERAGE)
Young's modulus (10^6 psi)	R.T.	2.99	0.74
	600	3.12	0.75
	1000	3.25	0.77
	1400	3.47	0.84
	2000	3.86	1.02
	2400	4.03	1.16
	2800	3.76	1.12
Tensile strength (psi)	R.T.	4405	1530
	1000	5090	1750
	1500	6270	1925
	2000	8540	2150
	2500	10075	2445
	2700	7660	2390
Flexural strength (psi)	R.T.	5605	2350
Compressive strength (psi)	R.T.	7085	12,600
CTE, 20–100°C (10^{-6}/°C)		0.62	9.00
Thermal Conductivity (cal/cm/sec/cm^2/°K)	R.T.	0.517	0.208
Specific Resistance (10^{-4}ohm-cm)	R.T.	6.97	21.87

While Table 2-1 is derived from data upon bulk graphite, and not carbon fibers, the data is obtained under temperature conditions where an oxidative atmosphere is not as drastic in effect as upon fine fibers. Furthermore, comparisons with data obtained upon fibers will prove useful.

In addition to the above information, a series of graphite-refractory composites were prepared containing additions of transition metal borides, carbides, and silicides. Among the more promising composites was a grade made from a mixture of graphite, zirconium diboride, and silica. This grade had a superior oxidation resistance at 2100°C, as determined by weight loss, as compared to the more common grades of graphite. This was attributed to the self-healing characteristics of the induced oxide coating. In addition, flexural strength (grade JTA) was 27,000 psi at 1500°C, as compared with 5605 psi reported for the ZTA bulk material.

The AFML 1965 report briefly commented upon the high strength to weight ratios, low thermal conductivity, and good thermal shock resistance of some of the early fibrous graphite composites.

PROPERTIES OF HIGH STRENGTH WHISKERS

High strength whiskers offer promise as very high strength, short fiber reinforcements because they are small (submicron to several microns in diameter) with a high degree of crystalline perfection. This quality arises from low dislocation density, low void content, low surface imperfections, and the absence of grain boundaries.[3] They are considered experimental, and other than Exxon's silicon carbide whiskers, are not available commercially. An earlier submicron diameter potassium titanate fiber ("Fybex" by Du Pont) was withdrawn from the market, as it did not have the desired super-strength and toughness. Highest tensile strengths and elastic moduli are realized in whiskers. Considerable encouragement was forthcoming from several government agencies, notably the Air Force Materials Laboratory at Wright Field.[4]

The theoretical and ultimate strength of candidate whiskers were reported by Sterry and Lachman for monocrystalline fibers or whiskers, which are not necessarily representative of commercial fibers.[5]

Table 2-2, showing the strength and moduli of certain whiskers, underscores the reasons for basic interest in carbon and related fibers. The above referenced paper presented significant data on some of the ceramic fibers, showing marked reduction in tensile strength of zirconia fibers as the size increased from 1 mil (25 microns) to 2, 3, and 4 mils. High strength carbon whiskers, measuring 1 to 5 microns, were grown in an electric arc under argon pressure. Authors noted that pyrolysis of organic fibers represented another promising technique.

High strength whiskers and their preparation appeared in a 1966 NASA report.[6] Table 1 of this report shows the physical properties of a number of inorganic materials with potentials as reinforcements. Unfortunately, a low strength carbon fiber was used in the comparative data. More recent data on graphite fibers with moduli of elasticity over 100×10^6 psi, would have placed this material at the top of the list in modulus to density ratio. Carbides of silicon and boron also demonstrate high modulus to weight ratios. This information warrants attention, because if impurities appear in the parent subject or fiber, or if modifications are introduced, they might as well be in a direction that would enhance performance. Absorbed, as well as internal, trace impurities have an

Table 2-2. Tensile Properties of Strong Whiskers.[5]

MATERIAL	MODULUS OF ELASTICITY (psi)	ULTIMATE TENSILE STRENGTH (psi)*
Silicon	23,000,000	550,000
Carbon	150,000,000	3,000,000
Iron	29,000,000	1,900,000
Quartz	11,000,000	600,000
Zirconia	62,000,000	600,000

Estimated tensile strength: $T_s = (2VE/A)^{1/2} \cong E/10$.

T_s = Estimated tensile strength (approximately $\frac{1}{10}E$)

V = Specific surface energy

E = Elastic modulus

A = Spacing between two planes

*Authors based this data on experimental values.

important role in the growth of whiskers. In some species, compositional differences at the whisker tip and side surfaces are theoretically and practically possible. The examination of the surfaces of these whiskers by optical and electron microscopy reveal dislocations, geometric patterns, localized defects due to impurities and voids, surface corrosion, oxidation, etc.

About the same time that the early NASA publication appeared on inorganic fibers and whiskers, there was another report by the Society of Automotive Engineers.[7] Steg reviewed some of the then available high strength whiskers and fibers in resin and metal matrices, showing a preference for the aluminum oxides over graphite in metal matrices, because of their better chemical inertness to the metal. It was also apparent in his report that not only the modulus/density of graphite whiskers was superior to all other candidate materials at that time, but that the specific modulus of C/G resin composites was the highest. In specific strength, boron/resin, aluminum reinforced with boron carbide whiskers, and aluminum reinforced with Al_2O_3 whiskers showed much promise. In the light of recent developments in advanced composites, one might be inclined to view these earlier reports as outdated. However, they contain early data which have influenced trends in advanced composite materials. Some of the relevant earlier data are given along with more recent observations to enable the reader to draw meaningful conclusions.

Schmidt and Hawkins[8] discussed early fibrous carbon developments with considerable foresight. The state of the art as known at that time (1965) was carefully reviewed. Also reviewed were silica coated carbon and graphite fibers which represented improvements over non-treated C/G fibers. A large improvement in oxidation resistance was reported. Pyrolitic graphite was deposited on C/G fibers to improve high temperature strength. Other refractory compounds imparted chemical and physical benefits to the fibers. Applications were principally to ablative heat shields for re-entry space vehicles.

Figure 2-1 shows the relationships between specific tensile strength and specific modulus for commercial filamentary materials. Glass fibers are strong and account for their prolific applications in many reinforced plastics. The C/G fibers owe their structural acceptance to their stiffness qualities (high elastic modulus), though these advantages are not necessarily accompanied by an increase in strength. As the strength properties of advanced composites are examined in a later chapter, it will be noted that the higher moduli C/G fibers suffer loss in interlaminar shear strength, necessitating greater attention to preparation and treatment of fiber surfaces.

A good status review by Dr. Economy[9] called attention to commercially available high strength fibers. Attention was directed to the advantages of boron nitride, as an experimental high strength fiber. Data in Table 2-3 were taken in part from his paper. The graphite fibers were from polyacrylonitrile (PAN) precursors.

It should be noted that, during the early 1960's, the necessity was recognized

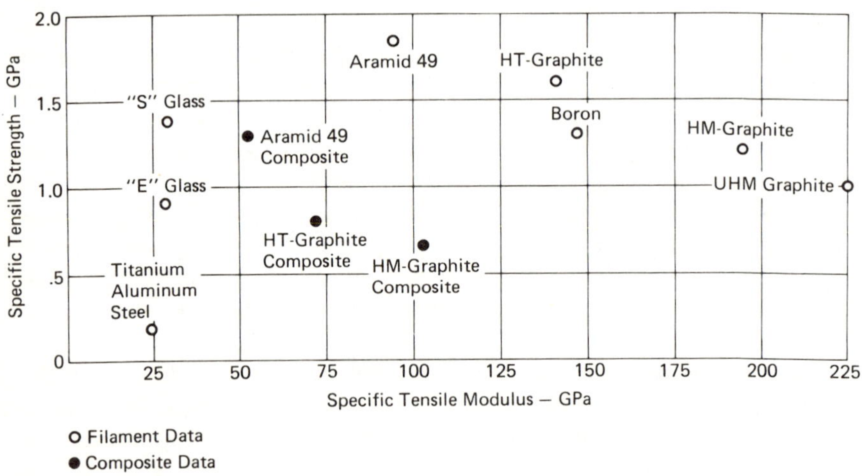

Fig. 2-1. Commercial filaments and composites relationships between specific tensile and specific modulus properties.

Table 2-3. Commercially Available High Strength Fibers.

FIBER TYPE	DENSITY (g/cc)	TENSILE STRENGTH (ksi)	(MPa)	TENSILE MODULUS (msi)	(GPa)
Boron (4 mils)*	2.65	495	3410	56	386
Boron (8 mils)*	2.38	540	3720	56	386
Graphite (A-S)	1.75	410	2820	30	208
Graphite (HT-S)	1.77	410	2820	34–37	234–253
Graphite (HM-S)	1.91	340	2340	50–55	345–375
Graphite (UHM-S) (GY-70)	1.96	270	1860	70–75	485–517
Aramid ("Kevlar") 49	1.45	400	2760	20	138
Glass (E)	2.54	500	3440	10.5	72.5

*Tungsten core.

to understand micromechanics of composites and failure modes. Also acknowledged were the need for reliable engineering data and design specifications, the importance of fabricating composites with predictable properties, and the necessity for procedures for adequate joining techniques. At the time of writing this book, good engineering data is being developed, as well as the influence of environmental extremes on service performance (subjects of subsequent chapters).

CHARACTERISTICS OF C/G CRYSTALS

In the graphitization of carbon, a layered hexagonal crystal structure is formed under applied heat and/or pressure (see Fig. 2-2, which illustrates the hexagonal structure of crystalline graphite and the ABABA layering, and the *c* and *a* axes). Bernal [10] proposed the stable hexagonal lattice as the structure of ideal graphite. A system of infinite layers of fused hexagons was visualized. Within each layer, the C-C bond distance has been determined as 1.415 Angstroms (Å). X-ray diffraction has accurately determined the spacing of the parallel layers as 3.3538 Å.[11]

Pauling discusses the structure of the graphite crystal as consisting of hexagonal layers of molecules separated by a distance so large that there can be no covalent bonds between them. He visualized each of the layers as a giant molecule, and the superimposed layer molecules held together by weak van der Waals forces. The four valences of each carbon atom are used to form bonds with its three neighbors; the later molecule resonates among many valence-bond structures, so that each C–C bond achieves one-third double bond character.[12]

Ideal monocrystalline graphite is composed of planar layers of hexagonal networks stacked in either an ABAB sequence (hexagonal lattice) or in an ABCABC

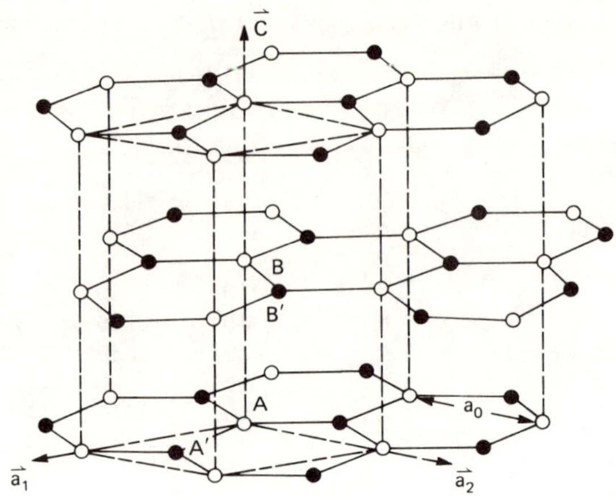

Fig. 2-2. Hexagonal structure of crystalline graphite showing ABABA layering, c and a axes.[10]

sequence (rhombohedral). The planar networks are comprised of hexagonal rings of carbon atoms bonded by the stabilizing, hybrid sp^2=pi bonds (1.415Å spacing) according to Ubbeholde and Lewis.[13] In the diamond structure, the carbon atoms are tetrahedrally bonded with sp^3 bonds (1.555 Å) in a nonplanar hexagonal network. The *c*-axis interlayer separation determined by X-ray diffraction is 3.3538 Å for the hexagonal or graphite lattice. Theoretical density for the ideal structure is 2.265 g/ml. The hexagonal lattice is considered more stable, and the rhombohedral form is converted to the hexagonal at 2000° to 3000°C. Edge defects which result from incomplete C-C bonding establish finite limits to the layers. They may form bonds with H, OH, =O, or other foreign atoms at the edge defects. X-ray diffraction also shows a high degree of three dimensional crystalline order in graphite whiskers, which are essentially crystals.

Processes for the formation of various carbons from solid or liquid hydrocarbons have been studied for many years.[14] In a recent publication, Woollman[15] described a number of ordered and partially ordered carbons, including the non-graphitizable glassy or vitreous carbons which do not achieve overall alignment by heating.

The distinction between carbon fibers and graphite fibers is primarily one of degree of crystalline order. Physical differences, such as density of fibers and elastic moduli, will also be apparent. Graphite fibers may exhibit more gloss or sheen than the amorphous carbon fibers–though sizing agents will also have an influence. The original fiber precursors will have an influence on the ordering of the carbons in the graphite fiber, as well as tensile stresses on the fiber during graphitization. Graphite fibers do not comprise 100% pure graphite crystals and retain a certain amount of the amorphous form of carbon.

C/G fibers are in fact heterogeneous materials which reflect on the precursors used for their preparation, as well as the heat treating processes employed in their manufacture. This was best expressed by Tokarsky and Diefendorf[16] in a recent paper: "A carbon fiber must not be thought of as a homogeneous anisotropic entity, but rather as a collection of anisotropic units linked such that the fiber may be viewed as a composite system. The moduli of the individual fibers will depend on the details of the axial preferred orientation in each fiber structure while the strength will be a function not only of surface and volume flaws, but of the axial as well as radial textures and gradients present." It was also pointed out that tensile strengths of 7 to 8 GPa (over 1,000,000 psi) were obtained on C/G fibers when data was extrapolated down to very low test gage length.

ORGANIC FIBER PRECURSORS FOR C/G FIBERS

The decision to utilize the cellulosic fibers as precursors for carbon fibers was an important step forward in the technology of this industry. They were available as continuous filaments possessing high initial strength. While cotton, cellulose acetate, flax, and hemp were carbonized with varying degrees of success, commercial interest focused upon the regenerated cellulose from viscose rayon. In the middle 1960's, commercial carbon fibers from Union Carbide, such as "Thornel" 25, 50, and 75, became available and hence appear in early reports on the advanced composites. The C/G fibers from PAN precursors have received more acceptance in recent years. In fact, while stocks of carbon fibers from rayon precursors may still be available, new interest in structures have favored some of the higher modulus fibers from PAN precursors and the more recent pitch fibers.

Cross-sections of some of the carbon fibers will give a clue as to their origin. The crenulated circumference of some carbon fibers may well reflect very rapid drying of their organic fiber precursor during the wet spinning process. Cross-sections in general vary from circular to bilobal to oval shape, reflecting the rate of removal of volatiles during the spinning stage. C/G fibers from pitch have more uniform cross-sections, as the pitch products have had significant processing to remove volatile matter before the extrusion operation—yielding fibers of higher density and carbon content.

Most of the commercial C/G fibers fall in the range of 6 to 11 microns in diameter, with the pitch fibers averaging 10 microns. There has not been sufficient availability of assorted diameters of good quality carbon fibers to determine optimum size of fibrous reinforcement. There is good reason to speculate that even smaller diameters will yield better physical properties in the composites—though the manufacturing costs would increase significantly in producing such fibers. For record purposes, the statistical diameter profile of a

Table 2-4. Diameter Distribution Statistics for High Modulus Samples (HMS).

CARBON FIBER SAMPLE	
Number of measurements	500
Mean (microns)	6.89
STD Ser.	0.24
% C.V.	3.5

FREQUENCY DISTRIBUTION

INTERVAL (MICRONS)	FREQUENCY
4.00–5.99	23
6.00–7.99	471
8.00–9.99	5
10.00–11.99	1

typical batch of PAN graphite fibers performed for Delsen Laboratories (Glendale, CA) is shown in Table 2-4.

Data shown on the distribution of diameter were taken on a sample of high modulus C/G fiber from a PAN precursor, using a procedure developed by Barker.[17] Short sections of fibers were cut at random, distributed onto microscope slides, and scanned with the image analyzer.

As the data demonstrate, the distribution of fiber diameters was very narrow, indicating that the variability among the diameters of the carbon fibers was small. Data have also been reported for pitch-based carbon fibers (VSB-32-0).[18] Characteristics for three lots of VSB-32-0 pitch fibers were:

LOT NUMBER	TENSILE STRENGTH ($\times 10^{-3}$ psi)	YOUNG'S MODULUS ($\times 10^{-6}$ psi)	MEAN DIAMETER (MICRONS)	STANDARD DEVIATION (MICRONS)
107F	198.6	56.9	10.27	0.68
110F	184.9	50.6	10.39	0.73
112F	184.1	55.0	10.18	1.03

Figure 2-3 depicts the percentage frequency distribution of diameters of pitch-based carbon fibers of Lot 110F, showing a slightly wider variability than reported for the PAN precursor fibers illustrated above.

As brought out in the earlier review of process patents for the manufacture of carbon fibers, pyrolysis of regenerated cellulose is generally carried out in an inert (nitrogen) atmosphere in the absence of air. As the starting fibrous material is heated, thermal decomposition commences in the 300° to 400°F range. Occluded gases and water vapor are liberated, as the fiber shrinks in dimension.

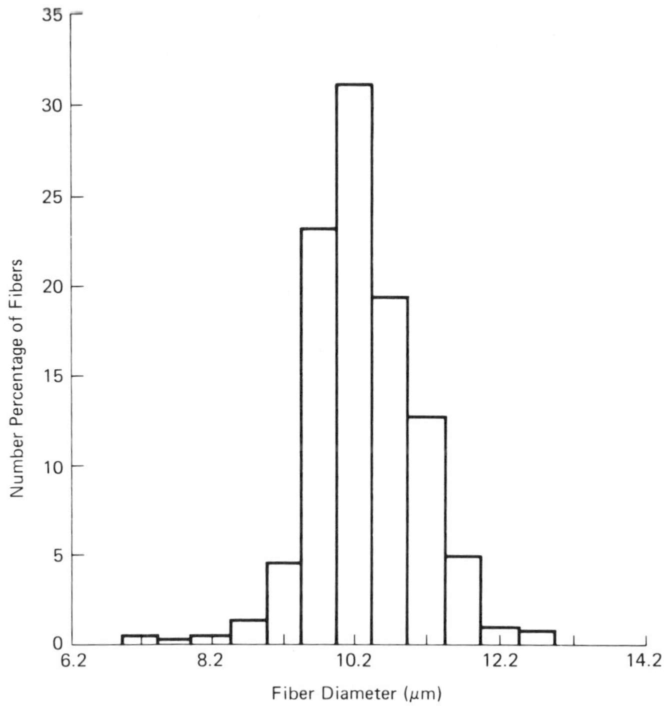

Fig. 2-3. Frequency distribution of diameters of pitch-based carbon fibers, VSB-32-0.[18]

Subsequent thermal degradation involves some chain scission, and at the same time chain condensation reactions in which broken side chains are reformed. Further heating produces a carbonaceous char. The amount formed depends upon the rate of polymer degradation, the presence of foreign substances in the product, and the heating chamber environment and the degree of cross-linking.

Slow carbonization has the advantage of increasing the carbon yield, though it is counterproductive to high production rates. As noted in the patent abstracts, this problem receives considerable attention. As a consequence, variations in the carbonization process result in carbon fibers having widely different properties. As the basics are better understood, and they have been in more recent years, fiber uniformity has been considerably better. It is appropriate to point out that earlier test results are not always easy to compare with later findings, because of some question on the origin and processing of carbon fibers.

The preparation of graphite fibers requires an additional thermal treatment in the 2500° to 3000°C, range. During graphitization, there may be shrinkage of

3 to 6%; a typical weight loss of 5 to 15% and an increase in the density of the fiber though the amount of change depends on precursor and process. By the time graphitization is completed, virtually all of the original oxygen and hydrogen have been eliminated, and the final carbon is over 99.0%. A knowledge of the presence and nature of the functional groups on the surface of the final fiber is useful in choosing appropriate matrix materials and determining surface modification procedures for enhancing fiber-matrix interactions. Samples were heated in vacuum at linear temperature increases of 0.5° to 2.5°C/minute from 135 to 1500°C. Gas chromatography was designed to separate and measure major gases evolved. The pyrolytic volatile analyses of different carbon types by Perkins, Smith, and Rivin[19] are shown in Table 2-5.

Studies on the graphitization of carbon fibers from PAN precursors have examined crystallite orientation through measurements of magneto-resistance. It has also been observed that carbon fibers bonded with furfuryl alcohol resins will yield graphite components from the resin matrix with uniaxial crystallite distributions parallel to the fiber axis.[20]

The fibers prepared from synthetic organic polymers have received considerable attention as possible precursors for carbon fibers. The techniques have usually involved "setting" the fibers under oxidative conditions at some moderately low temperatures to convert thermoplastic (meltable) fibers into thermosets, leaving blackened, unidirectional tow of filaments or a woven fabric. This has been followed by high temperature carbonization in an inert atmosphere, followed by exposure to still higher temperatures, achieving graphitization. Among the polymer fibers reported in the literature as having been studied as precursors for carbon fibers are aromatic polyamides, polyvinyl

Table 2-5. Pyrolytic Volatile Analyses of Carbon Types.[19]

CLASS	TYPE	CO_2	CO (MICROMOLES/g C)	H_2
Pyrolyzed coal	Austin Black	113	980	19,940
Carbon blacks	Channel	1130	4504	2098
	Thermal A	29	47	2520
	Furnace	248	372	884
	Acetylene	81	156	372
Carbon fibers	Kureha KCF	462	885	1403
	Hercules HM-S	30	66	101
	Morganite II	64	34	141
	Morganite I	43	4	63
	Thornel 50	45	32	106

alcohol, polyphenylenes, polyacetylenes based on diethynyl benzene, benzamidazole, polyvinyl chloride, polyoxadiazoles, and others. No dramatic advantages have been apparent so far from these fibers over those from PAN copolymers or regenerated cellulose. Hence, continued research has concentrated upon the PAN precursors which at present are dominating carbon fiber composite applications in the aerospace industry.

In the first chapter, the carbonization of rayon fibers was described in U.S. Patents 3,305, 315 and 3, 235,323. Typical processing details for carbonization of rayon appear in Fig. 2-4, taken from U.S. Patent 3,305,315. In order to reduce the time of exposure to the oxidizing atmosphere, chemical treatments of the fiber before entering the oxidizing furnace have been known to help. For example, immersion in 7% by volume of phosphoric acid-denatured alcohol solution for woven fabrics and 3% phosphoric acid/ethanol for filaments, will permit lower temperature thermal treatments at 100° to 350°C than at the 180° to 350°C required without acid pretreatment. However, as discussed later in this

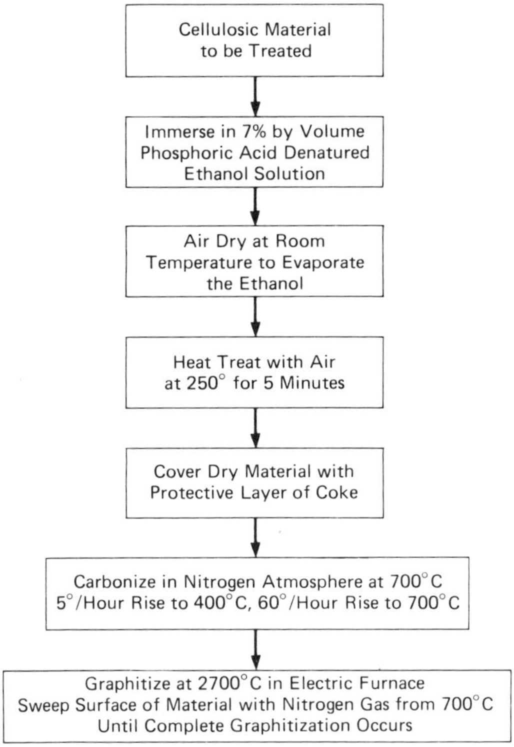

Fig. 2-4. Flowchart for bulk processing of carbon fiber textiles. (U.S. Patent 3,305,315.)

chapter, there is a growing pre-occupation with elimination of traces of ionic impurities because of their deleterious effects on high temperature performance, and some of the earlier art may be obsoleted. Heat treatment of the cellulosic fibers in an oxidizing atmosphere will drive off volatiles to yield a permanently dehydrated material. This product is substantially nonflammable, non-fusible, strong, and flexible, ready for further processing. When no acid pretreatment is used, the time to convert a cellulosic textile material to a heat treated material in an oxidizing atmosphere is as follows:

At 180°C–time in excess of 48 hours
At 250°C–about 10 hours
At 350°C–less than 1 minute.

From patent 3,305,315, Bacon *et al.* offer the following data:

TYPICAL VALUES AFTER HEAT TREATMENT (RAYON PRECURSORS)

TREATMENT	PERCENT VOLATILES	PERCENT LINEAR SHRINKAGE	TEMPERATURE FOR 1% WEIGHT LOSS IN 8 HOURS
Heat treatment (150° to 300°C) dehydration	50	18	100°C
Carbonization in inert gas (900° to 1200°C)	15	6	350°C
Graphitization in inert gas (2000°C)	0	0	500°C

Carbon fibers based upon a rayon precursor are irregular in appearance (crenulated along the circumference). Individual fibers, depending upon the spinning orifice and drying rates for wet spinning processes, may cover a broad range for several products (5 to 50 microns in diameter). The structures of PAN- and cellulosic (rayon)-based carbon fibers were investigated optically using various microscopic techniques. Commercial PAN fibers were circumferentially oriented on the outside and radially oriented on the inside. "Thornel" 50 fibers (rayon presursor) have oriented crystallites parallel to the fiber surface with the structure penetrating further toward the fiber center in high modulus fibers.[21] Some of the PAN based carbon fibers are bilobal in cross-section (see, for example, the fibers of Great Lakes Carbon Co.).

Novel methods for achieving graphitization temperatures have included passing the carbonaceous fibers between spaced electrical contacts (in an inert gas), which carry electrical current through the carbonaceous fibers to raise their temperature (see Cranch, U.S. patent 3,313, 597). Prospects of breakage of fibers and non-uniform heating are increased. By applying stress to the yarn during a final high temperature processing, the resultant carbon strand developed a high modulus and high strength capability. Representative of the first high modulus C/G fibers was "Thornel" 50, from regenerated cellulose. Typical properties are tensile strength 315,000 psi and modulus of elasticity 57×10^6 psi for the graphite fiber.

PAN FIBER PRECURSORS

The PAN fiber precursors have become the most important source for C/G fibers. The production of acrylonitrile reached about one billion (10^9) pounds in the United States in 1978, and about one-half of this went into the production of fibers of PAN and its copolymers. Most of this material found its way into textile products. A special product is generally used as the starting point for C/G fiber production. In the United States, the PAN fibers are produced principally by E.I. du Pont de Nemours Co. under the tradename of "Orlon," while in England, Courtaulds has been the principal supplier, and in Japan, Toray has produced most of this fiber. The fact that a preference has been shown by the principal suppliers of C/G fibers in the United States for the PAN precursors from overseas sources (Hercules has close ties with the English sources and Celanese and Union Carbide have close ties with Japanese sources) suggests that there are technical differences in the PAN fibers that are advantageous to the production of C/G fibers, or that production techniques developed about one fiber source are not readily transferable to fibers from another source. The fact that major suppliers in the United States have declared their intentions to invest in their own facilities in this country within the next few years indicates a changing picture. In reviewing the activity in PAN precursors, there are references to studies of various copolymers of PAN in an attempt to improve processing and enhance the performance of the ultimate C/G fibers in resin composites.

Fitzer[22] reported on the influence of copolymer content of the PAN precursor in the range of 1 to 5% methyl acrylate, using systematic studies of fibers prepared with varied copolymer content and consistent fabrication conditions. Length changes of the PAN precursors during oxidation and during carbonization were shown as dependent upon load on the fiber, as well as upon copolymer content. Improvements in modulus and tensile strength of carbon fibers were illustrated in a series of curves.

Processing of PAN fibers beyond the requirements of ordinary textile practices began with the work of Houtz.[23] Oxidation of PAN fibers under controlled temperatures of 220° to 250°C for several hours rendered them infusible (see also the U.S. patents 2,789,915; 2,913,802, Barrett; 3,027,222, Wilkinson; and 3,125,404, Crawley, abstracted in the first chapter). The fibers acquired stability by oxidative cross-linking to enable them to undergo processing at higher temperatures.

Oxidized PAN fibers were converted commercially to carbon fibers in the early 1960's, as reported by Shindo.[24] Details were revealed in the Tsunoda U.S. patent 3,285,696 (1966), which covered copolymers of PAN, which served as the principal polymer (over 85 mole percent); and two Japanese patents: Showa 37-4405 (1962) and Showa 38-12375 (1963). These were licensed to Tokai Electrode Co. and the Nippon Carbon Co.

In was shown that direct heating of fibers of PAN and some of its copolymers at 1000°C in an inert atmosphere (nitrogen) resulted in a brittle, low strength product. However, with the prior oxidation step noted above at 220° to 225°C, the way was paved for a tough, much stronger fiber when processed further at 1000°C (see also the German patent 1,923,622, February 12, 1970, which described commercial processes that heated PAN fibers for 7 hours at 220°C to promote complete diffusion of the oxygen into the materials, followed by carbonization for 4 hours at 1000°C). Stabilized acrylic fibers are very hygroscopic and rapidly equilibriate with a gain in moisture of about 8%, and concurrently elongation at about 1.5%. Data has been shown relating the gain in moisture with the time of oxidative heat treatment at 255°C.[25] After 2 hours at 255°C, the moisture gain was 6%, and after 4 hours, the gain in moisture was 8%.

Further improvements were attained by applying various stresses at different stages of the process to align the PAN molecules, orienting the fibers and preventing spatial disorganization and weakening caused by shrinkage and release of volatile substances during heat treatment. Carbon and graphite fibers of improved strength and higher modulus were realized. It was also noted that by limiting the heat treatment temperature, there were improvements in tensile strength in some instances, though usually at a sacrifice of modulus of elasticity. Prior stretching of the yarn in a steam chamber prior to oxidation also led to significant improvements according to another process by Great Lakes Carbon Corp. (U.S. patent 3,533,743). Good quality, high modulus C/G fibers were prepared in this manner.

PAN precursors possessing a continuous "backbone" of carbon atoms may be stretched in the early stages of heating, because strengths are markedly lowered by pyrolysis. With rayon precursors, the necessity for removing oxygen linkages in the polymer chain precludes early application of tension. During pyrolysis, though, they may be stretched during latter stages of heating.

The successful conversion of PAN fibers to high strength, high modulus fibers depends in part upon understanding of the thermal and oxidative reactions. Gaulin and McDonald[26] abandoned the theory of intramolecular cyclization in favor of a series of elimination reactions, followed by intermolecular cross-linking and cyclization, induced by dipole interactions between pendant nitrile groups. The full potential of high modulus C/G fibers is achieved by changes in the surface morphology (surface area, roughness, pore size distribution, etc.) or chemical functionality. These studies are significant to the enhancement of adhesion, and will be discussed in the chapter on surface treatments. PAN precursor fibers consist of carbonized fibrils (250 to 1000 Å). Each carbon fibril consists of turbostatic graphite crystallites about 50 Å in size, arranged with their basal planes oriented along the axis of filaments. This degree of orientation determines modulus. Modulus improves during the course of heat treatment, so that when filaments have a modulus of 60 msi, the mean deviation angle between the basal plane and the filament axis is only 5°.

While on the subject of the graphitization of carbon and graphite fibers, the work of Ezekiel should be noted. In his research upon the high tensile strength and high modulus graphite fibers from synthetic organic polymers and high molecular weight natural pitch, he observed the effect of the presence of boron upon the fibers during graphitization. Not only were higher values of modulus obtained, but for some precursors, such as the acrylonitrile polymers, graphitization is lowered by about 250° to 350°C (see Fig. 2-5) to obtain the same modulus attained without boron, or, at a specific temperature, a higher modulus is obtained.[27] The boron appears to facilitate bond migration during graphitization and the formation of a graphite network at lower temperatures than those normally required. PAN fibers which had been stabilized by isothermal air oxidation for several hours at 275°C were wound upon tubing under constant tension (100 g) and then treated with an aqueous solution (5% H_3BO_3) of boric acid for 26 hours. The attainment of moduli of elasticity in excess of 100 million psi after graphitization is quite exciting. Earlier studies on the doping of bulk graphite with boron have also been reported.[28] Boron was diffused into specimens placed in boron powder in an argon atmosphere at 2900°C. Changes in mechanical properties of the bulk graphite were relatively small.

Deposition of boron during the graphitization of carbon fibers increases both the strength and modulus, according to a more recent report.[29] An atmosphere consisting of argon/hydrogen/methane/triethylborane was used to deposit borocarbon coatings on the fibers using an arc plasma furnace. The healing of surface and internal flaws was thought to be the major factor giving strength to improvements.

In a further reference on preparation of high modulus graphite fibers from PAN, Ezekiel[30] noted that a critical step in processing the PAN yarn was the

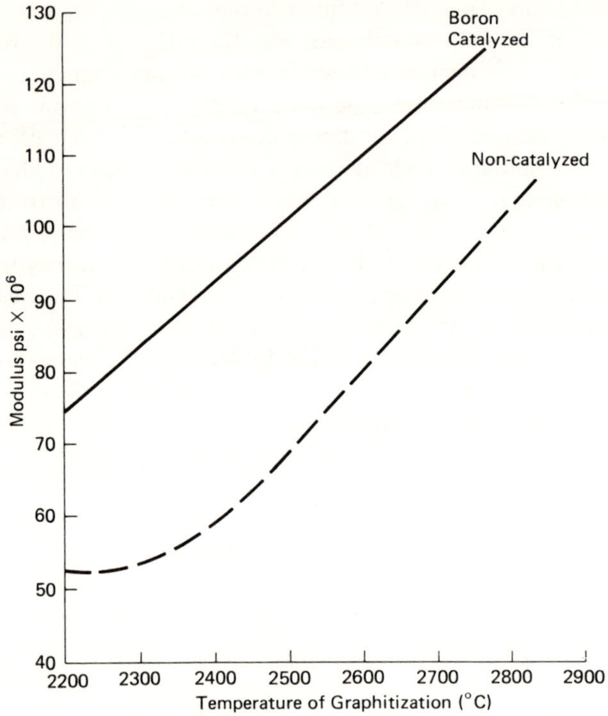

Fig. 2-5. Temperature of graphitization versus modulus.[27]

complete removal of the spinning size agent prior to the stabilization oxidative treatment.

In the preparation of high tensile strength carbon fibers, Toray (Japan) oxidizes the PAN precursor by contacting the fibers intermittently (less than 1 second) on multiple heated rolls at a surface temperature range of 200° to 400°C.[31] The temperature of oxidizing atmosphere is lower than heated rolls. The oxidized fiber is then carbonized in a non-oxidizing atmosphere above 800°C. In another process modification by Toho Belson (Japan), preoxidized PAN fibers are fed continuously through a double chamber. The first chamber is vertical and operating at the 500° to 1000°C range. To the inert gases are added the decomposition products of higher temperature processing, taking place in a second horizontal chamber at 800° to 2000°C (British patent 1,531,915, November 15, 1978).

The relationships of processing temperatures to the properties of C/G fibers are shown in Fig. 2-6. While there are a number of process variables which affect

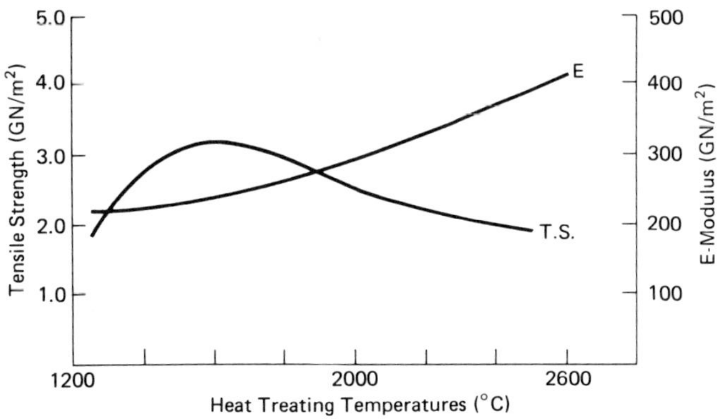

Fig. 2-6. Typical influence of processing temperatures on properties of C/G fibers.

the actual curve, the trend towards lower tensile strength and higher moduli at the higher temperatures of processing are clearly indicated.

British patent 1,032,542, granted to Union Carbide in 1966, described a process for the continuous graphitizing of textile threads by stretching at a temperature of at least 2200°C. British patent 1,110,791 emphasized the use of PAN precursors and the advantages gained by stretching during oxidative treatment in attaining high modulus after graphitization. Internal flaws often originate in the organic precursor, though they are amplified during the release of volatiles during heat treatment—a point to be noted because of the necessity of a controlled time and temperature procedure to minimize defects in the fiber. The presence of such flaws detracts from theoretical strengths which may be realized in the fibers. Based upon 140×10^6 psi modulus obtained in graphite whiskers, a level of tensile strength of one-tenth of this figure has been anticipated.

It has been shown that the tensile strength of PAN-based fibers increases with decreasing gage length—due to reduction of available flaws in the gage length, which lower tensile strength. Diefendorf[32] has extrapolated such data and concluded that zero flaw carbon fibers would display tensile strengths in the range of 1.5×10^6 psi. Strength limiting features seem to arise from stresses associated with the decrease in crystallite orientation from the surface of the fiber to the core. Consequently, there are differences in coefficient of thermal expansion between the material forming the core and the material forming the surface, creating stresses during the cooling process. It should also be recognized that inasmuch as the oxidation of the PAN fibers is diffusion controlled, the surface materials see a different oxidizing environment than the core material. Thus, the carbonization and graphitization results at the fiber surface and the core would be expected to be different.

Indicative of the research upon the synthesis of modified polyacrylonitriles and their conversion into carbon fibers is the report issuing from the Royal Aircraft Establishment in England.[33] The process developed at RAE for the conversion of PAN fibers into high modulus carbon fibers requires an initial oxidation of the PAN fibers in air between 200° and 300°C, followed by carbonization and then graphitization in an inert atmosphere up to 3000°C. The rate of oxidation of PAN is diffusion controlled and the time required is proportional to the square of the fiber diameter (at 220°C, it is about 5 hours for 13 micron diameter fibers). In developing copolymers of PAN with acrylic acid and with acrylamide, attempts were made to synthesize soluble, fiber forming, oxygen containing acrylonitrile copolymers which could be converted by heat treatment below 300°C into structures similar to oxidized PAN. In addition, attempts were made to speed up the cyclization process through appropriate chemical modifications. Solutions of a 50% aqueous sodium thiocyanate were prepared for spinning the copolymers. As indicated later in this chapter, the use of sodium salts is now discouraged, because of adverse influence of sodium ions on high operating temperatures of C/G fibers. Fibers spun from a stainless steel spinneret nozzle into a coagulation bath were stretched in a steam tube to align molecules along the fiber axis prior to oxidation.[34] Great Lakes Carbon Corp. also discussed steam stretching of PAN precursors in their U.S. patent 3,533,743 (see Fig. 2-7).

Acrylonitrile/acrylic acid (8 to 1) copolymer fibers 20 microns in diameter gave carbon fibers with a modulus of 25×10^6 psi (172 GPa) after 2 hours

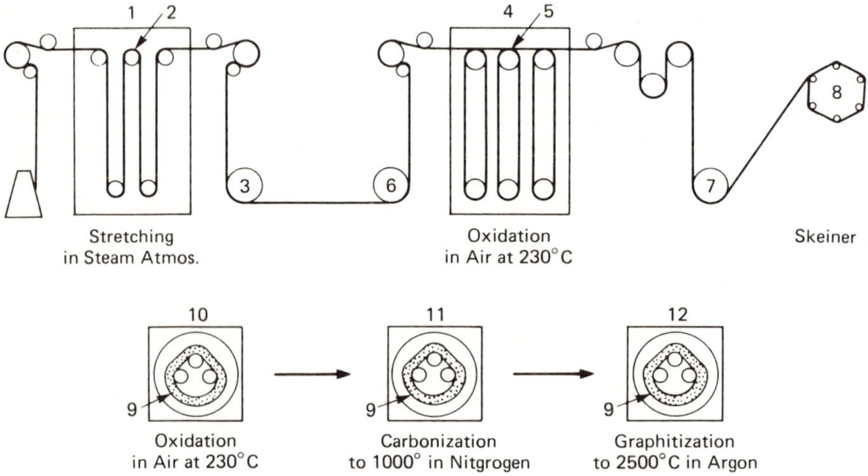

Fig. 2-7. Processing of PAN fibers according to U.S. 3,533,743 patent.

oxidation at 220°C and carbonizing to 1000°C, only 7% lower than the values obtained upon the carbonization of oxidized commercial PAN fibers. This represents a significant reduction of processing time of the carbon fibers from PAN. Celanese Corp. reports that, in developing its high modulus graphite fibers from PAN precursors, graphitization temperatures in excess of 3000°C are used. Elastic moduli of 70 to 75 MSI are obtained on the filaments, which have a density of about 1.96 g/cc (GY-70 fibers).

Some good data on GY-70 high modulus fibers in epoxy molding compounds have been reported by Kalnin.[35] Using these fibers at 40% volume in an epoxy molding compound, he established the influence of fiber length on physical properties as shown in Table 2-6 (compounds were solvent blended).

Table 2-6. Effect of GY-70 Graphite Fiber Length on Mechanical Properties of Molding Compound.[35]

		FLEXURAL		
FIBER LENGTH (inches)	BULK DENSITY (g/cc)	STRENGTH (ksi)	MODULUS (msi)	IZOD IMPACT (ft • lb/inch notch)
1/4	1.49	14.4	3.1	0.66
1/2	1.48	18.3	4.1	0.56
1	1.48	19.3	4.9	0.50

PITCH PRECURSORS FOR C/G FIBERS

It is generally recognized that, while ordinary pitch may yield carbon fibers by extrusion processes, these fibers are usually low in strength and low in elastic modulus. Even when they are carbonized to high temperatures, the increase in strength and modulus does not take place, unless there is hot stretching at very high temperatures, between 2700° and 3000°C, a very costly and impractical process. As pointed out, the ordinary pitches are isotropic in character, and new processes had to be developed to convert ordinary pitch into a mesophase or nematic liquid crystal.[36] Union Carbide Corp. has developed both coal- and petroleum-based precursors as satisfactory materials for the preparation of high modulus fibers. Commercial pitches are complex mixtures of aromatic compounds which, when heated to temperatures of 350°C or higher, undergo dehydrogenation condensation reactions to form planar aromatic molecules which aggregate into a liquid crystalline phase (mesophase). When the mesophase content reaches about 40%, a phase inversion occurs and this anisotropic material becomes the continuous phase. Further details are discussed in the basic Singer patent.[37]

The utilization of low cost commercial pitches without the costly stretching of the fiber at high temperatures was reported by Singer *et al.*[38] While isotropic carbon fibers had been produced for several years prior to 1975 by Otani and coworkers in Japan, the efforts of Singer *et al.* led to the preparation of high modulus C/G fibers. The degree of preferred orientation obtained in raw mesophase pitch fibers is high, as shown by the X-ray diffraction patterns of collimated bundles of as-spun pitch fibers. The influence of the pyridine insoluble components derived from aromatic pyrolysis tars upon carbon fiber quality is brought out in the subject paper. High strength carbon fibers from mesophase pitch have also been reported by Barr and Didchenko in the order of 4 to 5 GPa at low gage lengths for tensile strength.

Improved techniques have been developed for converting the mesophase pitch to high modulus graphite fibers without sacrifice of tensile strength. Barr *et al.*[39] pointed out that a defect-dependent mechanism can be recognized by an inverse tensile strength-gage length for carbon fibers produced from mesophase pitch (also previously reported for PAN-based fibers). Significant improvements in strength have been demonstrated for multifilament carbon fibers spun and processed on production equipment by taking care to avoid contamination of the pitch.

When discussions of low cost carbon fibers ensue, the usual reaction is to look upon pitch as a low cost raw material—and it is. However, the refinements required in processing are costly and the quality control measures are sophisticated, with the consequence that processing costs far outstrip raw materials costs. It is estimated that, as full industrial potentials are realized, the C/G fibers from pitches will be significantly lower (say 30%) than those from PAN precursors, though this subject is pure speculation at this time.

The utilization of pitch in carbon and graphite products has been evident for many years, as one examines developments in carbon products, such as graphite electrodes, electrical resistors, anodes for cathodic protection of underground pipe lines, moderators in nuclear power plants, and packing and gasketing materials. The new technology in preparing C/G fibers of high strength and high modulus has appeared only within the last ten years. The reader can obtain more complete information on the patent references noted herewith by examining the patent abstracts in Chapter 1. Otani (U.S. patent 3,392,216) discloses heating of organic materials in an inert atmosphere at 300° to 500°C, extruding them under pressure to shape the fibers, exposing them to an oxidizing atmosphere prior to the final carbonization at higher temperatures. He discussed the application to polyvinyl chloride powders and PAN, as well as coal pitch and petroleum pitch. Values of 230,000, 190,000, and 105,000 psi tensile strength were reported for fibers of 8.5, 10, 15 microns diameter, respectively. Applications of pitch-based carbon fibers by Kureha Chemicals in Japan were discussed by Okuda.[40] General purpose grades for 1) heat insulation for vacuum furnaces, 2) packings for pumps, and 3) reinforcements for thermoplastic engineering plastics were described. The author felt that sooner or later asbestos will be replaced by carbon fibers and, furthermore, that carbon fibers demonstrated better resistance to alkali attack in portland cement than presently used glass fibers.

Although recognition of coal tar pitch as a source of carbon fibers has been made by manufacturers of pitch-based carbon fibers, Dr. Romey has concentrated on this raw material and has reported on some promising results in recent publications.[41, 42] In early work, he utilized cleaned pitch extract, whereby tar pitches from bituminous coal are solubilized in quinoline to remove insoluble components and inorganic matter. After the solvent removal, the cleaned pitch is used to produce carbon fibers.

During pressure gasification of coal, char tars with high solids content are obtained in the temperature range of 500° to 600°C. After separation of insoluble constituents in quinoline, the base products have proven suitable for making carbon fibers. Fibers produced at a spinning temperature of 230°C were in the range of 5 to 20 microns in diameter. They were dusted with activated carbon to minimize sticking during spinning. Oxidation in air up to 400°C was followed by subsequent carbonization in nitrogen. Char tars oxidized by air blasting yielded the highest tensile strength carbon fibers in the range of 130,000 to 150,000 psi. More recent work has led to the removal of quinoline insolubles from hard coal tar pitches without the use of solvents.

A patent of Otani (U.S. patent 3,629,379) specifically covered the production of carbon filaments from low-priced pitch. He recognized the importance of the starting material possessing the capability of easy melt-spinning and its adaptability to subsequent oxidation to achieve infusibility. While there are many sources of carbonaceous materials from asphalt, petroleum pitches, and coal tar pitches, which permit ease of spinnability, infusibility treatment after fiber formation is difficult—depending upon the origin and pretreatment of the material. Selection focused upon materials which had a carbon content of 92 to 96% and a mean molecular weight of 400 to 2000. Pitches which did not lend themselves to infusibility of their extruded fibers, after oxidation, were subjected to dry distillation by bubbling nitrogen gas through the pitch at 380°C and applying vacuum to remove volatiles and low molecular weight fractions in order to achieve the molecular weight range and desired carbon content. Filament strengths of 60,000 to 130,000 psi were reported after carbonization of the fibers, in several examples.

As further interest developed for the low cost carbon fibers from pitch, attention was given to identification of the mesophase content of the pitch, and the rheology of the fibers. Singer (U.S. patent 4,005,183) used pitches containing 45 to 65% mesophase content by weight (liquid crystal or so-called mesophase state). They exhibit vicosities of 30 to 60 poises at temperatures of 340° to 380°C. The fibers which are extruded at rates of 100 up to 1000 ft/minute will have diameters of 10 to 20 microns. The pitch must be non-thixotropic and exhibit Newtonian or plastic flow during the spinning of the fibers. Fibers having a carbon content greater than 98% are obtained by carbonization in an inert atmosphere at temperatures in excess of 1000°C (usually 1500° to 1700°C for 1 to 5 minutes). Carbon fibers produced in this manner have a highly oriented structure characterized by the presence of carbon crystallites aligned parallel to the fiber axis. Further graphitization at temperatures above 2800°C develops the three dimensional order characteristic of polycrystalline graphite. Typical characteristics are shown in Table 2-7 (data from Singer patent).

Some work has been reported on classifying carbon fibers from mesophase pitch (PM), fibers from isotropic pitch (P), and fibers from a PAN precursor. Fischbach and Gilbert[43] suggested that the order of magnetic behavior indicated the order of graphitizability of the precursor, listing PM, PAN, and P fibers in that order.

Table 2-7. Characteristics of Carbon Fiber Filaments From Mesophase Pitch.[37]

	AFTER CARBONIZATION (1500° to 1700°C)	AFTER GRAPHITIZATION (2800°C)
Interlayer spacing of carbon crystallites	3.40–3.43 Å	3.36–3.37 Å
Density (g/cc)	2.1- 2.2	2.1–2.2
Tensile strength of filaments	140–160 ksi	250–350 ksi
Young's modulus	25–35 msi	75–120 msi
Electrical resistivity	800–1200 $\times$ 10^{-6} ohm-cm	150–200 $\times$ 10^{-6} ohm-cm

The Singer patent[37] is also noteworthy by the inclusion of numerous X-ray diffraction patterns and polarized light micrographs to identify the degrees of crystallite orientation and the influence upon the properties of the C/G fibers. These fibers are being directed towards advanced composites in aerospace applications, as well as in the manufacture of pressure vessels.

When drawing comparisons between the manufacturing methods for PAN-based carbon fibers and pitch-based carbon fibers, there are advantages in favor of the latter. The carbon yield from mesophase pitch is about 75 to 85%, whereas the yield from PAN is only about 50%. Because the as-spun pitch fibers are highly anisotropic, there is no need to apply tensioning during the oxidative treatment, and carbonization may be carried out at a rapid rate. During oxidative treatment to develop thermoset qualities (250° to 400°C), there is an oxygen pick-up. Early in the production of pitch-based fibers by Union Carbide, a hairline crack developed along the fiber, which contributed to a radial crystal structure and less efficient use of the fiber in applications to composites. This problem has been resolved in more recent manufacture. "Thornel" P is now available as a continuous twisted yarn (20 turns/m), 2000 filaments, with an epoxy compatible finish. The average filament diameter is 10 to 11 microns.

It was an early objective of the manufacturers of pitch fibers to offer grades which would be compatible to the PAN-based C/G fibers. Applications and early uses were described by Chambers,[44] with emphasis on the transportation industry and the sporting goods industry.

If one were to identify one of the shortcomings of C/G materials in the early 1970's, fiber property variations would be cited. However, later materials from leading producers have been very uniform. As manufacturing problems for C/G fibers have been resolved, as the process variables have been ascertained, as basic research establishes the criteria for optimum fiber properties, and as the influences of sizing agents and the resin matrices with carbon fibers are better understood, superior materials began to appear, and current C/G fibers are better than older ones in uniformity and properties. The problem is posed as to the reliability of the earlier data in published reports; one should place greater reliance upon more recent reports and current materials.

There are recent literature references which at this time show preference for the PAN-based C/G fibers over those from pitch. Bell and Perry[45] evaluated composites using C/G fibers from pitch and PAN sources. Tests included mechanical properties in dry and wet conditions at room and elevated temperatures. They concluded that epoxy composites prepared with pitch/graphite fibers were suitable for stiffness critical applications, but flexural, tensile, and compressive strengths were lower than for those composites prepared from HM-S fibers from PAN precursors.

Commercially available C/G fibers derived from pitch and from PAN are compared in Table 2-8. While the new pitch-based fibers are not as strong as the low modulus PAN fibers, the ease with which they can be processed to high moduli makes them attractive to stiffness critical applications.[46]

HM-S graphite fibers yielded higher flexural strengths and higher interlaminar shear strengths in the composites than pitch origin fibers. While pitch-based graphite fibers have attained moduli in excess of 100×10^6 psi, this has been accomplished at some sacrifice in tensile strength and impact resistance. Comparisons between PAN-based fibers and pitch fibers in Nylon 66 molding compounds favored the PAN-based carbon fibers on the basis of better physical

Table 2-8. Basic PAN Fiber and Pitch Fiber Comparison.[46]

	PAN FIBERS		PITCH FIBERS	
	"THORNEL" 50 (UNION CARBIDE)	GY-70 (CELANESE)	TYPE P-VSB-32 (UNION CARBIDE)	TYPE P-VS-0053 (UNION CARBIDE)
Fiber				
Tensile modulus (10^6 psi)	50	70	50	75
Tensile strength (ksi)	350	300	300	300
Composite				
Tensile modulus (10^6 psi)	32	44	32	44
Tensile strength (ksi)	175	110	150	150
Interlaminar shear (ksi)	10.0	7.0	4.5	3.0

properties. In this example, non-oriented carbon fibers were dispersed into thermoplastic molding compounds–30% by weight.[47]

However, the higher moduli fibers of graphite appear to be more effective in reducing thermal coefficients of expansion and in improving resistance to fatigue of structural laminates and molding compounds.

LARGE DIAMETER C/G FILAMENTS

Special efforts have been directed towards the development of large diameter C/G filaments–usually by uniting fine carbon filaments with a carbonizable and graphitizable binder. The method consists of impregnating a bundle of carbon fibers with a high char yield resin, such as phenolic, and after heat curing, subjecting the composite fiber to controlled carbonization temperatures. Variable cross-sections and shapes are produced in this manner.[48]

Work at Lockheed indicated that best strengths were obtained from the combination of "Thornel" 75 fibers and resin precursors and subjecting the combination to copyrolysis. Methods are described to reduce, within the carbonaceous matrix, pyrolysis flaws which lower composite strength.[49]

A process was developed at United Aircraft Corp. for preparing large diameter carbon-boron monofilaments.[50] A carbon fiber substrate received a deposit of carbon-boron alloy from a chemical vapor mixture of $BCl_3/CH_4/H_2$. In further studies, the gas flows and deposition temperatures were studied in their effect on tensile strength of the monofilaments. The carbon fiber base has been used to replace the heavier tungsten core of boron monofilaments (see the chapter on matrices over 300°C, which includes a section on graphite fiber/metal matrix composites).

Carbon fibers of a somewhat intermediate size (25 microns) were prepared from reactive polyacetylenes based on diethynyl benzene ($HC \equiv C -$ ⌬ $- C \equiv CH$). Polymerization by oxidation yields polymers of 250 to 1000 degrees of polymerization. In the process described by General Electric, the fibers are drawn at 120° to 150°C and elongated up to several hundred percent.[51] They will rapidly cross-link at 250° to 500°C in the presence of oxygen. Further carbonization at 1100°C in an inert atmosphere, followed by graphitization and stretching at 2800° to 3000°C, yields highly oriented C/G monofilaments with tensile strengths of 300 ksi and moduli of 50 msi for fibers 25 microns in diameter.

While large diameter carbon monofilaments have not been prominent in the structural advanced composites, larger diameter monofilaments of boron fiber have been used in epoxy or polyester impregnated tapes.

Applications in rocket nozzles and in atmospheric re-entry vehicles require the heat resistant C/G fiber composites. The techniques of impregnating C/G filamentary structures with matrices that are co-pyrolized—perhaps several times after subsequent reimpregnations—are used to prepare the outer surfaces of space shuttles, as the temperature resistance of the pryolyzed materials and their strength at re-entry temperatures are exceptionally high. In the chapters on high temperature (above 300°C) matrices for advanced composites, attention will be given to C/G fibers with carbon matrices. Illustrated herewith are examples of early carbon fiber nose cones for rockets—selected for heat resistance and light weight (Fig. 2-8).

Also noteworthy is the early paper on carbon-based fiber reinforced plastics by Schmidt and Jones.[52] Attention at that time (1961-1962) was directed towards thermally stable ablative plastic materials. The authors recognized the many inherent advantages of composites made with carbon fibers, such as high thermal stability, light weight, chemical inertness, low thermal expansion, nonflammability, and semi-electrical conductivity. They were clearly in advance of their times in recognizing these products—and, as a measure of the advances of the last 20 years, their early tables are reproduced here based upon carbon fiber filled molding compounds (Table 2-9) and carbon fiber laminates (Table 2-10).

Among other carbonaceous materials examined as binders for C/G fibers is mesophase, a liquid crystal phase from pitch, which molds under pressure at temperatures at 350° to 500°C. This process yielded composites with reduced void, though subsequent carbonization and graphitization caused considerable

Fig. 2-8. Early uses of carbon fiber in rocket nose cones. (*Courtesy Union Carbide Corp.*)

shrinkage.[53] The application of pyrolytic graphite through chemical vapor deposition has also aided the production of carbon/carbon composites.

HYBRIDS

The preparation of hybrid constructions is becoming more important, as it is unlikely that any one fibrous reinforcement will answer all the technical and economical requirements of automotive, aerospace, or other industries. Combinations of fibers, referred to as "hybrids," such as glass and carbon fibers; carbon and graphite fibers; boron and graphite fibers; aramid and graphite fibers; glass and aramid fibers; and others are finding engineering and economical reasons for their utilization. Aramid fibers are desirable for combining with graphite fibers to reduce composite costs and improve on impact strength. (Average price of "Kevlar" in 1980 was $10.00/lb). (See Fig. 2-9, illustrating a cross-section of a hybrid of C/G and aramid fibers, and differences in filament diameter. The smaller cross-section is carbon and the larger diameter fiber is aramid.)

Table 2-9. Strength Properties of Carbon-based Fabric Reinforced Plastics.[52]

MATERIAL	DENSITY (lb/ft^3)	TENSILE STRENGTH (lb/in^2)	FLEXURAL STRENGTH (lb/in^2)	FLEXURAL MODULUS ($10^6 lb/in^2$)	COMPRESSIVE STRENGTH (lb/in^2)	COMPRESSIVE MODULUS ($10^6 lb/in^2$)
Phenolic[a] 40% Carbon fabric[b] 60%	89.8	16,600	26,500	2.27	40,200	2.47
Phenolic[a] 35% Carbon fabric[c] 65%	82.0	9160	18,300	0.92	23,800	1.50
Phenolic[a] 46% Carbon fabric[c] 54%	85.2	7250	14,800	1.01	27,100	1.22
Epoxy novolac[d] 27% Carbon fabric[c] 73%	78.7	9310	16,000	0.93	16,700	1.19
Phenolic[e] 27% Graphite fabric[f] 73%	82.6	11,600	20,100	1.69	13,200	2.13
Phenolic[e] 37% Graphite fabric[f] 63%	86.7	9970	17,900	1.17	10,900	1.75
Phenyl silane[g] 30% Graphite fabric[f] 70%	83.2	11,200	19,900	1.58	11,500	1.95

[a] Monsanto "SC 1008."
[b] "CCA-1."
[c] "VCA."
[d] Dow Chemical "D.E.N. 438."
[e] C.T.L. "91LD."
[f] "WCB."
[g] C.T.L. "37-9X."

Table 2-10. High-temperature Flexural Properties of Carbon-based Plastic Laminates.[52]

MATERIAL	DENSITY (lb/ft^3)	FLEXURAL STRENGTH (lb/in^2) 75°F	400°F	700°F	FLEXURAL MODULUS ($10^6 lb/in^2$) 75°F	400°F	700°F
Phenolic[a] 40% Carbon fabric[b] 60%	89.8	26,500	26,500	21,200	2.27	1.49	1.31
Phenolic[a] 35% Carbon fabric[c] 65%	82.0	18,300	17,500	13,600	0.92	1.06	0.82
Phenyl silane[d] 32% Carbon fabric[c] 68%	83.5	18,600	19,600	14,600	1.19	1.10	0.89
Phenolic[a] 30% Graphite fabric[e] 70%	82.6	20,200	17,600	8160	1.64	1.40	0.65
Phenyl silane[d] 30% Graphite fabric[e] 70%	83.2	19,900	17,200	10,600	1.58	1.51	1.04

[a]Monsanto "SC 1008."
[b]"CCA-1."
[c]"VCA."
[d]C.T.L. "37-9X."
[e]"WCB."

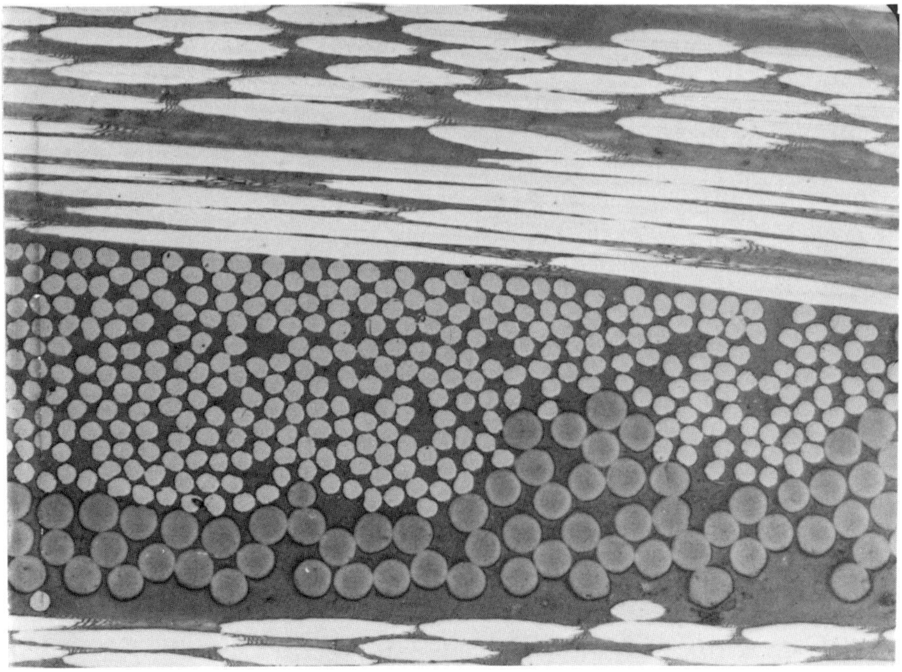

Fig. 2-9. Cross-section of hybrid composite aramid fibers (large diameters) and carbon fibers (small–7 micron–diameter).

There are several ways of combining the fibers to form the "hybrids":

1. Strands or tows of different fibrous yarns may be collimated together to form specialty tapes, which are usually impregnated with a resin matrix of epoxy, phenolic, polyimide, or polyester. The fiber proportions would be determined by the properties expected of the tape. Presumably, such tapes would be a specialty item at this time and prepared only for custom orders. In practice, alternate tapes of different fibers are placed in molds, as required by specifications.
2. Fabrics may be woven in whatever combinations will meet the requirements of this application. For example, some fabrics of graphite have been augmented with other fibers to enhance properties such as interlaminar shear strength.
3. Woven fabrics or mats of C/G fibers may be intercalated with woven fabrics or mats of other materials, such as glass or aramid, so as to concentrate those fibers in specific areas where they will perform the best. The binder

or resin matrix of the two fibrous structures should be the same. A case in point is the insertion of a graphite fiber near the surface of a composite, where the better electrical conductivity of the graphite may serve a useful purpose, as to reduce static charges.

4. Non-woven mats of random fiber materials of any form and combination may be prepared in aqueous or solvent slurries, or in extrusion processes (with thermoplastics) to prepare molding compounds for specialty purposes.
5. Graphite fibers in place of tungsten wire have also functioned as a repository or carrier for boron fibers or whiskers—to assist a manufacturing operation.[54] The cost of the boron filament would be materially reduced if the tungsten core were replaced by C/G filaments. Another unusual combination is the impregnation of cellulosic continuous yarns with PAN. Co-pyrolysis and heat treatment have provided interesting carbon fiber precursors (U.S. patent 3,395,970).

In most cases, the role of the graphite fibers has been to lend stiffness, to reduce the coefficient of thermal expansion, to improve the modulus of elasticity of the composite structure, and to enhance fatigue strength. In molded parts, the presence of chopped C/G fibers not only increases modulus, but also reduces shrinkage, lowers the expansion coefficient, improves lubricity, and reduces electrical resistivity.

The problems associated with the preparation of yarns from various hybrids were reviewed by Riley.[55] A single slurry vortex spinning apparatus was developed to form a multi-component (hybrid) yarn. The apparatus was based on developments in the textile industry to produce yarn by means of open-end spinning. A slurry of dispersed fibers was transferred to the spinning zone, and as they passed through a rapidly rotating tube, the hydrodynamic action twisted the fibers into a yarn. Integral mixtures of high modulus graphite fibers and lower synthetic fibers were prepared. This reference also contains a summary and illustrations of open-end spinning systems, described in the literature. The vortex generator developed is shown in Fig. 2-10, with its rotating tube for spinning the fiber hybrids together.

In reviewing the commercial fabrics involving hybrids, cost savings are mentioned most frequently when glass fibers and aramid fibers are blended with graphite fibers. In one example of a horizontal stabilizer for Sikorsky's new S-76 commercial helicopter, the trends in mixed composites are evident. Unidirectional graphite/epoxy spar caps; aramid/epoxy spar webs, skins and torque tubes; and aramid paper and aluminum honeycomb cores are described.

When "Kevlar" 29 or "Kevlar" 49 aramids are combined with graphite fibers

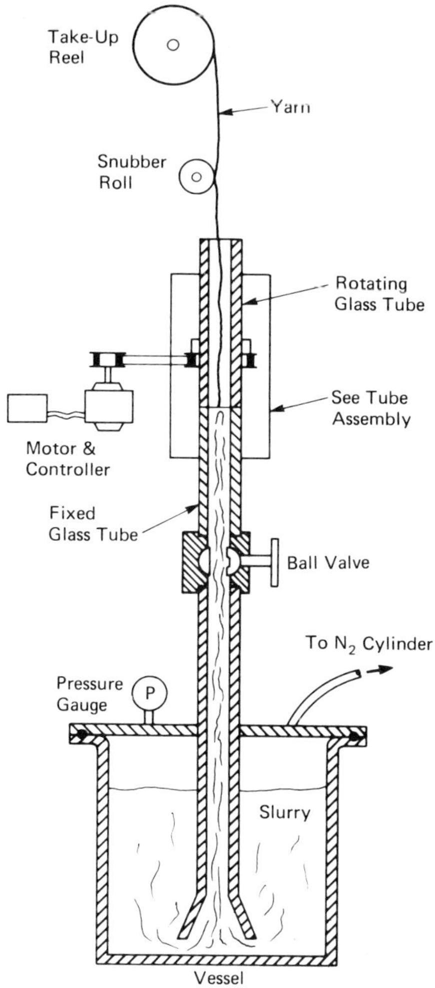

Fig. 2-10. Schematic of open-end spinning apparatus.[55]

to prepare prepreg tapes, lower cost laminates and higher impact laminates are obtained. Data are presented for a number or combination of aramid fibers and graphite fibers. Table 2-11 shows the tensile properties obtained on composites using an epoxy resin matrix as a binder. Commercially, these combinations are being used in the manufacture of golf shafts which acquire stiffness from the graphite fibers and improved shear strength from the presence of glass or aramid.[56]

Table 2-11. Tensile Properties of Unidirectional Composites Made from Hybrid Combinations of Aramid Fibers and Graphite Fibers, 0° Orientation. Data Normalized to a 60 v/o Total Fiber Content.[56]

		TENSILE PROPERTIES		
FIBER	v/o FIBER	ULTIMATE STRENGTH (10^3 psi)	MODULUS (10^6 psi)	STRAIN TO FAIL (%)
"Kevlar" 29	60	196	7.25	2.70
"Kevlar" 49	60	200	11.0	1.80
A-S Graphite	60	147	15.2	0.97
"Thornel" 300	60	185	19.2	0.96
HM-S Graphite	60	110	29.0	0.44
"Kevlar" 49	20.9	61.2	24.5	0.25
HM-S Graphite	39.1			
"Kevlar" 49	26.1	143	14.1	1.01
AS Graphite	33.9			
"Kevlar" 49	29.4	153	15.5	0.99
T-300	30.6			
"Kevlar" 29	13.0	162	17.6	0.92
T-300	47.0			
"Kevlar" 29	31.2	144	13.8	1.04
T-300	28.8			

In the development of hybrids for advanced composites, the trend appears to utilize the aramid 49 in conjunction with C/G fibers. Not only is there a marked improvement in fracture strength and impact resistance, but also the lower cost of the aramids and their lower specific gravity (1.38) are desirable advantages. Aramid 29 appears to be utilized to a greater extent in markets for ropes, cables, and coated fabrics.[57] In the chart shown in Fig. 2-11, the stress-strain relationships of unidirectional composites of graphite fiber, glass fiber, and aramid fiber are compared. From the shape of the aramid fiber composite curve, one would expect improved impact strength and improved vibrational dampening characteristics. In combination with C/G fibers, it has been possible to design advanced structures of optimum efficiency.

Metals, composites, and hybrid structural components are being examined to a greater degree.[58] High modulus composites from hybrid materials are shown to offer attractive alternates to metal structures. The application and proliferation of carbon fiber hybrid composites are growing. Short and Summerscales[59] reviewed some of the activity and manufacturing methods. Many of the recent

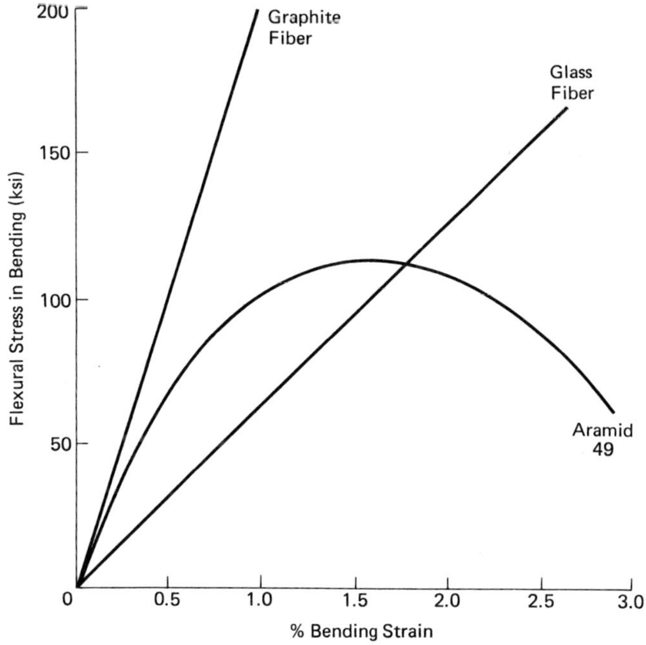

Fig. 2-11. Unidirectional composite (epoxy resin) stress-strain in bending.

examples cited were glass/carbon fiber hybrids of reinforced plastics among manufacturers in England and Europe. It was pointed out that instead of designing to a material, the material is tailored to the design, often resulting in fewer components. Also noted is that limit strain is a more relevant condition of failure than stress analysis. Significant progress is also being made in areas of discontinuous and intimately mixed carbon/glass fiber hybrids.

Hybrid design methodology has been adapted to tubular structures such as golf clubs and bicycle frames. High strength and high modulus graphite fibers are used with glass fibers and aramid fibers, the latter for cost reduction. A series of charts were developed to demonstrate torsional rigidity and stiffness for different hybrid laminates.[60]

To improve interlaminar shear strength of high modulus graphite fibers, silicon carbide has been deposited on C/G fiber substrates. Moduli of elasticity approached 100 msi.[61] This represents still another approach in hybridization. Silicon/glass/graphite hybrids have been studied on helicopter rotor blades. Physical and structural properties were determined from these hybrids, prepared as solid laminates, tubular specimens, and sandwich beams.[62]

OXIDATIVE RESISTANCE

When making comparisons among the various types of carbon and graphite fibers, there is no one criterion which may be used in making selections. Basically, the lower modulus fibers and those possessing the highest strain at the time of failure may be expected to yield the higher impact composites, while those with the highest moduli offer the stiffness factors not obtained with other materials of construction. Hybrids, to an increasing degree, will permit compromises to be made, as seen from the brief review of these combinations. The true measure of the success lies in the application and the cost effectiveness of the product in meeting its requirements. As laminates, composites, or molding compounds, C/G fibers are used with many types of matrices, which form the other half of the partnership, and for this reason the types of resin matrices which are most likely to be used will be discussed in the subsequent chapters.

It is also a fact of life that at the time of this publication there is a considerable maturity in C/G fibers. This maturity now allows us to look objectively at results which have appeared in many reports over the past 10 to 15 years. Fibers have been improved and, in consequence, changes have been made, of necessity obsoleting earlier products and earlier published results that may have dealt with fibers suspect in character. At the time the C/G fibers were first manufactured, cleaning and activating treatments, or possibly the heat treatments themselves, were not fully comprehended and lot to lot variations were extant in fiber properties. As service tests became more demanding, particularly for aerospace use and environmental extremes, the weaknesses of some early fibers became apparent—sometimes in the guise of trace impurities on the fiber surface. More precise test methods and quality assurance measures have been adopted. These are the elements of maturity. Test methods on fibers and their composites are the subject of a separate chapter in this book, though at this time, a particular thermal test of the oxidative stability is selected for making comparisons among the C/G fibers for two important reasons: 1) most of the C/G fiber precursors undergo oxidative treatments before carbonization to stabilize the fibers, and 2) identification of the more oxidatively stable fibers paves the way for their selection in high temperature applications of the aerospace activities, currently the largest market. This picture changes dramatically as the automotive applications begin to grow and as the demand for precision molded parts with high accuracy makes itself felt. Hence, for comparative purposes, the selectivity afforded by high temperature oxidative studies on C/G fibers has merit. The work of McMahon and Gibbs in this area is particularly noteworthy, and some of their findings are reproduced from references in the following paragraphs. This chapter will be closed with a summary of trade names and manufacturers, as available at the time of publication. Some of the earlier manufacturers have withdrawn or have become divisions of other organizations, and identifications which are now obsolete will not necessarily be reproduced.

Increasing attention has been given to the oxidative resistance of various commercial carbon fibers and their composites, by measuring weight loss for different time intervals at elevated temperature. Such data give insight into the heat treatment and processing treatments of the fibers during manufacture. Gibbs, Wendt, and Wilson[63] studied the thermal-oxidative stability of several available carbon fibers at 600°F (589°K). Substantial differences were observed on the weight loss behavior in air aging studies run on bare fibers. These data bore a direct relation to the stability at high temperatures of the composites prepared from the fibers. In addition, trace amounts of certain chemicals, such as sodium sulfate, were found to contribute to the fiber instability in the presence of air at higher temperatures. Furthermore, lot to lot variations of certain fibers were large, suggesting the need for employing high temperature oxidation of fiber lots as a quality control measure prior to laminating.

Data reproduced in Table 2-12 show the wide spread of weight loss for commercial fibers of certain designated lots at the time tests were performed. One can be confident that manufacturers of such fibers have been alerted to these discrepancies, and where necessary are able to improve their oxidative stability. In reporting their results, the investigators subtracted the two-hour reading from the total, to account for residual moisture or other volatiles on the bare fiber.

Table 2-12. Oxidative Stability of Selected Carbon Fibers According to Percent Weight Loss After 700 Hours in Air at 600°F (589°K).[63]

CARBON FIBER	LOT NUMBER	PERCENT WEIGHT LOSS
"Fortafil" 5	R128(B)C	0
"Fortafil" 3 U	2-1093 W/8C	97
"Fortafil" 3	RS494(B)B	0.18
"Modmor" II-S	C266/12	0.08
"Modmor" II-S	265-2	0.38
"Magnamite" HMS	34-1	0.08
"Magnamite" HTS	66-7	1.2
"Magnamite" HTS	54-4	49
"Magnamite" AS	36-3	67
"Celion" 6000-Sized	6711	3.5
"Celion" 6000 Unsized	6Y11	5.5
"Thornel" P	–	0.90
"Thornel" 300	93.2	58.0

All of the product identifications in quotations are United States registered trade names. The "Celion" products consisted of 6000 fiber tow and the "Fortafil" products were reported as consisting of 40,000 fiber tow. In most instances, the carbon fiber densities lay between 1.62 and 1.82 g/ml. Typical results are shown for weight loss in the accompanying curves from the Gibbs *et al.* paper (see Fig. 2-12). The author's conclusions emphasized the following points:

1. Carbon fiber stability affected the stability and performance of high temperature composites, such as those prepared with polyimide matrices.
2. A direct relationship exists between carbon fiber stability and sodium

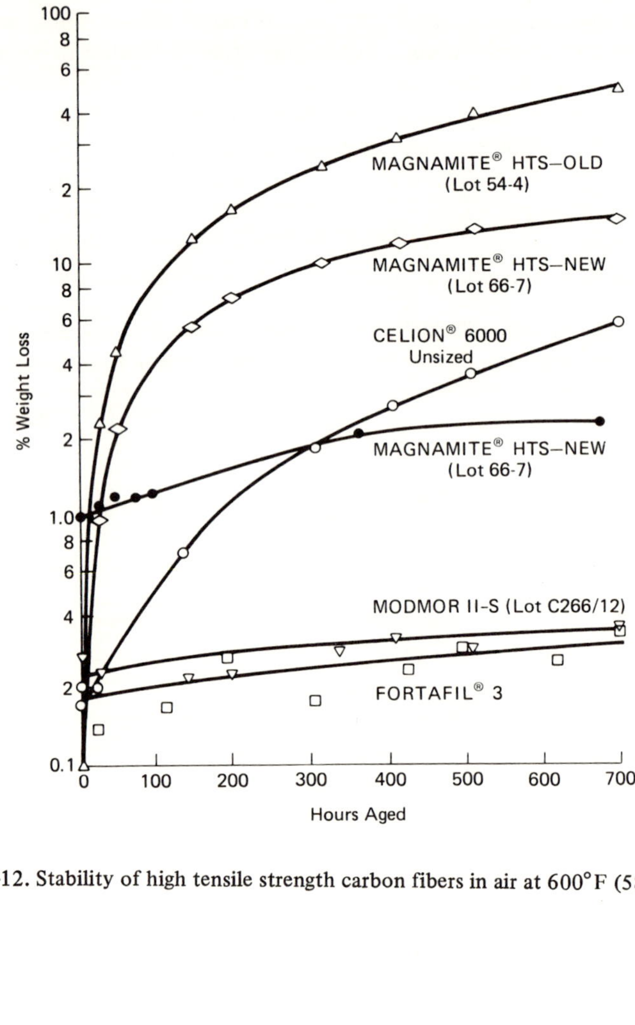

Fig. 2-12. Stability of high tensile strength carbon fibers in air at 600°F (589°K).[63]

content. The higher the sodium content, the poorer the stability of the carbon fiber.

These findings are noteworthy as well as timely, and manufacturers of carbon fibers and fabricators of advanced composites have shown increased concern for impurities which may mitigate against performance of their products. Additional data were also forthcoming from McMahon[64] in 1978. He reported that good oxidative stability in the carbon fibers contributed to good property retention after 500 hours at 600°F (589°K) for polyimide composites. He pointed out that high modulus carbon fibers (345 GN/m^2) have good resistance to thermal oxidation in air, and that the more economically produced lower modulus materials (240 GN/m^2) have poorer oxidative resistance at high temperatures.

McMahon brings out in his report the effect of processing temperature on the oxidative stability of the carbon fibers, citing significant difference in the weight loss of "Courtauld-based" fibers and "Celion" fibers from PAN precursors. Some of his data are reproduced for different temperatures on three commercial fibers: "Celion" (Celanese), Hercules HTS, and U.C.C.'s "Thornel" 300 in Table 2-13.

Table 2-13. Percent Weight Loss of Carbon Fibers After High Temperature Exposure to Air.[64]

		AT 533°K	AT 588°K		AT 648°K
CARBON FIBER–	MODULUS (GN/m^2)	500 HR	500 HR	1000 HR	24 HR
"Celion" 6000 with finish	241	1.5%	3.3%	7.3%	1.9%
"Hercules HTS" Batch 66-7	276	0.6%	1.2%	–	3.4%
"Thornel" 300, several lots	241	4.2%	22.8%	49.7%	14.9%

Table 2-13 suggests that more rapid evaluations of carbon fiber weight loss at elevated temperatures may be expected by heating at higher temperatures. In fact, McMahon shows some data suggesting that TGA weight loss in air under a uniform heat-up rate of 20°K/minute would yield useful comparisons. The inference, of course, is that the more thermally stable carbon fibers would attain higher temperatures in a TGA analysis before losing significant weight.

McMahon also used accelerated aging with three-hour exposure in air at 773°K (932°F) to classify several commercial grades of carbon fibers. His data is reproduced in part in Table 2-14.

Table 2–14. Weight Loss of Graphite Fibers After Three Hours at 773°K (932°F) in Air.[64]

FIBER	PERCENT CARBON	INITIAL MODULUS (GN/m^2)	PERCENT WEIGHT LOSS
GY-70 (Celanese)	99.3	517	0.1
VS-Pitchbase (UCC)	99.0	345	0.6
HMS (Batch 42–2)	–	345	3.3
"Celion" 3000/6000	93.0	241	3.5
"Thornel" 50	–	345	9.6
Great Lakes 4T	–	276	9.9
Hercules HT-S	–	276	5 to 30
Hercules AS	93.9	241	98.7
"Panex" 30A	–	207	98.8
"Thornel" 300	91.3	241	99.3

It should be noted that the elemental analysis of the carbon fibers has a bearing on weight loss at high temperature. "Thornel" 300 and AS fibers tested showed over 1000 ppm of sodium. (Commercial fiber grades are available with low sodium.)

The subject of fiber oxidation has been examined in considerable detail because of its influence upon the ultimate properties of the C/G composites. The interlaminar shear strength of graphite fiber-epoxy composites have been significantly improved by various treatments, though the reasons for the improvement are complex and not fully understood. It is believed that the improvement in composite shear strength is related to an increase in fiber surface area and to the alteration of the nature of the surface of the carbon fibers, through the establishment of oxygen-containing functional groups.[65] The methods of enhancing the properties of composites through the prior surface treatments of the C/G fibers are the subject of a separate chapter.

REVIEW OF INDUSTRY ACTIVITY IN THE PRODUCTION OF C/G FIBERS

This review of industry activity in C/G fibers is dated as of early 1980. Production of the C/G fibers is relatively small as compared to other industrial fibers, though its potential and, in particular, its impact on aircraft design and on automotive design is large. In retrospect, one is looking at a current manufacturing capacity of about 1,000,000 lb total (using United States and Japanese figures) of high performance C/G fibers, soon to grow to 2,500,000 lb. For comparison, current production of glass fibers is well in excess of 500,000,000 lb and current production of aramid fibers is about 15,000,000 lb.

How is the activity in C/G fibers distributed? In a growth industry, aside from the key markets, which have been identified as aerospace and sporting goods, there are many small segments, which are difficult to interpret, first because available data is limited or restricted, and secondly, because the processors of C/G fibers (such as the prepreggers) are usually engaged in handling other fibers, such as the aramids, glass, and boron. With the growing recognition of the value of hybrids which combine several fibers in the final product, the clear-cut delineation of C/G contribution is not always apparent. There are interrelationships between producers and distributors of carbon and graphite fibers, in which their respective roles have been changing, but in the paragraphs below, one can obtain a picture of the status of the activity—as defined above—as of early 1980.

PRINCIPAL PRODUCERS OF CARBONIZED FILAMENTS

Union Carbide Corp. U.C.C. has been identified with the manufacture and distribution of C/G fibers from the early 1960's (from rayon precursors, produced by textile groups such as Burlington and carbonized by U.C.C.), and from the 1970's the distribution of C/G fibers produced by Toray (Japan). In more recent years, U.C.C. has been producing high quality pitch fibers based upon their own patents, and has begun producing some of their own PAN-based fibers. An 800,000 lb capacity plant is expected in middle 1981. The U.C.C. products have been called by the trade name, "Thornel," which was associated in the late 1960's with rayon precursors (those having a 25×10^6 psi modulus were known as "Thornel" 25 and those with a 50×10^6 psi modulus as "Thornel" 50). More recently, there has been an identification of "Thornel" 50 PAN for the PAN-based C/G fibers, though for many years now, and on many specifications, the symbol T-300 has marked the material imported from Toray in Japan—the high strength C/G fiber—from PAN precursors. The pitch fibers have been described as Type P, with the added symbols VSB in various grades. Earlier in this chapter, the physical properties of the PAN-based fibers and pitch fibers were compared.

Hercules, Inc. Hercules, like U.C.C., has been distributing C/G fibers, though more recently the firm has been producing its own PAN-based C/G fibers. The announced plant capacity for the early 1980's is 600,000 lb—making it the largest United States producer. Hercules has been importing its C/G fibers from Courtaulds of England under an exclusive licensing arrangement for the past decade. It is expected that Sumitomo of Japan will begin selling Hercules fibers in Japan in the near future. At this time, Courtaulds will also be selling the products in the United States. The identification of the PAN precursors C/G fibers offered by Hercules has been high strain, low modulus (A-S), high tensile strength (HT-S), and high modulus (HM-S)–the "S" identification referring to surface treatment on the fiber, and the symbol "U" for unsized. Type AS, HT-S, and HM-S fibers have been extensively called out on government and aircraft specification. HT-S of Hercules and T-300 of U.C.C. are prominent in aerospace requirements. Hercules has also supplied "prepregs" of its different fibers, and will also carbonize filaments and woven products under contract. The C/G products of Hercules are now being identified under the trade name "Magnamite."

Celanese, Inc. Celanese has smaller distribution and manufacturing capacity than Hercules or U.C.C., though it does offer a unique, ultra-high modulus graphite fiber (70 to 75 $\times$ 10^6 psi), identified as GY-70. Toho-Belson Company of Japan, with a capacity of 180,000 lb, has been the principal source of the Celanese products. Increasing emphasis is being placed on the "Celion" 3000 and 6000, products from PAN precursors, the numerals indicating the number of filaments in the basic tows. The advantages of the high modulus fiber in aerospace are being extensively evaluated. Celanese now owns the experienced Narmco organization, which has pioneered many resin developments in the field of advanced composites. Narmco (formerly Whittaker Corp.) specializes in the prepregging of advanced fibers for industry. Narmco resin systems and C/G fiber specifications appear frequently on development programs.

Courtaulds (England). This English company has pioneered in the manufacture of PAN-based C/G fibers developed by the Royal Aircraft Establishment (RAE) in the late 1960's. Today their products are sold under the trade name of "Grafil." The fiber is supplied as a tow containing 10,000 individual parallel C/G filaments. Their production capacity in Europe is reported as 180,000 lb/year. Mitsubishi distributes their products in Japan. Their principal fibers are the Type AS, HT-S, and HM-S, the same designations as the Hercules fibers. The European activities in advanced composites follow the lead of United States aerospace manufacturers. With sub-contract work on advanced composites for commercial aircraft arising in Europe, the activity will increase significantly in this area. Absent from this tabulation is another English manufacturer, Morganite Co., which, during the early 1970's, supplied C/G fibers identified as "Modmor" I and "Modmor" II. They are no longer producing these materials.

Toray (Japan). This organization has pioneered in the development of PAN-based C/G fibers and, through U.C.C., has been supplying a substantial quantity of materials to this country. "Torayca" carbon fiber is manufactured from PAN precursor, and the high strength fibers are sold as T-300, available in 1000, 3000, and 6000 filament yarns. Grade M-40 is a high modulus fiber available in 3000 and 6000 filament yarns. Toray also offers a line of woven goods, braids, chopped fibers, and prepregs. More recent products include woven yarns of hybrids of glass fibers and carbon fibers.

Great Lakes Carbon Corp. One of the smaller manufacturers of C/G fibers, Great Lakes has products under the trade name of "Fortafil." The C/G fibers are offered in a 40,000 continuous tow for applications requiring high modulus and strength. The oxidative resistance of the fibers at high temperatures is excellent. The C/G precursor is an acrylic polymer, obtained from a United States source, and carbonized by Great Lakes to obtain an oriented graphite structure in the fiber.

Stackpole Fibers Co. Another of the smaller manufacturers of C/G fibers, Stackpole offers its 160,000 high strength, high modulus filaments in a continuous tow. Under the trade name, "Panex" 30, the fibers are treated specifically for various resin matrices such as those used in the manufacture of reinforced thermoplastic molding compounds. High modulus fabrics are made from a spun yarn for application to multiple curvatures in rocket nozzle and heat shields. A United States source of PAN precursor is used for manufacturing the C/G fibers.

Hitco. One of the latest entries into the C/G manufacturing activity is the Hitco organization, a subsidiary of Armco Steel Corp. Carbonized fabrics are offered for ablative insulation on solid rocket motor nozzles, as an alternate to the ablative fabrics manufactured from rayon precursors. Intermediate, high modulus filaments are also being offered in continuous filament carbon tows under the trade name of "Hi-Tex."

PREPREG

An important link in the manufacturing of advanced composites is the preparation of prepregs with the application of resin matrices to the C/G fibers. This activity requires a high degree of technical competence to apply sophisticated resin systems (usually of epoxy, polyimide, phenolic, and polyester) to the yarns or tows of unidirectional C/G fibers and to collimate them into tapes of specified widths and thicknesses, which are impregnated to closely controlled resin contents, volatiles, and flow. There are several manufacturers active in "prepregging" of tapes and woven fabrics of C/G fibers, and who are capable of maintaining the close quality control required by the fabricator. These include manufacturers such as the following:

1. Fiberite Corp. (Minnesota)
2. Ferro Corporation (Los Angeles)
3. Fiber Materials, Inc. (Maine)
4. Hexcel, Inc. (California)
5. Narmco Division of Celanese
6. U.S. Polymeric (Santa Ana, California)
7. Avco Products (Massachusetts)

8. Du Pont Division in Newhall, California
9. Forthergill and Harvey (England)
10. Prodesco (Pennsylvania)
11. Proform (Texas)
12. Minnesota Mining and Manufacturing Co.
13. Reliable Manufacturing Co. (Ciba Geigy)

Manufacturers of molding compounds purchase from the suppliers of C/G chopped C/G fibers–usually in the 1/4 to 1 inch range, for inclusion in materials requiring increased stiffness, improved fatigue resistance, lower shrinkage, lower coefficients of thermal expansion, and improved lubricity. Amoco Chemicals, Inc., U.S. Polymeric, Hexcel, Fiberite Corp., and LNP, Inc., are among these. The Fiberite and LNP groups are both divisions of Beatrice Foods, Inc.

REFERENCES

1. Reynolds, W.M., *Physical Properties of Graphite.* Elsevier Publishing Company, New York, 1968.
2. AFML Research Technology Division, Technical Report ML-TDR-65-114, Wright Patterson AFB, 105 pages, April 1965.
3. Milewski, J.V., "Short Fiber Reinforcements," *Plastics Compounding,* page 17, November/December 1979.
4. "Composite Materials and Composite Structures," *Proceedings of Sixth Sagamore Research Conference in New York,* pages 56-57, August 1959.
5. Sterry, J.P. and Lachman, W.L., "Ceramic Fibers," *A.I.Ch.E. Forty-eight National Meeting,* page 8, Denver, August 1962.
6. Harmon, C.G., "Non-Glassy Inorganic Fibers and Composites," NASA SP-5055, August 1966.
7. Steg, L., "Whisker and Short Fiber Reinforcements," *SAE Report 660640,* Aeronautic and Space Meeting, Los Angeles, October 1966.
8. Schmidt, D.L. and Hawkins, H.T., "Fibrous Carbon Reinforcements," *SPE, Retec, Conference,* page 111, July 1965, Seattle, Washington.
9. Economy, J., "Present Status and Future for High Strength Fibers," *SAMPE Journal,* Vol. 12, page 5, November/December 1976.
10. Bernal, *Proceedings of the Royal Society (London),* A-106, 749, 1924.
11. Walker, P.L., chapter on "Ultrafine Carbon," page 297, *Electrochemical Society, Symposium 1961.* John Wiley and Sons, New York, 1963.
12. Pauling, L., *Nature of the Chemical Bond,* Third Edition, pages 234-235. Cornell University Press, 1960.
13. Ubbeholde, A.R. and Lewis, R.A., *Graphite and Its Crystal Compounds.* Oxford University Press, London, 1960.
14. Blackman, L.C., Editor, *Modern Aspects of Graphite Technology.* Academic Press, New York, 1970.
15. Woollman, J., "Review of Electrical and Thermal Properties of Carbon and Graphite," *ACS Symposium No. 21 on Petroleum Derived Carbons,* 1975.

16. Tokarsky, E.W. and Diefendorf, F.J., "Modulus and Strength of Carbon Fibers," *Twelfth Biennial Conference on Carbon,* page 301, American Carbon Committee, Pittsburgh, Pennsylvania, July 1975; and *Polymer Engineering and Science,* Vol. 15, page 150, July 1975.
17. Barker, R.L., "Optical Scanning Technique for Measuring Fiber Size," *Journal of Engineering for Industry,* Vol. 100, page 1, February 1978.
18. Hsiung, P.L. and Rowe, C.R., "Statistical Characteristics of Diameters of Pitch-based Carbon Fibers," *Fourteenth Biennial Conference on Carbon,* page 226, American Carbon Committee, Penn State University, June 1979.
19. Perkins, J., Smith, D.S., and Rivin D., "Pyrolytic Analysis of Carbon Fibers," *Eleventh Biennial Conference on Carbon,* page 349, American Carbon Committee, Oak Ridge, Tennessee, June 1973.
20. Hishiyama, Y. and Yamada, S., "Graphitization of Carbon Fiber and Glassy Carbon Composite," *Eleventh Biennial Conference on Carbon,* page 326, American Carbon Committee, Oak Ridge, Tennessee, June 1973.
21. Larsen, J.V. and Smith, T.G., "Carbon Fiber Structures," NOLTR-71-166, October 1971.
22. Fitzer, E and Simitzis, J., "Influence of Copolymer Content of the PAN Precursor and Tensile Load During Oxidation and Carbonization on Formation of Carbon Fibers," *Twelfth Biennial Conference on Carbon,* page 291, American Carbon Committee, Pittsburgh, Pennsylvania, July 1975.
23. Houtz, R.C., "Orlon Acrylic Fibers—Chemistry and Properties," *Textile Research,* Vol. 20, page 796, November 1950.
24. Shindo, A., Government Ind. Institute, Report #317 (1961), Osaka, Japan.
25. Warner, S.S., Peebles, L.H., and Uhlmann, D.R., "Oxidative Stability of Acrylic Fibers—Moisture Sensitivity," *Journal of Material Science,* Vol. 14, No. 11, page 2764, November 1979.
26. Gaulin, C.A. and McDonald, W.R., *Polyacrylonitrile Precursor Fibers.* (Aerospace Corp.) N.T.I.S. Report 755-350, May 1970.
27. Ezekiel, H.M., "Effect of Boron Catalysts During Graphitization," *Eleventh Biennial Conference on Carbon,* page 267, American Carbon Committee, Oak Ridge, Tennessee, June 1973; and British Patent 1,295,289, March 1973.
28. Eslick, M. and Bacon, D.J., "Room Temperature Mechanical Properties of Boron Doped Graphite," *Thirteenth Biennial Conference on Carbon,* page 439, American Carbon Committee, Irvine, California, July 1977.
29. Nelson, D.C., Pepper, R.T., and Goldfein, S., "Improvement in Strength of Carbon Fibers by Alloying with Boron." *SAMPE National Technical Conference,* Vol. 11, page 358, Boston, Massachusetts, November 1979.
30. Ezekiel, H.M. (AFML), "Formation of High Modulus Graphite Fibers from PAN Yarn," Report AD766-327/1-NTIS, November 1972.
31. Kinoshita, Y. (Toray), U.S. Patent 4,065,549, December 27, 1977.
32. Diefendorf, F.J., Division of Petroleum Chemical Preprints, Vol. 20, No. 2, page 444, 1975.
33. McLaughlin, V.C. and Moreton, R., NTIS-AD-A026-743, December 1974.
34. Moreton, R., "Spinning of PAN Fibers for the Production of Carbon Fibers," RAE Technical Report 70156, 1970; and *Fiber Science Technology,* Vol. 1, page 273, 1969.
35. Kalnin, I.L., "Graphite Fiber Reinforced Bulk Molding Compound," *Twenty-eighth Technical Conference,* SPI, RP/C Institute, Section 19B, Washington, D.C., February 1973.

36. Volk, H.F., "High Modulus Pitch-based Carbon Fibers," *Symposium on Carbon-fiber-reinforced Plastics,* Bamberg, West Germany, May 11, 1977.
37. Singer, L., "High Modulus, High Strength Carbon Fibers Produced from Mesophase Pitch," U.S. Patent 4,005,183, January 25, 1977.
38. Singer, L.S., Didchenko, R.D., Barr, A.B., Chwastiak, S., Lewis, I.C., and Lewis, R.T., "High Modulus Carbon Fibers from Mesophase Pitches," *Twelfth Biennial Conference on Carbon,* page 329, American Carbon Committee, Pittsburgh, Pennsylvania, July 1975; and page 96, *Thirteenth Biennial Conference on Carbon,* American Carbon Committee, Irvine, California, July 1977.
39. Barr, J.B., Jones, J.B., and Smith, R.E., "Analysis of Flaws in High Strength Carbon Fibers from Pitch," *Fourteenth Biennial Conference on Carbon,* page 224, American Carbon Committee, Penn State University, June 1979.
40. Okuda, L., "Pitch-based Carbon Fibers as Reinforcements," *Conference on Advanced Composite Technology.* Technology Conference Association, El Segundo, California, March 1978.
41. Romey, I., "Carbon from Purified Coal Tar Pitch," *Twelfth Biennial Conference on Carbon,* page 79, American Carbon Committee, Pittsburgh, Pennsylvania, July 1979.
42. Romey, I., "Production of Carbon Fibers from Char Tars," *Thirteenth Biennial Conference on Carbon,* page 94, American Carbon Committe, Irvine, California, July 1977; and U.S. Patent 3,997,654.
43. Fischbach, D. and Gilbert, W., "Diamagnetic Characteristics of Carbon Fibers," *Journal of Material Science*, Vol. 14, page 1587, July 1979.
44. Chambers, W.E., "Low Cost High Performance Carbon Fibers," *Mechanical Engineering,* December 1975.
45. Bell, J.E. and Perry, J.S. (Boeing), "Pitch Fiber Composite Design Data," *SAMPE, Tenth National Technical Conference,* Vol. 10, page 651, New York, October 17-19, 1978.
46. Shepler, R.E., Towne, M.K., and Saylor, D.K. (Union Carbide), "Carbon-Fiber Composites," *Machine Design*, May 1979.
47. Anon., *Plastics Technology,* page 101, July 1978.
48. Cooper, G.A., Gladman, D.G., Sillwood, D.M., and Sims, G.D., "Development of C-Fibers of Large Diameter," NPL-IMS-18, November 1972.
49. Bradshaw, W.G., Pinoli, P.C., and Vidoz, A.D., "Development of Process for Large Diameter Composite Monofilaments," NASA-CR-120973, 112 pages, 1972.
50. Jacob, B. and Neltri, R.D., "Large Diameter Carbon Monofilament," NASA-CR-121229, 40 pages, January 1973.
51. Krutchen, C.N., Flom, D.G., Gorowitz, B., and Roberts, B., *Eleventh Biennial Conference on Carbon,* American Carbon Committee, Oak Ridge, Tennessee, June 1973.
52. Schmidt, P.L. and Jones, W.C., "Carbon-based fiber reinforced Plastics," *Chemical Engineering Progress,* Vol. 58, No. 10, page 42, October 1962.
53. Gilliam, H.K. and Whittaker, M.P., "Mesophase Graphite," *Eleventh Biennial Conference on Carbon,* page 211, American Carbon Committee, Tennessee, June 1973.
54. Welhart, E.K. *et al.* (McDonnell Douglas), NTG-23360-NTIS, March 1976.
55. Riley, M.R. (Monsanto), "Development of Method of Concentrating High Modulus, Open End Fibers into a Yarn," Monsanto/Washington University Report HPC-72-147, April 1972.
56. Riewald, P. G. and Zweben, C. (DuPont), "Kevlar," *49-Hybrid Composites for Commercial and Aerospace Applications, Thirtieth Technical Conference,* SPI Institute, Section 14B, Washington, D.C., 1975.

57. Langston, P. (DuPont), "Kevlar in Plastic Reinforcement," *SAMPE, Fourth Annual Seminar,* Newport, California, October 1979.
58. Chatterjee, S. and Kulkarn, S., "Evaluation of Metal, Composite, and Hybrid Structural Members," *Thirty-Third Annual Technical Conference*, SPI, RP/C Institute, Section 12F, Washington, D.C., February 1978.
59. Short, D., and Summerscales, J. "Hybrids–A Review." Composites, Vol. 10, page 215, October 1979.
60. Kliger, H.S. (Celanese). "Ultra High Modulus Graphite Hybrids," *Thirty-First Technical Conference,* SPI, RP/C Institute, Section 12A, February 1976, Washington, D.C.
61. Bourdeau *et al.* (United Aircraft Research Laboratories), "Development of High Modulus C-Fibers," NTIS-AD 765 497/3, May 1973; and Simon, R.A., "Silicon Carbide Whiskers on Carbon," NTIS-AD757-659, January 1973.
62. Pinckney, R.L. and Freeman, R.B., "Physical and Structural Properties of Mixed Modulus Composite Materials," NTIS-AD-732-489, June 1971.
63. Gibbs, H.H., Wendt, R.C., and Wilson, F.C., "Carbon Fiber Structure and Stability Studies," *Thirty-Third Technical Conference,* SPI, RP/C Institute, Section 24F, February 1978.
64. McMahon, P.E., "Thermal Oxidative Resistance of Carbon Fibers and Their Composites," *SAMPE Symposium,* Vol. 23, page 150, Anaheim, California, 1978; and ASTM-STP-658, J.R. Vinson, Editor, pages 254–266, 1978.
65. Goan, J.C., "Graphite Fiber Oxidation," NOLTR 69-153, November 1969.

3
Synthetic Resin Matrices for Service to 200°C

Carbon and graphite fiber technology has made enormous strides in recent years in combination with synthetic resin matrices, and much attention will be given in this chapter to those applications which may reach 200°C, and in the following chapter to matrices for serving the higher service temperatures up to 300°C. Because we will examine those features which promote high strength advanced composites, emphasis will be placed in Chapter 3 upon those thermosetting resins (such as epoxies) which are most frequently identified with advanced composites. Chapter 4 includes resins which have thermosetting qualities as well as thermoplastic characteristics (such as polyimides). Attention will be given in Chapter 5 to the group of C/G fiber products which are combined principally with thermoplastic resin matrices for use at more moderate temperatures up to 200°C. Thermoplastic/graphite fiber composites hold promise for large scale industrial applications such as moldable, lightweight, high strength and heat resistant components. For comparison purposes, attention will be directed to the considerable work performed on advanced composites using glass fibers. These experiences have proven invaluable to carbon fiber technology.

One must not overlook the promise of C/G fibers in combination with metal matrices, and while experiences and knowledge of these products are not as voluminous as with the synthetic resins, there are reports which will be reviewed that underscore this growing interest. This subject is examined in a later chapter.

In deciding between the types of resin matrices at this time, the reader will soon recognize that, in achieving optimum performance and cost benefits, compromises may be necessary and that, in the true sense of the word, materials engineering will be used in selecting the best materials for the purposes intended. In consequence, acceptance of hybrid constructions will be anticipated, because there will be combinations of other fiber types, such as glass with carbon and graphite fibers, as well as the possibility of multiple resin matrices. The reasons for such decisions will be examined, and by example will suggest materials and processes to overcome the shortcomings of a specific combination.

Prominent among the thermosetting resin binders for advanced structural composites are epoxy resin systems and, more recently, the polyimides. There are other promising polymer types and these will be referenced. To draw meaningful conclusions in the present state of technology of C/G composites, attention is directed to those current materials and products, their strengths and their weaknesses.

EPOXY RESIN SYSTEMS

Developments in epoxy resin systems have spanned at least 30 years, starting with the Ciba Swiss patents in the late 1940's and the Shell Chemical Company patents in the early 1950's. In their application to C/G composites, they have been notably successful because of their excellent adhesion characteristics; a lower order of shrinkage during cure than with other competitive thermosetting materials; and a good balance of physical and electrical properties. When epoxy resin matrices are used in combination with graphite and carbon fibers in advanced composites for the aerospace industries, additional requirements of improved heat resistance and stability under environmental extremes have brought about improvements in the performance of the materials. The graphite/epoxy composites enjoy the advantage of several years of successful service applications—which, of course, make them more difficult to replace with newer materials unless there is a dramatic improvement in some properties. There are compelling reasons for considering other materials, such as cost advantages: the polyesters and some thermoplastics have an edge and are in use in experimental automotive applications, where performance requirements are not as high as on aircraft. Furthermore, the higher temperature capabilities of some of the polyimides have prompted experimental field adaptations of these resin matrices in advanced composites, which will be discussed after the epoxy systems, in the next chapter.

Discussions of epoxy resin systems entail examination of the basic epoxy polymers on the one hand, and the curing agents on the other hand. A wide variety of structures can be obtained depending upon the chemical combinations. Lee's book on epoxy resins[1] gives a background of the available materials and systems, though for the purpose of advanced composites, attention will focus upon those resins and curing agents used in this industry. Not only is their performance as cured parts important, but also their performance during processing, because if the C/G resin system cannot be easily applied, there is little to be gained by hoping for production of what can only be achieved in a small, limited capacity in an experimental study.

In establishing criteria for the selection of epoxy resins for use in C/G composites, it is appropriate to start with heat resistant systems. In general, not only do the heat resistant resins fulfill the requirements of the aerospace industry, which has been the most active in fostering such products, but well cured heat resistant resin systems may also be expected to have good environmental stability. There are many other factors which must be weighed in considering the choice of the resin systems, such as toughness, handling qualities during application to fibers, cure conditions, an awareness of problems related to carcinogeneity and flammability of cured products, and the influence of physical and thermal stresses and moisture upon the cured structure and the "aged" structure which has been in service. Recognition of some of these factors has been accorded in various reports. The long-range success of advanced composites will be influenced by the care with which the candidate systems are examined and by the attention given to earlier work which may have revealed some shortcomings. To make a point, there are reports in recent years which have extolled the virtues of a "new" epoxy resin or polyimide resin at high temperatures, overlooking the fact that ten years or so ago, industry had considered such products but rejected them because of carcinogenic tendencies.

Final decisions on the epoxy resin system will be based upon the performance of the C/G composite—but selection can be aided by examination of the "neat" resin alone, rather than having results obfuscated by the presence of a high volume of fillers or reinforcements. Filler loadings must stand on their own merits and contribute value, as well as resin systems which must respond similarly. Advanced composites owe their appeal to the synergism arising from reliable fibers and reliable resin components.

HIGH HEAT RESISTANT EPOXY RESINS

Epoxy resin systems are available for good high temperature performance (up to 200°C). The epoxy systems are the ones most widely used as thermosetting resin matrices for C/G fibers. More than ten years experience has been vouchsafed for these combinations in aerospace, and a growing confidence has been developed in C/G/epoxy composites. Published data on these materials will be presented further in this chapter, including interim reports as the carbon fiber technology was developing. The epoxy resin systems most frequently identified with advanced composites are listed below, together with their basic chemical structures. The curing agents most frequently used with epoxy composite structures are listed afterwards. High temperature stability depends not only on the resin used, but also on the curing agent, accompanied by the proper times and temperature of cure, as well as processing measures to minimize the presence of voids in the composites.

Diglycidyl Ethers of Bisphenol (Standard Epoxy)

By far the most widely used epoxy resins are the diglycidyl ethers of bisphenol A (see Fig. 3-1). They do not form the basis of the most heat resistant types, but they find large volume applications in protective coatings, and significant uses as adhesive, sealants, impregnants, bonding, and laminating materials. The low molecular weight standard epoxy resins are liquids, and the high molecular weight resins are solids at room temperature. They are sometimes used in conjunction with other epoxy resin types to improve upon some performance or processing qualities. Epoxy resin precondensates, which entail partial reaction with a portion of aromatic polyamines, are also being used in Europe with

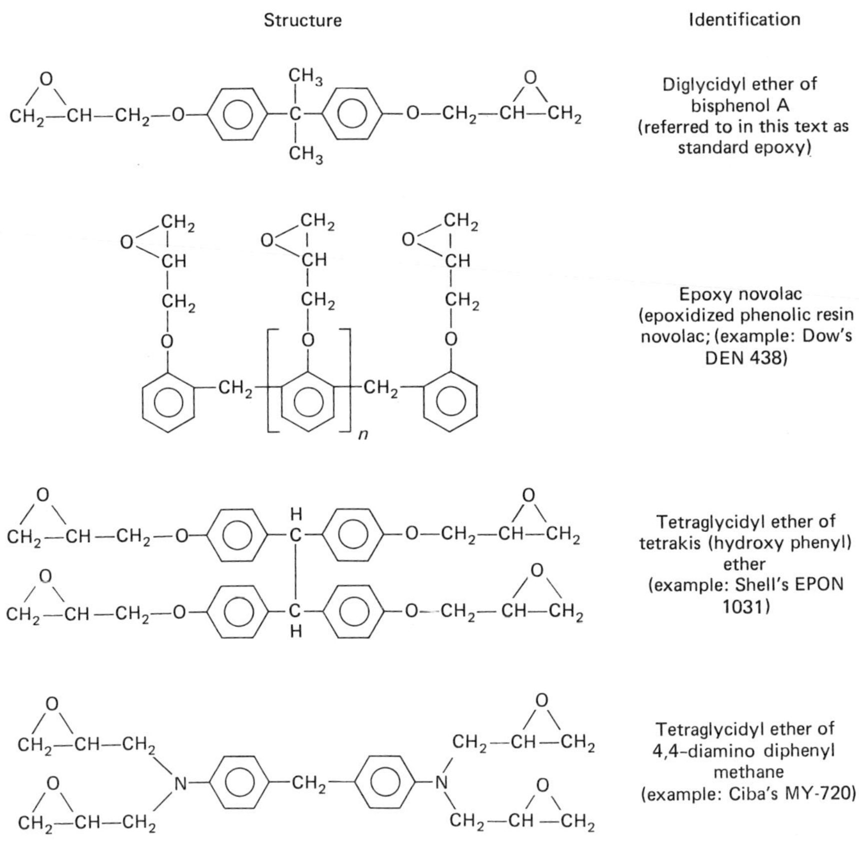

Fig. 3-1. Basic epoxy resins used in advanced composites.

graphite fibers to prepare composites (for example, "Epikote" 210[2]). Cured with BF_3MEA, graphite fiber composites demonstrated good water resistance on long-term aging, and better thermal resistance than the "standard" epoxy resin matrix.

Epoxidized Phenolic Novolacs

(See Fig. 3-1.) Like the standard epoxies, these resins are available in liquid and solid states, depending upon the molecular weight. In general, they react more rapidly with the same type of curing agent than do the standard epoxy systems. Epoxy novolacs brought a good measure of improved heat stability to epoxy resin systems, especially with reduced weight loss of cured materials when exposed to high temperatures for prolonged periods of time. Blending with liquid cycloaliphatics will produce lower viscosity without sacrifice of heat resistance. Cycloaliphatic epoxy resins are to be preferred over the monoglycidyl ethers frequently suggested as diluents. Phenolic resin blends with epoxidized novolacs offer processing advantages in manufacture of high temperature molding materials. The phenolic resins will react with and cure the epoxy resins–a system of potential value to composites.

Tetraglycidyl Ether of Tetrakis (Hydroxyphenyl) Ether ("Epon" 1031)

(See Fig. 3-1.) This epoxy resin makes available greater cross-linking density and hence better heat stability when cured. To achieve greater fluidity for preparing advanced composites, the solid "Epon" 1031 has been used in conjunction with liquid nadic methyl anhydride (see cure agents). This resin system was a favorite among the high temperature epoxy systems used with glass fibers in the early 1960's and was the subject of many reports. It was used to modify standard epoxy resins.

Tetraglycidyl Ether of 4, 4'-Diaminodiphenyl Methane (Ciba's MY-720)

This tetrafunctional glycidyl ether (see Fig. 3-1) has been used in preparing C/G composites. At room temperature, it is a viscous liquid (viscosity–15,000 cps at 25°C) and for laminating purposes, it is reacted with liquid anhydride curing agents, sometimes in conjunction with fluid cycloaliphatic epoxies. It is cured after 8 hours at 160°C or 3 hours at 180°C, followed by post-cure at 200°C for higher heat distortion. With NMA cure agent, ASTM heat distortion temperatures of 200°C are reached and higher values obtained on post-cure. Mixtures of

MY-720 and anhydrides are hygroscopic, and during application and lay-up of structural laminates, precautions must be taken to minimize water pick-up before cure, to attain good properties. Morgan and O'Neal[3] reported upon the cure of MY-720 with diamine diphenyl sulfone, which attained a peak T_g in the range of 25 to 30 parts of DADS per hundred of resin. They commented on the effects of absorbed moisture in crazing structures of this system.

Cycloaliphatics

The cycloaliphatic resins brought to the epoxy industry a whole new group of reactive polymers, several of which achieved prominence in the advanced composite field during the early 1970's. While both liquid (usually a mixture of isomers) and solid types of cycloaliphatic epoxy resins have been prepared, greater attention has been given to the liquid systems. Application as binders to printed circuit boards and as cast outside insulators for the electrical power industry have been prominent applications. ERLB-4617 has appeared frequently in developmental work on advanced composites (see Fig. 3-2).

The liquid cycloaliphatic epoxy resins made possible new developments in the field of epoxy resins by enhancing the physical properties of the products cast from them, as well as the composites which were prepared from them. Tensile stress-strain curves of epoxy cast resin systems are shown in Fig. 3-3.[4] Results in cycloaliphatics were due in a large measure to the pioneering efforts of Union Carbide Corp., though, in more recent years, Ciba-Geigy Corp. became more active in the distribution of some of the products. The manufacturing effort has been faced with problems due to the ease with which some of the cycloaliphatic epoxies would polymerize—sometimes with explosive violence. Commercial products have also been derived from blends with other epoxy resins, which have been less reactive, and at higher viscosities which would be better suited to the manufacture of prepregs. As a group, the cycloaliphatic epoxies have yielded higher heat distortion, higher strength, and better weather resistant products than the usual standard epoxies.

Those cycloaliphatic epoxies which have been important in the development of structural composites as resin matrices are given in Fig. 3-2 with their principal physical characteristics. (Data from bulletins prepared by Union Carbide Corp.)

A new epoxy resin system with improved processing qualities is based upon the triglycidyl ether of para-aminophennol, and cured with a tetrafunctional anhydride BTDA, which is readily soluble in this epoxy. The optimum molar ratio of the anhydride to the epoxy is 0.8. Three hours cure at 125°C is reported to yield a heat distortion temperature of 230°C.[5] Typical properties of composites prepared with this epoxy resin are as follows:

FIBER TYPE	S-GLASS	ARAMID	HERCULES AS-CARBON
Fiber content (%)	45	53	46
Tensile strength (ksi)	126	158	147
Tensile modulus (msi)	3.9	9.4	13.0
Flexural Strength (ksi)	142.0	61.0	177.0

CURING AGENTS FOR EPOXY RESINS

Among the various curing agents for epoxy resins are the chemical types which are listed below. In general, the aromatic diamines and aliphatic primary polyamines listed are used in stoichiometric proportions to the epoxies with which they co-react. The acid anhydrides, on the other hand, are usually proportioned 0.8 mole to 1 mole of epoxy, though there are exceptions which are specified by the manufacturer of the resin. Optimum heat resistance generally occurs when the curing agents are employed slightly under the correct stoichiometric ratio, and longer cure times and higher temperatures are practiced. Excessive curing agents will lower heat distortion and lead to poorer chemical resistance. Insufficient curing agents contribute to poor thermal stability and inadequate chemical resistance.

Aromatic Amines

These cure agents, which require elevated temperature cures, are widely used in laminating and filament winding formulations. The more common diamines are meta-phenylene diamine (m-PDA), 4-4′-methylenedianiline (MDA), and diaminodiphenyl sulfone (DADPS). All of these aromatic amines are solid at ambient temperature and require melting prior to mixing with the resin. Liquid versions are common with m-PDA and MDA which lend themselves to useful eutectic mixtures or solutions, or special derivatives (such as Hn 9231),[6] which offers improved heat resistance (over 50°F, higher heat distortion than the aromatic diamines). Co-accelerators are sometimes used to reduce cure time or to modify properties.

Aliphatic Primary Amines

Most polyamines in this group are reactive with the glycidyl ether resins at ambient temperatures. The principal examples are diethylenetriamine (DETA), triethylenetetramine (TETA), and tetraethylenepentamine (TEPA). These are liquid chemicals and are strong irritants to the skin. They must be used under good ventilation with protection for the hands. Their irritating qualities are

Structure and U.C.C. Designation		Name	Density (g/cm^3)	Viscosity (CPS at 25°C)	Epoxy Equivalent Weight (g/g•mol)
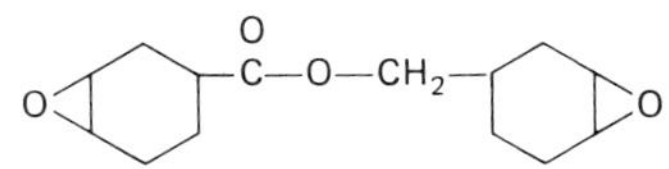	ERL-4221	3,4,-Epoxycyclohexylmethyl-3,4-epoxycyclohexane carboxylate	1.175	350-450	131-143
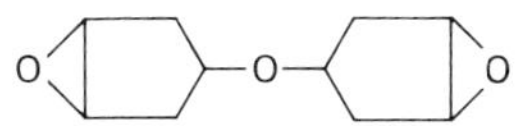	ERL-4205	Isomeric mixtures of bis (2,3-epoxycyclopentyl) ether (discontinued in favor of ERL B-4617)	1.16-1.18	< 100 (45°C)	91-102
ERL-4205 and $HOCH_2CH_2OH$	ERLB-4617	Copolymer ERL-4205 and ethylene glycol	1.17-1.18	80-150	110-120
ERL 4205 and standard epoxy liquid resin	ERL-2256	Approximately 50/50 blend	1.15-1.17	500-900	140-150
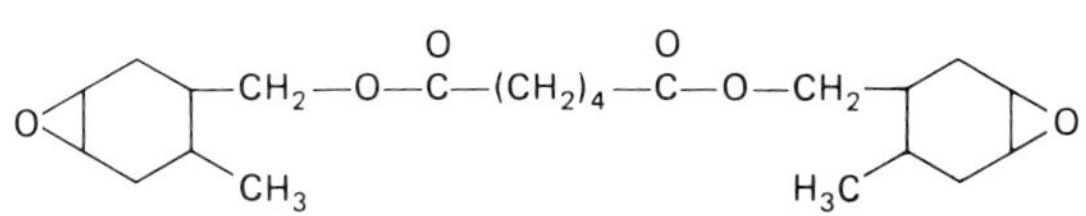	ERLA-4289	Bis (3,4-epoxy-6-methyl cyclohexylmethyl) adipate	1.124	500-1000	205-216
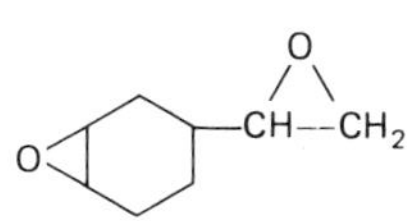	ERL-4206	Vinyl cyclohexene dioxide (class 6; SPI irritant: handle with care)	1.08-1.10	< 16.	227

Fig. 3-2.

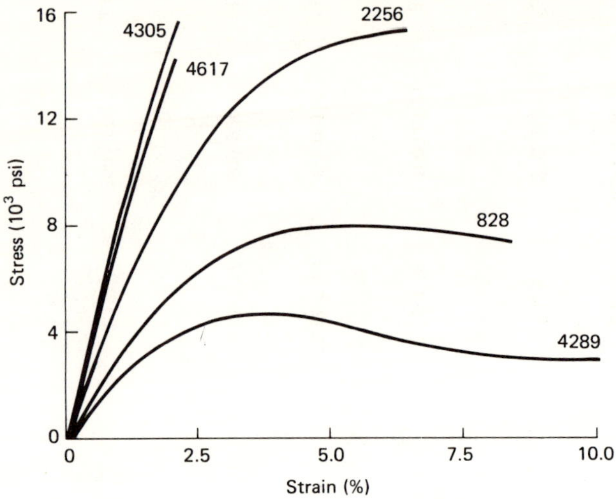

Fig. 3-3. Tensile stress-strain curves of cast resin matrix systems.[4]

reduced to some extent by the formation of resin adducts. Because they will react, for the most part, at room temperature, they are not used in prepreg systems and are used chiefly on formulations for the assembly or repair of aircraft components.

Polyamides and Imidazolines

Various commercial curing agents prepared from aliphatic primary amines and dibasic acids are available as relatively "safe" handling curing agents for ambient temperature cure and are employed in the adhesive and sealant formulations for industrial and aircraft components. Dicyandiamide (DICY) warrants special attention because of its extensive use as a catalyst for epoxy resins. It is latent at ambient temperature, though will cure at higher temperatures–above 150°C. DICY is generally used with a co-catalyst to reduce cure temperature and cure time, forming the basis of "250°F" epoxy cure systems. It is particularly valuable in prepreg formulations and in one-component adhesive formulations, though it will not yield the higher heat distortion temperatures of the aromatic amine curing agents.

Tertiary Amines

Tertiary amines function as catalysts or accelerators. They are employed principally as accelerators with epoxy/anhydride resin systems, such as are used in laminating formulations. Benzyldimethylamine (BDMA) and Tri(dimethylaminomethyl) phenol (DMP-30) are typical of the tertiary amines. They are usually present to the amount of 1/4 to 2% of the resin in the formulation of the epoxy matrix. Imidazoles are also fulfilling a similar role as catalysts in room temperature stable epoxy formulations.

Lewis Acids

Boron trifluoromonoethylamine ($BF_3 \cdot MEA$) is the most common cure agent for epoxies of this class. Lewis acids react very slowly at room temperature (1 to 3% by weight is used) and hence find usefulness in room temperature adhesive or laminating systems, which are cured with heat. The corrosiveness of the gaseous BF_3 during cure must be considered.

Acid Anhydrides

Most anhydrides require elevated temperature cures and post-cures with glycidyl ether resins, and are generally used in conjunction with a small amount of tertiary amine as a catalyst. The structural properties of laminates prepared from anhydride/epoxy resin systems are influenced a great deal by the chemistry of the materials used. Highest heat resistant systems are developed from the "Nadic" methyl anhydride (NMA), pyromellitic dianhydride (PMDA), and benzophenone tetracarboxylic acid dianhydride (BTDA) systems. Only the NMA anhydride of this group is liquid. Other acid anhydrides (such as dodecylsuccinic anhydride) will impart some flexibility to a system when this feature is required in the advanced composite, and will be discussed in another chapter. Chlorendic anhydride (HET) has been used in flame retardant formulations, though because of copious smoke evolution at the time of decomposition, its use has been minimized. The basic work on NMA cured epoxies has been performed by Delmonte and Hirosawa.[7, 8]

There are many other anhydride curing agents, besides those discussed above. Some, as trimellitic anhydride, are unstable, particularly in the presence of moisture, and it is difficult to establish process controls for them. Others, such as hexa-hydrophthalic anhydride, are fluid, though like phthalic anhydride do not contribute much to higher heat performance (although they may have value as co-anhydride cure agents). Others, because of their chemical structure, will increase flexibility and lower heat distortion temperatures. In addition, there are benefits to some epoxy/anhydride systems from the additions of small amounts of polyols (less than 2%) to minimize small incipient crack formations in such systems when they are exposed continuously to temperatures of 200°C for hundreds of hours.

In summarizing the review of the curing agents used with epoxy resins, it is only appropriate to remind the reader that manufacturers of the epoxy resin systems (including cure agents, modifiers, and fillers) have taken steps to offer optimum materials which will cure and perform as expected under well defined conditions. From the viewpoint of the manufacturer of the advanced composite, alterations in the materials provided (such as additions to lower viscosity) or in the cure procedure recommended may negate the otherwise good qualities of the product. If the desired results are not attained, discussions with technical representatives of the material supplier are the obvious measures to follow. While these comments will be viewed as redundant, designers of advanced composites must recognize that they are not dealing with a sheet of uniform material, but rather with a chemical system, properly cured and applied, which in combination with high strength fibers will yield structures superior to those obtainable from other sources.

PROCESSING MEASURES FOR THERMOSETTING RESIN SYSTEMS

Designers of advanced composites find that knowledge of the processing methods and variables will enable them to determine adequate design limits. To identify the basic problems that may arise, one must have some grasp of the processing measures used to develop prepregs of C/G fibers and resin matrices, and of the cure cycles which follow. These comments are applicable to the thermosetting epoxy, polyimide, and polyester systems. The comments which follow are general summaries, and interested readers are urged to refer to the considerable literature available on the subject of processing of reinforced plastics.[9] While these references have been developed principally around glass fiber reinforcements, there are advantages from perusing experiences in other fields of activity.

Most C/G composites for structures utilize continuous filaments of fibers or woven fabrics. The resin matrix must be capable of wetting and penetrating into the bundle of fibers, providing the reinforcement, replacing the dead air spaces therein, and providing those physical characteristics capable of enhancing the performance of the fibers. There is a considerable effort to minimize voids in the final laminate, as this will improve strength and fatigue resistance and enhance resistance to environment. Recognizing that substantially non-flowing resin matrices must be present in the composite at the time of lay-up in the tools (flow may occur later during heat cure under pressure to an infusible state in the cases of thermosetting resins; i.e., phenolics, epoxy, polyimides, and polyesters), one must recognize that the preliminary penetration into the fibers, as part of processing by the prepreggers, was accomplished in one of the ways described below.

1. *Solid or semi-solid resins and their curing agents were solubilized in solvents to permit easy penetration into the fiber system.* In this method, the solvents are substantially removed after application to the fibers, but not completely—for compelling reasons: 1) a small amount of solvent will improve drapability and tack of the fibers and their fabrics, valuable to laminate lay-up; 2) any attempts to remove all of the solvent, as by heat, will advance or polymerize the resin system so that it will be difficult to form or shape during processing; and 3) resin proportions added to the fibers may be more readily controlled with low viscosity systems. Complete removal of solvents from components is essential during cure (which may be staged at increasing temperatures and pressures).
2. *Solid or semi-solid resins and their curing agent were hot-melted into and upon the fabrics.* While this process eliminates the retention of solvents, compounding may be necessary to include low molecular weight components to achieve drape and flexibility. Great care must be exercised to avoid advancement of cure of the resin system to the point where it would not attain good impregnation. As low a viscosity as attainable with solvent systems is not realized. However, for systems that will experience a rapid through-put, as in the case of some filament winding and pultrusion operations (we will examine examples of these processes for C/G composites in a later chapter), liquid or hot melted resin systems with a rapid cure may be adapted with success.
3. *Inert carrier processes (used to a lesser degree) permitted a large variation of fiber/resin matrix to be attained.* The resin components and the fiber components are added to a rapidly stirred inert base (water or hexane, for example). When all components have been accurately metered into the mixture (which is not solubilized by the carrier), the inert carrier is filtered away through screens, sometimes shaped to approximate the final product. Traces of inert carrier are removed by heat and vacuum, leaving behind a

blend of fibers and resin systems in the desired proportions. These preparations do not have the fiber orientation required by the high strength, high modulus composites, but they do provide a technique of achieving polyblends of unlimited proportions. They are particularly useful for thermoplastic composites with non-oriented fibers. This process has been used to disperse epoxy resins and BF_3MEA cure agents with 20% carbon fibers to yield products with a uniform modulus of 7×10^6 psi. The same product prepared by intensive mechnical mixing, which damaged carbon fibers, had a 40% lower modulus.[10]

4. *A liquid system was used.* There are thermosetting resin matrices which are completely liquid at room temperature. If they are compounded to cure rapidly, they may be used for lay-up purposes in fabricating advanced composites. Work must proceed expeditiously, as the resin system continues to cure during application. It is for this reason that pre-distributed resin matrices applied and functioning as "dry" prepregs permit much greater latitude in lay-up, in that the resin matrix will not flow from place, as from vertical walls. Liquid or semi-solid systems for producing smaller components and for repair and maintenance in the field, have fabricating advantages, when manual measures are used to apply them, as pressures required for shaping are minimum.
5. *A mechanical process was used.* Conventional mechanical mixing devices, such as intensive internal mixers, ball mills, three-roll mills, and extrusion blenders, are available for blending fibers and resin matrices, though they are designed for non-aligned fiber formulations. Furthermore, the fibers are broken down in size during such mixing, and the available time is short for thermosetting resin matrices, which tend to react with the curing agents that have been added to the mixing operations.

These processes have been developed to handle resin matrices and their curing agents, and a wide choice of materials and procedures must be available to accommodate both small and large parts and materials with different cure requirements. Designers should have recognition of some of the chemical limitations which accompany these methods. As noted earlier, there are publications concerned with modern methods of preparing prepregs and composites (particularly for glass fibers), which should be consulted for further information. Prepreg systems, particularly if they consist of mixed and partially cured resin and cure agents deposited on C/G fibers or fabrics, must be protected from moisture in the atmosphere. For some thermosetting resin components, moisture will react with the resin components and negate otherwise good properties in the final cured product. Not only should protective films be applied at the time of preparing the prepregs, but also steps must be taken to avoid contact of moist air with tapes or fabrics of prepregs freshly removed from refrigerative storage, which are subject to condensation of moisture from the atmosphere. The burden is shifted to the user of the material to maintain such prepregs under the manufacturer's recommendation for refrigeration for matrix stability until ready for application and cure at high temperature. The reader is reminded of these details once again, but in examining reports on advanced composites, it is not always evident that proper storage and controls have been exercised upon the resin matrix system, and variable performance may sometimes be traced to these oversights. These comments are likewise applicable to most thermosetting resin matrices.

Most C/G resin tape and fabrics are pre-impregnated with resin systems and stored until ready for fabrication. Because time available for room temperature storage is limited, some refrigeration is required (the low temperature depending upon the manufacturer's recommendation, influenced by the reactivity of the cure agents used). Commercial practice is to ship such goods in a refrigerated truck, if the prepregs are prepared at a location different from the actual fabrication area. The one receiving such merchandise must promptly store it at recommended low temperatures until ready for use. Time of removal from the refrigerator before lamination and use is carefully monitored to assure the material is in a proper state of cure for the subsequent heat and pressure application. Resin suppliers work towards developing systems with long "working lives" to compensate for fabrication delays, without sacrificing processability or physical properties.

While production oriented engineers may plan in terms of rapid turnover of tools for fabricating C/G prepregs, present activity in high performance composites leans towards optimization of stength and quality. Staged cures and controlled rates of heating are more the rule for production. It is not uncommon for some of the heat resistant systems to require cure temperatures up to 600°F (317°C) and higher for post-cure in air. The processing advantages of one-component prepregs are obvious in that the fabricator or the advanced composites is not concerned with the stoichiometry of the resin systems that he is using as a matrix. The proper combinations are already provided by the supplier of the matrix system to effect cure. This advantage is also evident when using prepregs which already have the resin and curing agents applied to the C/G filaments or fabrics.

Recognition and acceptance of thermosetting resin matrix binders for C/G fiber composites appeared in the 1960's and onward. A review of some of the early papers draws attention to these valuable contributions, which have led to the advanced technology now being applied. In an early study of liquid epoxy resins and C/G fibers, Spain and Schwartz[11] conducted tests upon Union Carbide's "Thornel" 25 graphite fibers from a regenerated cellulose precursor. Resin matrix consisted of ERLA 2256 resin (cycloaliphatic epoxy in combination with a bisphenol A glycidyl ether). Fiber characteristics shown in this 1966 report were:

Average tensile strength–	184 ksi
Average modulus–	24.1×10^6 psi
Density–	1.5 g/cc.

Composites were prepared with fiber volumes ranging from 45.7 to 64%. There was a definite trend in increased flexural strength with higher fiber volume, though tensile modulus and flexural modulus showed no definite trend for the samples prepared in this early experiment. While a polyvinyl alcohol finish was used on the graphite fibers, no definite conclusion was drawn at this time relative to its effectiveness.

Also noteworthy is the early paper on carbon-based fiber reinforced plastics by Schmidt and Jones,[12] discussed in Chapter 2. (Tables 2-9 and 2-10 contain the authors' early data for reference; examination of these data indicates that Schmidt and Jones were clearly ahead of their times.)

The evidence for the advantages of high heat distortion temperature resins over lower heat distortion matrices of the same class, comes from many sources. In studies over a five-year period on epoxy systems, Schorr and Cianciarulo[13] reported on various epoxies, including the cycloaliphatic resins and their stability in outside weather exposure. High temperature cure systems, high heat distortion resins, and high temperature epoxy/anhydride resins were reported as the most stable.

In selecting epoxy resin systems for high temperature service, Delmonte[14] used weight retention and strength retention after exposure for prolonged periods at elevated temperatures. A standard diglycidyl ether of bisphenol A was cured with several curing agents and then exposed to 450° and 500°F for different intervals of time. Table 3-1 shows the results.

Above 400°F continuous service, most epoxy systems commence thermal degradation. Some of the cycloaliphatic epoxies and combinations with epoxidized phenolic novolacs offered improvements and began to appear in the late 1960's (re "Epocast" 8265 and "Epocast" 37, developed by Furane Plastics[15] in Los Angeles).

Another valuable paper was presented by Christie,[16] on the effect of resin parameters on the properties of cured epoxy resins. Using statistically designed experiments, many tests were performed to ascertain the effects of curing agent concentration, the weight per epoxy equivalent of the resin, and cure conditions on the mechanical properties of cured epoxy systems. He concludes that by increasing the cross-link density of the polymer through the use of curing agents with increased functionality, by utilizing aromatic amine or anhydride curing agents, and by formulating at slightly above 100% of stoichiometry for amines,

Table 3-1. Various Resin Systems Cured With Diglycidyl Ether of Bisphenol A.

CURING AGENT (PARTS PER HUNDRED OF RESIN)	PERCENT WEIGHT LOSS OF SELECTED EPOXY SYSTEMS* AFTER 24 HOURS AT		AFTER 100 HOURS AT	
	450°F	500°F	450°F	500°F
30 Diaminodiphenyl sulfone	-0.84	-3.4	-1.91	*D*
80 Nadic methyl anhydride (1% BDMA)	-0.59	-1.5	- .87	-3.4
24 Methylene dianiline	-0.80	-5.0	-2.7	*D*
12.5 TETA	-2.5	*D*	-5.5	–
15 Meta-phenylene diamine	-1.9	*D*	-4.2	–
4 BF_3 Monoethanolamine	-3.5	*D*	-6.4	–

**D* = badly cracked.

one can maximize the tensile strength of the epoxy resin system. On the other hand, with reduced cross-linking through the use of linear aliphatic curing agents, with a deficiency of anhydride cure agents, or with the presence of long hydrocarbon chains between reactive groups in the polymer, there is an increase in impact strength.

Lee[17] has carried out some programmed thermogravimetric research on diphenylolpropane (bisphenol A) and novolac-based epoxies cured with nadic methyl anhydride and methylene dianiline. He confirmed the observations that the epoxidized novolac resins are more stable than the epoxidized diphenylolpropane resins, and that anhydrides give more stable resins than amine hardeners.

Studies of epoxy resins as castings have yielded valuable information, when considering them for applications to composites. A good comparison was initiated by Craven[18] in evaluating the performances of epoxidized phenolic novolacs (DEN 438) and glycidyl ethers of bisphenol A (DER 332). Two Dow epoxy resins were used—DER 332, a low molecular weight resin of 4500 cps at 25°C, and a semi-solid DER 438. The influences of cure temperature and time upon the tensile properties of the castings (Table 3-2) are interesting. (Data in Table 3-2 are from Craven's paper.)

The data indicate clearly that the epoxy resin systems were not cured completely at 300°F. The mechanical data therein reflect the amount of cure applied. The data also remind us that heat distortion temperatures alone are not sufficient criteria in selecting resin systems for high temperature.

With the objective of developing better resins, it is propitious to study those properties of thermosetting resins which are the most significant in affecting the

Table 3-2. Tensile Properties of Unfilled Epoxy Casting Cured With Meta-phenylene Diamine.[18]

RESIN SYSTEM	CURE	HEAT DISTORTION TEMPERATURE (°F)	TENSILE STRENGTH R.T. psi	TENSILE STRENGTH 300°F psi	TENSILE MODULUS R.T. psi	TENSILE MODULUS 300°F psi	PERCENT ELONGATION R.T.	PERCENT ELONGATION 300°F
DER 332	1	277	15,700	560	608,000	15,700	4.5	27.0
	2	361	8,900	3940	412,000	41,900	3.8	14.0
	3	441	8,500	3870	432,000	41,000	3.4	10.9
	4	453	7,000	3390	420,000	35,900	2.8	10.0
DER 438	1	342	12,700	3260	526,000	40,800	3.8	20.0
	2	430	8,100	3060	521,000	44,200	2.1	9.7
	3	459	7,900	2510	535,000	42,400	1.9	8.6
	4	477	9,000	1990	511,000	42,500	2.4	7.4

Cure 1–2 hours at 200°F + 2 hours at 300°F.
Cure 2–Cure 1 + 2 hours at 400°F.
Cure 3–Cure 1 + 6 hours at 400°F.
Cure 4–Cure 1 + 24 hours at 400°F.

structural properties of C/G fiber epoxy resin composites. This subject was intensively investigated by Mahieu, Weidner, and Kuhbander.[19] Composites were fabricated with several different epoxy resin systems. Tensile and compressive loadings were made longitudinally in the direction of the unidirectional graphite fiber selected, which was "Morganite II," manufactured in England. Transverse loadings produced principal tensile stresses at 90° to the fiber direction (tensile test) and at 45° to the fiber direction (torsional test). Cast epoxy resins were also evaluated without reinforcing fibers in an attempt to determine those qualities that most affected composite performance.

The resin systems selected by Mahieu *et al,* for comparison purposes (see Table 3-3) were as follows:

A. Cycloaliphatic epoxy resin ERLA 4289, cured with a combination of hexahydrophthalic anhydride and a dibasic acid ("Empol 1022"). This combination yielded a low modulus, high elongation resin. The cycloaliphatic epoxy is distinguished by the longer than usual chain between the epoxy terminal groups for added toughness.
B. Standard diglycidyl ether of bisphenol A, modified by the addition of a finely divided synthetic rubber "CTBN" (which, when added to adhesive compounds in the amount of 8 to 10% by weight, contributes to improved peel strength and toughness). The "CTBN" co-exists with the epoxy resin as a separate phase. The curing agent used was "Epon" D tris-ethyl hexoic acid salt of DMP-30 (4%, based on the weight of the epoxy resin).
C. The cycloaliphatic resin derivative, "ERL-2256," which consists of a blend of diglycidyl ether of bisphenol A and a cycloaliphatic epoxy, ERN-4205. The latter cycloaliphatic compound, while very promising in its ability to offer fine mechanical properties, polymerizes very rapidly exothermically and must be moderated for safe handling in castings or when applied to composites. The curing agent was a eutectic of meta-phenylene diamine (27 p.h.r.)
D. "ERLA-4617," chosen as one of the prime candidates for use because of a good balance of mechanical properties and processing characteristics with C/G composites. In the current study reported herein, 24 phr of meta-phenylene diamine was employed as the cure agent.

While four different types of epoxy resins and their performance are shown herein, it should also be noted that four different types of curing agents were used, and hence another variable was introduced. The resins are all liquids, and the cycloaliphatic resins used were all under 1000 centipoises viscosity—suitable for penetration and minimizing voids in composites. Cure schedules followed the recommendations of the manufacturers of the resin systems. In preparing

Table 3-3. Key Composite Properties for the Various Resins.[19]

RESIN MATRIX	4289	828*	2256	4617
Cast resin properties				
Initial modulus (10^6 psi)	0.24	0.32	0.60	0.78
Proportional limit (10^3 psi)	1.7	3.2	2.4	7.2
Ultimate elongation (%)	81.0	8.1	6.5	2.2
Ultimate strength (10^3 psi)	5.4	8.0	15.2	14.8
Longitudinal composite properties				
Tensile strength (10^3 psi)	116.0	205.0	207.0	150.0
Tensile modulus (10^6 psi)	13.6	22.4	21.0	19.5
Proportional limit (10^3 psi)	38.1	80.7	70.4	56.7
Compressive strength (10^3 psi)	12.4	134.0	149.0	150.0
Compressive modulus (10^6 psi)	20.0	19.5	19.1	20.9
Proportional limit (10^3 psi)	7.8	44.3	76.2	36.0
Fiber content (% by Vol.)	67.8	62.0	62.7	61.3
Transverse composite properties				
Tensile strength (10^3 psi)	1.2	3.4	3.9	4.3
Tensile modulus (10^6 psi)	0.19	1.18	1.33	1.49
Proportional limit (10^3 psi)	0.08	1.13	2.3	4.0
Torsional stress (10^3 psi)	0.77	3.7	6.9	7.9
Torsional modulus (10^6 psi)	0.054	0.68	0.81	0.96
Proportional limit (10^3 psi)	0.13	1.2	1.7	3.2
Fiber content (% by Vol.)	64.8	68.2	66.7	61.2

*Modified with "CTBN".

the composites, however, methyl ethyl ketone solvents were used in all instances to facilitate impregnation of the fibers. The composites were press cured at 100 psi, starting at temperatures of 200° to 220°F, and finally up to 320°F.

Regarding the Morganite type II graphite fiber roving that was used for preparing composites, it consisted of a bundle of collimated graphite filaments, comprising approximately 10,000 filaments, each having a diameter of about 7 microns. Manufacturer's tensile strengths for the fibers were reported as 350,000 to 450,000 psi. Although kinks were reported in the bundles of fibers, which required removal of some samples from the test, and though there is a significant variation in percent volume of fibers in the composites, the comparison is interesting in that the graphite fibers exert the dominating influence upon the composite performance. The large variations in the ultimate elongation of the unreinforced base resin system are noteworthy. The added toughness expected from the low modulus, high elongation resin is more evident in high impact tests, as high impact, lower modulus systems are more prone to inhibit micro-crack formations under impact.

Representative epoxy resin systems for advanced composites are reviewed in the Mil Handbook, "Plastics for Aerospace Vehicles."[20] Materials selected have appeared prominently in development activities for advanced composites. A chart comparing candidate epoxy resins for advanced composites is shown in Fig. 3-4. The higher modulus and the higher strength of ERLA-4617/IM PDA system is evident.

Among the early recognition of the value of epoxy cycloaliphatics as high performance resins for advanced composites was a paper by Soldatos *et al.*[21] Two of the newly developed materials were singled out—ERLA-4305 and ERLA-4617—both of which utilized bis(2, 3-epoxycyclopentyl) ether. These two resins catalyzed with benzyl dimenthylamine exhibited excellent properties as cast resins, and furthermore imparted high performance to glass reinforced composites. In particular, the ERLA-4617 was utilized in latter work in the development of C/G fiber composites.

Further comparisons of the high performance epoxy resins were made by Soldatos and Burhans.[22] Data taken from Table 11 of this paper shows properties of the cast resins:

CURE AGENT	STANDARD EPOXY M-PDA	STANDARD EPOXY/ "EPON" 1031 MNA/BDMA	ERLA-4617 M-PDA
Tensile modulus (ksi)	458.0	507.0	783.0
Tensile strength (ksi)	13.0	9.1	19.2
Flexural modulus (ksi)	462.0	597.0	815.0
Flexural strength (ksi)	17.5	15.4	31.0
Heat distortion temperature (°C)	158.0	143.0	175.0

The ERLA-4617 epoxy resin system became the subject of further investigation in the preparation of C/G fiber epoxy composites. A report by Soldatos *et al.*[23] emphasized aromatic amine curing agents, directing attention to one which had consistently yielded high heat distortion temperatures for most epoxy resin systems. The curing agent or hardener which was singled out is a proprietary polymeric amine hardener manufactured by Furane Plastics. The trade designation of this cure agent is HN-9231. It is a liquid, and when blended with liquid epoxy resin systems, it offers a much longer working life at room temperatures, and better stability, than the usual meta-phenylenediamine or its liquid eutectics used with epoxy resins. Approximately 1.5% $BF_3 \cdot$ MEA will not only accelerate

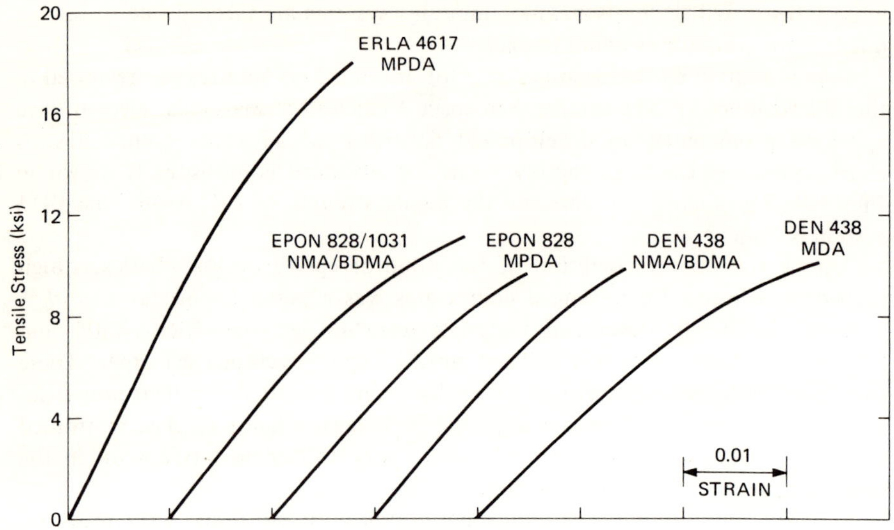

Fig. 3-4. Tensile stress-strain for epoxy resins.[20]

the cure of ERLA-4617 cycloaliphatic resin with graphite fibers, but also will raise the heat distortion temperature even further (see Table 3-4).

Table 3-4. Heat Distortion Temperature.*[23]

CAST RESIN SYSTEM		ASTM D-648
ERLA-4617/9231	100/24	205°C
ERLA-4617/m-PDA	100/24	175°C
ERLA-4617/9231/BF_3·MEA	100/24/1.5	235°C
ERLA-4617/m-PDA/BF_3·MEA	100/24/1.5	190°C
ERLB-4617/9231/BF_3·MEA	100/22/1.5	255°C
Bis A Epoxy/m-PDA	100/14.5	161°C
Bis A Epoxy/MDA	100/27	167°C

*Cure: 3 hours at 100°C/2 hours at 120°C/16 hours at 170°C.

With the cast cycloaliphatic ERLA-4617 resins shown in Table 3-4, the typical physical properties developed were:

Compressive modulus (psi)	800,000-900,000
Compressive yield strength (psi)	32,000- 32,500

Tensile modulus (psi)	800,000–900,000
Tensile strength (psi)	16,000– 20,000

As reported by Soldatos *et al.*,[23] the effects of heat aging on "Thornel" 50S/ERLB-4617/9231 unidirectional graphite composites are as given in Table 3-5 (fiber volume was 60% graphite).

Table 3-5. Composites of ERLB-4617/9231/Graphite Fiber.[23]

	AGING TIME AT 177°C (HOURS)	TEST TEMPERATURE R.T.	177°C
Longitudinal compressive strength (psi × 10^3)	None	88	82
	200	90	73
	1000	83	53
	3000	43	16
Longitudinal flexural strength (psi × 10^3)	None	105	131
	200	107	118
	1000	133	133
	3000	130	118
Short-beam shear strength (psi × 10^3)	None	8.3	5.3
	200	7.9	5.2
	1000	8.4	4.9
	3000	7.1	5.4

Further data on ERLA-4617 epoxy resin systems in conjunction with aromatic amines and hardener 9231 are presented in the chapter on the effect of moisture on composites. With HN-9231, there is a significant improvement in strength retention after humidity cycling.

Using an epoxy binder and basing comparisons on a 60% fiber volume, Freeman and Kuebler[24] reported on mechanical properties of advanced composites (see Table 3-6.) Note that A, HT, HM, and UHM are graphite fiber designations—refer to Chapter 2.

In reviewing the results, the authors noted that the first failure in any ply was analogous to the yield point in metal testing. They also felt that the elastic properties of oriented laminated materials can be predicted from measured lamina elastic constants. In drawing comparisons of the data shown in the previous table with other materials of construction, the low specific gravities of the advanced composites should be noted.

Among some recent published results concerning graphite/epoxy composites were data reported by Katz[25] upon several graphite fibers from different sources. Fiber volumes differed in some instances (see Table 3-7).

Table 3-6. Orthotropic Mechanical Properties of Advanced Composites.[24]

PROPERTY	A	HT	HM	UHM	PRD-49	S-GLASS	BORON
Tensile strength (ksi)							
0°	220	200	130	200	220	260	230
90°	7–12	6–12	5–10	2–3	3–4	5–9	9–12
Compressive strength (ksi)							
0°	220	200	130	97	45	100	400
90°	40	35	30	27	25	25	40
Tensile modulus ($\times 10^6$ psi)							
0°	18	22	27	45	11	8.0	30
90°	1.5	1.3	1.3	0.9	0.8	2.0	3.0
Ultimate tensile strain (%)							
0°	1.2	0.9	0.5	0.5	2.0	3.0	0.7
90°	0.5–0.9	0.5–0.9	0.5–0.8	.2–.3	.3–.5	.3–.5	.3–.5
Composite density (g/cc)	1.55	1.58	1.63	1.61	1.39	2.0	2.1

One does get a feel for the order of magnitude of values that may be expected from epoxy/graphite advanced composites from the data in Table 3-7.

Graphite/epoxy laminates were thermally cycled to determine the effect on the tensile properties and the thermal coefficient of expansion of the laminates. Tensile strength, tensile moduli, and expansion coefficients were all lowered by the thermal cycling. Property degradation was attributed to the progressive development of matrix cracks, whose location depended upon the lay-up orientation of the laminate.[26]

A paper by Yokota[27] of General Dynamics establishes clearly relationships between the void content and short beam interlaminar shear strength for epoxy/graphite composites. The subject of his tests was a Hercules AS/3501-5 graphite/epoxy prepreg system. In reproducing data related to the processing of these materials, the relationship between void content and short beam shear strengths was illustrated (see Fig. 3-5.) If pressure is applied too early, a high void results because of high early flow of the resin, leaving a starved laminate, while if pressure is applied too late, a high void results from insufficient consolidation of the advanced resin. Good translation of fiber properties into composites of continuous tows of graphite fiber has been reported by Dauksys.[28]

Table 3-7. Unidirectional Composite Properties of Commercial C/G Prepregs.[25]

	CELANESE	GREAT LAKES CARBON	HERCULES			STACKPOLE FIBERS	UNION CARBIDE
	CELION 3000	FORTAFIL 3	A	HT	HM	PANEX 30	THORNEL 300
Fiber Volume (%)	62	60	62	62	62	60	65
Tensile strength (ksi)	246	190	270	225	165	225	220
Compressive strength (ksi)	–	190	260	–	–	180	230
Tensile modulus (msi)	20.2	–	21	24	30	18	20
Ultimate tensile strain (%)	1.2	–	1.2	0.9	0.55	1.2	–
Flexural strength (ksi)	268	215	260	220	140	250	280
Interlaminar shear strength (ksi)	14.2	12	18	15	8	17	18
Density (g/cc)	1.57	–	1.55	1.58	1.63	1.58	1.57

NOTE: ksi = psi $\times 10^3$; msi = psi $\times 10^6$.

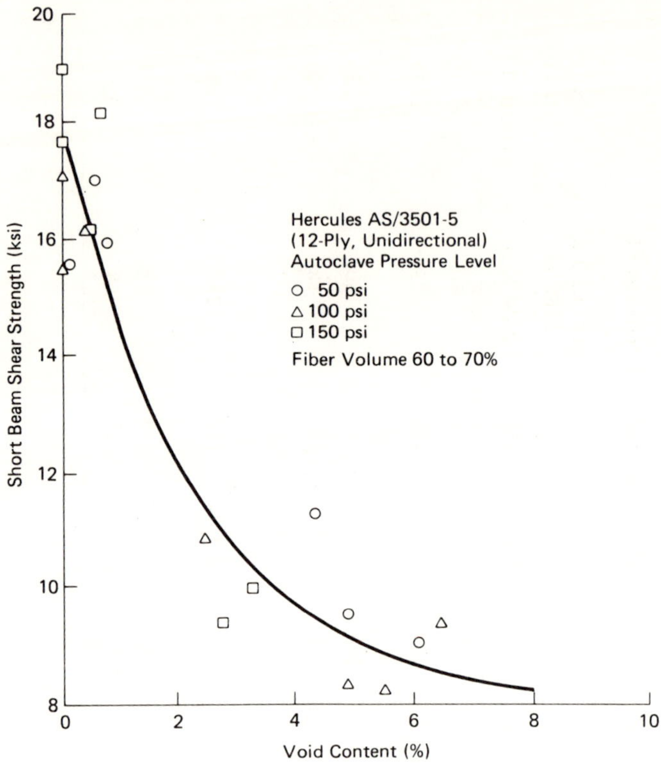

Fig. 3-5. Shear strength vs. void content.[27]

Ten contemporary graphite/epoxy materials were exposed to 350°F for up to 9000 hours in a circulating air oven. Mechanical properties of specimens were periodically evaluated.[29] Typical test results are shown in Table 3-8. T300 graphite and AS graphite were used in preparing the composites. As observed at 350°F (176°C) for up to 9000 hours, thermal aging badly degraded seven of the nine graphite/epoxy composites. The epoxy resin systems were not clearly identified, though it should be noted that, of the two specimens that fared well, one was a cycloaliphatic epoxy resin. It was concluded that 2000 to 3000 hours aging at 350°F (176°C) was the maximum period the majority of the epoxy/graphite materials could withstand without serious damage.

Morley[2] presented data on several graphite/epoxy composites prepared from surface treated "Grafil" HT-S carbon fibers. The composites, which were aged for up to 20,000 hours at 150°C, showed good stability. The emphasis was upon commercial applications of C/G epoxies which did not encounter the

Table 3-8. Average of Three Test Values for Test Specimens Aged at 350° F.[29] (Greer)

	FLEXURE				TENSION				COMPRESSION				SB SHEAR	
AGING (hours)	F ut R.T.	ksi 350°F	E 10^6 R.T.	psi 350°F	T ut R.T.	ksi 350°F	E 10^6 R.T.	psi 350°F	C ut R.T.	ksi 350°F	E 10^6 R.T.	psi 350°F	ksi R.T.	ksi 350°F
S/N 1														
Control	127.5	107.3	9.9	8.9	67.1	52.3	8.1	7.7	83.6	55.2	7.3	6.8	14.6	11.4
1500	125.3	145.0	12.7	11.7	60.9	52.6	7.6	7.4	52.7	53.5	6.9	6.9	9.0	8.1
3000	72.5	86.6	10.6	10.0	50.1	46.5	8.0	7.5	46.0	38.2	7.0	6.8	7.7	5.2
4500	42.1	39.3	11.6	10.3	48.4	43.7	7.6	7.3	27.7	24.2	7.6	7.3	3.4	2.0
6000	20.2	15.2	8.8	7.7	32.0	29.8	7.1	7.2	19.2	13.5	8.3	5.2	1.9	1.5
S/N 2														
Control	118.5	92.8	9.0	9.2	64.9	46.2	7.2	6.9	65.9	42.0	7.2	6.6	9.7	7.1
1500	138.1	118.8	11.1	10.6	58.5	56.2	7.3	6.9	61.8	46.9	6.6	6.3	7.6	6.0
3000	143.8	119.4	13.2	12.9	54.1	54.9	7.5	7.1	65.2	48.5	6.6	6.2	7.3	5.7
4500	124.3	103.4	10.7	9.5	60.9	55.9	7.5	7.3	52.7	48.7	6.6	6.6	6.5	5.7
6000	138.0	122.6	14.2	14.0	57.1	51.0	7.7	7.4	59.4	39.8	6.6	6.3	6.0	5.3
S/N 3														
Control	116.6	81.7	8.9	8.3	60.6	41.2	7.0	6.9	76.3	45.1	6.9	6.3	9.1	7.1
1500	122.0	109.7	11.1	10.8	52.0	46.3	6.8	6.4	56.3	45.9	6.6	6.5	7.5	6.5
3000	106.3	85.1	10.5	9.9	46.1	44.5	6.7	6.6	51.1	37.1	6.5	6.1	6.0	3.8
4500	63.3	51.5	10.3	8.8	37.6	37.4	6.6	6.5	28.3	20.4	7.5	6.8	4.7	3.4
6000	35.5	22.7	9.6	8.7	29.4	23.0	6.4	6.4	14.0	10.6	6.2	6.3	3.3	1.8
S/N 4														
Control	134.3	114.4	12.1	10.4	54.7	46.3	7.7	7.6	75.7	57.7	7.5	7.1	8.4	7.1
1500	125.4	119.0	11.1	10.2	62.4	58.3	8.0	7.6	58.1	48.1	7.3	7.3	8.4	6.3
3000	79.6	82.5	12.0	12.7	31.8	39.0	7.5	7.6	36.9	30.1	7.6	7.2	4.4	3.7
4500	49.5	37.1	9.7	9.2	31.3	28.0	7.1	7.0	17.0	14.2	7.7	6.9	3.7	2.3
6000	24.6	17.8	11.2	9.9	28.2	28.5	6.7	7.9	12.9	9.7	6.2	6.6	2.9	1.8
S/N 5														
Control	131.2	108.5	10.5	10.0	65.1	60.6	6.9	6.3	71.5	58.1	6.5	6.2	9.7	7.1
1500	128.5	109.9	10.5	8.6	64.4	55.0	7.0	6.7	64.3	54.9	6.5	6.1	7.6	6.7

Table 3-8. Average of Three Test Values for Test Specimens Aged at 350° F.[29]

	FLEXURE				TENSION				COMPRESSION				SB SHEAR	
AGING (hours)	F ut R.T.	ksi 350°F	E 10^6 R.T.	psi 350°F	T ut R.T.	ksi 350°F	E 10^6 R.T.	psi 350°F	C ut R.T.	ksi 350°F	E 10^6 R.T.	psi 350°F	ksi R.T.	350°F
S/N 1														
3000	111.6	105.4	9.0	11.5	47.1	46.0	7.4	7.0	61.2	49.2	6.4	6.3	6.8	5.3
4500	90.4	69.9	9.8	9.6	47.7	46.3	6.9	6.7	40.4	31.8	6.6	6.4	4.1	2.4
6000	51.9	39.4	8.7	8.8	33.9	34.8	6.4	6.4	25.4	22.1	5.9	5.7	4.1	2.6
S/N 6														
Control	130.0	89.6	11.6	9.2	67.0	56.8	7.5	7.1	66.0	45.4	6.2	5.7	9.3	6.9
1500	119.6	94.9	11.2	9.6	59.7	56.8	7.2	6.9	57.4	45.3	6.0	6.0	7.5	6.0
3000	100.9	98.4	9.3	11.5	53.5	49.1	7.5	7.7	42.9	42.6	6.3	6.0	5.9	4.9
4500	84.4	80.4	9.9	10.8	50.4	46.8	7.1	7.0	37.9	29.1	6.9	6.0	3.5	2.4
6000	60.8	49.5	9.2	9.1	38.5	40.3	7.0	7.1	20.2	18.4	5.9	5.8	3.0	2.3
S/N 7														
Control	129.2	32.8	8.3	5.3	68.6	54.8	6.4	5.2	65.5	8.8	5.6	4.4	10.1	4.4
1500	124.6	66.3	9.0	6.0	60.9	45.0	6.4	6.1	57.6	22.6	5.3	5.4	8.5	4.8
3000	101.0	73.5	7.8	5.9	59.0	44.2	6.9	6.4	55.2	29.9	5.8	4.8	8.5	5.2
4500	103.6	104.3	8.3	8.7	52.8	48.1	6.9	5.7	59.8	35.9	5.8	5.9	5.7	4.1
6000	124.7	107.7	11.0	10.6	53.4	47.0	6.3	6.0	53.1	35.7	6.2	5.1	5.8	4.5
S/N 8														
Control	148.7	104.1	9.7	9.5	74.8	66.5	6.6	5.9	89.0	58.3	6.6	6.4	14.1	7.7
1500	162.3	139.1	10.7	10.4	74.6	69.2	7.0	6.5	75.6	57.3	6.7	6.1	8.8	7.4
3000	115.3	101.7	9.8	8.9	71.6	64.5	6.8	6.6	55.2	42.5	6.3	6.1	7.0	5.7
4500	102.0	83.6	9.4	9.4	61.5	56.9	6.6	6.5	43.5	34.1	6.2	6.0	10.2	6.4
6000	59.7	52.6	9.6	10.5	47.3	44.3	6.2	6.3	30.7	26.9	6.4	6.2	4.4	3.5
S/N 9														
Control	107.8	–	10.7	–	57.7	–	7.6	–	80.7	–	7.5	–	11.1	–
1500	150.8	129.5	10.4	11.0	67.3	62.5	7.3	6.8	56.1	47.7	6.7	7.3	7.2	6.1
3000	110.5	79.7	11.1	11.6	66.9	60.3	7.1	6.5	38.7	31.8	6.9	6.9	4.0	5.1
4500	54.3	49.7	10.3	10.0	53.4	45.7	6.6	6.6	22.0	20.8	7.6	6.0	3.5	2.5
6000	42.1	24.9	10.5	10.5	41.0	37.0	6.0	6.2	16.8	17.7	7.4	6.0	1.6	1.2

higher temperatures of some aerospace applications. Some of the data reported in this paper are shown in Table 3-9.

Table 3-9. Results of Thermal Aging on Graphite Fiber/Epoxy Composites.[2]

RESIN SYSTEM	FLEXURAL PROPERTIES STRENGTH (MN/m^2)	FLEXURAL PROPERTIES MODULUS (GN/m^2)	INTERLAMINAR SHEAR STRENGTH (MN/m^2)
Standard epoxy and dicyandiamide cure agent			
Initial test at 23°C	1480	131	79
At 100°C	1000	115	55
After 20,000 hours at 70°C			
Test at 23°C	1500	125	90
Test at 70°C	1600	120	–
Precondensate epoxy and BF_3•MEA			
Initial test at 23°C	1380	130	79
At 100°C	1300	130	65
After 20,000 hours at 150°C			
Test at 23°C	1300	120	85
Test at 150°C	1000	120	–

PHENOLIC RESINS

Among the better known condensation-polymerization polymers are those derived from the reactions of phenols with aldehydes. They have been widely used in many industrial applications for well over 50 years, and are identified, for example, in the manufacture of cloth and paper laminates, prepared under high pressures and temperature. Lower in cost than epoxies or polyimides, phenolic resins still continue to serve important markets as binders, adhesives, molding, coatings, and laminating compounds. In the areas of advanced composites, phenolic resins have yielded to the epoxy resins because of the better adhesion, the lower cure shrinkage, and the absence of volatiles (hence voids) formed during polymerization processes. However, there are significant roles still being fulfilled by the phenol formaldehyde resins (phenolic resins). These are summarized below, though they are discussed throughout the book in connection with specific applications.

1. The high char strength of phenolic resins (as well as furfuryl alcohol resins) has been recognized and reported upon. For applications at temperatures which degrade most organic resins, such as ablative shields which must withstand the searing temperatures of atmospheric re-entry, as with orbiting vehicles, or in rocket nozzles, the special capabilities of phenolics are still recognized. Because controlled porosity is acceptable under such conditions, where physical integrity of the charred structure is important we have acceptance and use of phenolic resins and furfuryl alcohol resins, as binders for carbon fiber structures.
2. Phenolic resins function as cure agents for some of the epoxies, through their hydroxyl group, as well as serving as raw materials for epoxidation. As described earlier, the epoxidized phenolic-novolacs are available as candidate materials for advanced composites. The presence of phenol or thio-phenol in epoxy resin polymerization serves to accelerate the rate of reaction.
3. Recent developments in phenolic laminating resins have yielded binding resins of a low order of flammability and minimum smoke evolution during severe external fires. These are desirable qualities, particularly when one considers the alternatives, such as halogenated additives to the resin systems, which may lessen the probability of fires starting, though which, under external fire stimulus, yield noxious fumes and dense smoke.

POLYESTER RESINS

For many years, unsaturated polyesters have dominated the markets for reinforced glass fiber products. Major areas for reinforced polyesters include appliance, business equipment, consumer goods, construction, corrosion resistant products, electrical insulation, and transportation. The unsaturated polyesters are formed by reactions between: 1) dibasic acids composed of one or more saturated acids and/or an unsaturated acid (such as maleic anhydride or fumaric acid), and 2) glycols, such as propylene glycol, ethylene glycol, or diethylene glycol. The polymers normally are solid at ambient temperatures and, as such, are relatively stable and inert. To facilitate handling and impregnation of fibrous reinforcements, the polyester polymers are dissolved in reactive monomers, such as styrene, vinyl-toluene, diallyl phthalate, and methyl methacrylate. For optimum physical properties, 1.5 to 2.0 moles of monomer/mole of maleate in the polyester is recommended. Addition of phthalic anhydride or adipic acid as partial replacement of the maleic anhydride will reduce the reactivity of the styrene-modified resin. Catalysts for the cross-linking reactions are the organic peroxides, with an activiation through cobalt napthenate and/or tertiary amines. For improved high temperature performance, para-oxybenzoyl copolyesters are available as aromatic polyesters.

In the chapter on thermoplastic resin matrices for C/G fibrous reinforcements, some of the thermoplastic polyesters based upon polyethylene terephthalate and polybutylene terephthalate will be reviewed briefly. C/G polyester composites have not enjoyed a prominent role in aerospace applications, principally because epoxy systems have lower cure shrinkage and superior adhesion, and because of the availability of the higher heat resistant polyimides. The epoxies and polyimides are not only higher cost resins, but they also require longer processing times at higher temperatures than the polyester.

With the escalating interest by the automotive groups in resin matrices that are inherently lower in cost and which will cure rapidly, polyesters are receiving much recent attention in C/G fiber composites. Property requirements are not expected to be as high for automotive uses as those required in aerospace. One may expect more hybrid constructions with glass fibers to reduce the amount of C/G fibers, as well as the use of the lower cost resins.

There are some drawbacks in the use of styrene modified unsaturated polyesters, such as the high shrinkage of styrene, increasing restrictions on styrene levels in the air, and high exotherm temperatures during cure. With these considerations in mind, Edelman and McMahon[30] have concentrated upon using diallylphthalate as a monomer in a high performance polyester to be used in preparing C/G prepregs (using "Celion" 6000) for automotive uses.

Table 3-10. Comparative Properties of "Celion" 6000/Molded Composites at 22°C.[30]

	RESIN MATRIX		
	STYRENE POLYESTER	DAP POLYESTER	250°-TYPE EPOXY
Tensile strength (ksi; MPa)	229 (1579)	220 (1517)	221 (1524)
Tensile modulus (msi; GPa)	20.1 (139)	20.2 (139)	20.0 (138)
Tensile elongation (%)	1.1	1.1	1.1
Flexural strength (ksi; MPa)	303 (2089)	272 (1875)	287 (1979)
Flexural modulus (msi; GPa)	20.8 (143)	17.7 (122)	19.3 (133)
Interlaminar shear strength (psi, MPa)	10,350 (71.4)	13,400 (92.4)	14,000 (96.5)

Considerable significance is attached to interlaminar shear, as it is a good indication of adhesive qualities of the resin system. All of the C/G composites shown in Table 3-10 were compression molded for 15 minutes at 177°C under 500 psi. Test results on the experimental DAP polyester, when molded for one minute at the same temperature and pressure, showed flexural properties within 10% of those obtained at the longer cure time. Authors point out that prepregs of their system will not require refrigeration for several months of storage. The more reactive epoxy systems usually have to be stored at refrigerated temperatures.

POLYURETHANES

Most engineers look upon polyurethanes as potential resins for highly elastic products or as good candidates for low density foams–ignoring their potential qualities as structural materials. There are potential polymers with excellent structural properties, which, when coupled with high strength fibers, will yield good candidates for high production products for automotive industries. The principal candidate polyurethane polymers may be prepared from PAPI isocyanates and high molecular weight polyols, with specific mineral catalysts. Such systems are characterized by low cost, production rates of less than one minute, low shrinkage, and low exothermic temperature rise during cure. The principal drawbacks in the molding of C/G composites would arise from developing new processing techniques, because the urethane thermosetting resin systems are liquid and highly fluid under heat and pressure–though capable of rapid conversion to an insoluble and infusible state. Processing must accommodate the reactivity of the isocyanate reactive groups, which rapidly react with traces of moisture–*before* they have polymerized. With the rapid, automatic concepts of automotive production, this problem is solvable. One further comment on polyurethane structural composites is that the performance of the cured materials is not as good as epoxies at temperatures above 250°F. This drawback has been recognized for several years and, at the moment, no immediate solution appears in the offing, other than through drastic modification of the base polymer.

In spite of the above mentioned drawback, one should look upon polyurethanes reinforced with graphite fibers as good condidates for the intermediate temperature automotive applications. Comparisons of engineering properties of urethanes versus epoxies may be obtained.[31]

The adaptability of urethanes to the RRIM process (reinforced reaction injection molding) has been promoted in recent years in the automotive field because of their adaptability to the manufacture of lightweight, complex parts, which resist corrosion and denting. Noteworthy is a recent paper by Mikulec[32] on fiber reinforced polyurethanes, involving reinforcements of milled and chopped glass fibers, 1/8 inch graphite fibers, wollastonite, talc, etc. These products are currently being studied in conjunction with the RRIM processing.

The aspect ratio (length to diameter) of glass fibers appears critical for urethane composites. Tarin *et al.*[33] recommend an aspect ratio of 15 for best impact and best packing to achieve optimum concentration of fibers. Silane treated glass fibers out-performed the untreated fibers.

Processing considerations are paramount, and changes in equipment design for handling fiber filled resin slurries that are recommended are as follows.

1. High pressure cylinders for material delivery.
2. The mixing chamber should be capable of easy flush-out or clean-out when there is an interruption in processing. Static mixing tubes with multiple baffles have been effective in processing filled urethane systems.
3. Wear resistant components for parts exposed to abrasive slurry.
4. Short delivery lines and minimization of elbows and connections.
5. Continual mixing of fillers in their liquid components (in the absence of moisture) to assure uniform product delivery.

Among the advantages of C/G fiber in filled molded urethane systems is the reduction in the coefficient of thermal expansion (CTE) from a range of 60 to 90×10^{-6} inches/inches/°F to 20 to 30×10^{-6} inches/inches/°F, bringing it closer to the CTE of the steel members with which the molded components will be coordinated.

Nomenclature for the fiber reinforced polyurethanes is given in Table 3-11, reproduced in part from Mikulec's paper.

Table 3-11 demonstrates the continuing emphasis upon low density parts for automotive weight reduction. This appears to be one of the major thrusts to meet United States legislative standards for fuel economy in the 1980's.

MISCELLANEOUS THERMOSET MATRICES FOR ADVANCED COMPOSITES

PSP Resins

Recent research on a new group of aromatic heterocyclic polymers has been reported by a French group for use in advanced composites.[34] These polymers are classified as polystyrylpyridines and are referred to as PSP resins. The aromatic heterocyclic structure is obtained from the condensation of aromatic aldehydes with methylate derivates of pyridine (see the structure below). One important feature is their ability to be cured at 250°C after B-staging at 200°C,

Table 3-11. RIM and RRIM Nomenclature and Classifications.[32]

NOMENCLATURE	REINFORCEMENT (%)	TYPICAL FLEXURAL MODULUS ($\times 10^3$ psi)	DENSITY (g/cc)	TYPICAL APPLICATIONS
Low modulus RIM	–	5–50	1.0	Automotive fascia
Reinforced low modulus RIM	5–30	10–80	1.1–1.3	Automotive fascia, spoiler
Semi-high modulus (RIM)	–	50–100	1.0	–
Reinforced high modulus RIM	5–30	100–160	1.1–1.3	Fender, quarter extension, spoiler
High modulus RIM	–	100–350	1.0	–
Reinforced high modulus RIM	5–30	200–750	1.1–1.3	Fender, deck lid, hood, grille panel

with very little evolution of volatiles, while retaining a low melt viscosity. The resin precursor is soluble in dimethyl formamide or n-methyl pyrrolidone. As prepregs, the product is claimed to be relatively insensitive to moisture. Most of the work with this resin system has been conducted with unidirectional HTS C/G fibers, as well as some AS and HMS fibers. It is claimed that after 1000 hours in boiling water the flexural strength of C/G composites, which was 240,000 psi at R.T. and 175,000 psi at 250°C, remained virtually unchanged. Mechanical properties reported by the authors are reproduced in Table 3-12.

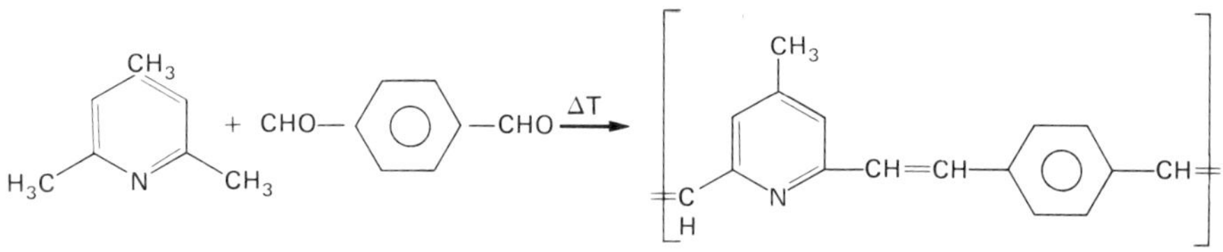

Polystyrylpyridine (PSP)

Table 3-12. Mechanical Properties of PSP-6030 Laminates Unidirectional Carbon Fiber.[34]

		HTS[a]	AS[a]	HMS[a]	T300[b]
Flexural strength	20°C	1700	1400	1100	1800
MPa	250°C	1250	1000	–	1480
Flexural modulus	20°C	110,000	90,000	150,000	134.000
MPa	250°C	110,000	90,000	–	121,000
Shear strength	20°C	90	100	50	90
MPa	250°C	70	70	40	62

NOTE: Cure cycle: 3 hours/200°C; 2 hours/250°C; pressure 1 MPa. Before test at 250°C–post cure 16 hours at 250°C.
[a]Courtauld fibers.
[b]Toray fibers.

A more recent report on the PSP resins[35] directs attention to the low flammability, low smoke evolution, and very low toxicity of pyrolysis products from composites of high char yield resin PSP-6022. It is also fortuitous that this product appears to resolve the problem of free-floating C/G filaments released under burning conditions from some composites made with epoxies.

Epoxy Terminated Polyimide

Another approach to high temperature polymers for use with C/G fibers was a series of epoxy-terminated imido structures by Ferro Corp.[36] In developing this resin, it was hoped that the high temperature performance of the polyimides would be retained, while gaining the process advantages of the epoxy. The resin that was synthesized was essentially N, N′-diglycidylbenzophenone tetracarboxy di-imide (DBTD). A peroxidation step on a diallyl intermediate attained the epoxy endcapping. Flexural strengths obtained from laminates molded from these novel resins demonstrated strength retention at 260°C (500°F) of the order of 70% of R.T. strengths, based on short-term exposure.

DGEBA-DGEBF Resins

Novel thermoset resins for fire resistant composites were described by Kourtides and Parker.[37] Using copolymer blends of the diglycidyl ether of bisphenol A (DGEBA) and diglycidyl ether of 9, 9-bis(4-hydroxyphenyl) fluorene (DGEBF), cure was effected with trimethoxyboroxine at 9.5 g/epoxy equivalent. High thermal stability is attributed to the fluorene group linked to the two phenyl groups in the polymer and to the trimethoxyboroxine catalyst, which enhances self extinguishing properties. This fire resistant resin system is considered by the authors to be an excellent candidate for thermally stable laminating resins. Their bibliography showed that DGEBF resin was first reported by the Soviet Union in 1971.

Bismaleimide resins prepared by reacting meta-maleimidobenzoic acid chloride with 4, 4′-diamino-diphenylmethane produced a high viscosity meltable polymer. The polymer, soluble in n-methylpyrrolidone and dimethylacetamide, was used to impregnate fibrous materials to achieve fire resistant laminates. It maintained flexural strength better than the usual epoxy laminates above 200°C. These resins were also described by Kourtides and Parker.[37]

Cyanate Resins

These polymers have been examined recently for possible use in aerospace applications to obtain improved interlaminar strength retention at 350°F for graphite composites after humid aging.[38] Some experimental cyanate systems utilizing "Thornel" T-300 graphite provided excellent 350°F strength retention after 1000 hour aging at 95% R.H. at 120°F. Cyanate resins are derived through reaction of bisphenol with cyanogen halide to provide a biscyanate ester, which upon thermal trimerization forms a meltable B-stage resin.

REFERENCES

1. Lee, H. and Neville, K., "Epoxy Resins," McGraw Hill Publ. Co. 1967.
2. Morley, J.M., "Role of Matrix in Preparation and Properties of Carbon Fiber Composites," *A.C.S. Organic Coatings and Plastics Chemistry,* Vol. 38, page 666, March 1978.

3. Morgan, R. and O'Neal, J., "Epoxies as Composite Matrices," *ACS Organic Coatings and Plastics Chemistry,* Vol. 38, page 485, March 1978.
4. Kuhbander, R.J. and Pagano, N.J., "Significant Properties of Epoxy Resins as Matrices in Graphite Composites," *NTIS* AD/A-026 028, March 1976.
5. Graham, J.A. and O'Connor, J.F., "Epoxy With Low Temperature Cure," *Adhesive Age,* page 20, July 1978.
6. Hirosawa, F., U.S. Patent 3,600,362, "Epoxy Resin Cure Agent," to Furane Plastics, August 17, 1971.
7. Delmonte, J., "High Temperature Anhydrides," *SPE Symposium,* page 10, Los Angeles, California, November 1956.
8. Delmonte, J. and Hirosawa, F., U.S. Patent 2,967,843, "Epoxy Resin Compositions Having High Heat Stability," January 10, 1961.
9. Lubin, G. (Ed.) "Handbook of Fiberglass and Advanced Plastics Composites," Van Nostrand Reinhold, 1969.
10. Yoshioka, N., Hakano, S., Takaku, A., and Sato, H., Chemical Abstracts 85-6713-1976.
11. Spain, R.G. and Schwartz, R.T., "Graphite Fiber Reinforced Composites," *SAMPE Symposium on Advanced Fibrous Reinforced Composites,* Vol. 10, page A21, San Diego, California, November 1961.
12. Schmidt, D.L. and Jones, W.C., "Carbon-base Fiber Reinforced Plastics," *Chem. Eng. Progress,* Vol. 58, No. 10, page 42, October, 1962.
13. Schorr, F. and Cianciarulo, A., *Polymer Eng. Sci.,* Vol. 5, page 166, 1965.
14. Delmonte, J., "Selecting Epoxy Resins," *Chem. Eng. Progress,* Vol. 58, No. 10, page 51, October, 1962.
15. McKague, E.L., "Graphite Filament/Epoxy Studies," *SAMPE Journal,* p. 49, April/May 1969.
16. Christie, S.H., "Effects of Resin Parameters on the Properties of Cured Epoxy Systems," *SPI Twentieth Annual Conference of Reinforced Plastics Division,* Chicago, Illinois, February 1965.
17. Lee, L.H., *Journal of Applied Polymer Science,* Part A-3, 859-1965.
18. Craven, P., "Epoxy Resin Properties and Performance as Castings, Laminates, and NOL Rings," *SPI Nineteenth Annual Conference of Reinforced Plastics Division,* Section 20A, Chicago, Illinois, February 1964.
19. Mahieu, W., Wiedner, J.C., and Kuhbander, "Significant Properties of Epoxy Resins as Matrices in Graphite Composites," Technical Report AFML-TR-70-305, page 82, October 1971.
20. "Plastics for Aerospace Vehicles," *MIL Handbook,* 17A Chapter 2, January 1971.
21. Soldatos, A.C., Burhans, A.S., and Cole, L.F., "Novel Cycloaliphatic High Performance Epoxy Resins for Reinforced Structures," *Twenty-third Annual Technical Conference,* SPI, RP/C Institute, Section 17A, February 1968.
22. Soldatos, A.C. and Burhans, A.S., "Novel High Performance Epoxy Resins," *IEC Product Research & Development,* Vol. 6, No. 4 page 205, December 1967.
23. Soldatos, A.C., Burhans, A.S., Eckstein, B.H., and Spence, G.B., "High Performance Epoxy Resin Systems with Improved Thermal Capability, Modulus, and Shelf Life for Graphite Fiber Composites," *Modern Plastics,* page 62, December 1971.
24. Freeman, W.T. and Kuebler, G.C., "Mechanical and Physical Properties of Advanced Composites," *Composite Materials–Testing & Design (Third Conference),* STP-546-(ASTM), pages 435–456, 1974.
25. Katz, H.S., "Carbon/Graphite Reinforcements," *Plastics Compounding,* Vol. 1, page 18, March/April 1979.

26. Sheldahl, R.E. and Wright, G.F., "Effect of Fabric Orientation on the Ablation Performance of Carbon/Phenolic," Report NTIS-Sandia-75-6088, 23 pages, 1975.
27. Yokota, M.J., "In Process Controlled Curing of Resin Matrix Composites," *SAMPE Journal,* Vol. 14, page 11, July/August 1978.
28. Dauksys, R.J., "Graphite Fiber Tow, Epoxy Matrix Composite," AFML-NTIS Report AD872-699/4ST, 43 pages, June 1970.
29. Greer, R.H., "Thermal Aging of Contemporary Graphite/Epoxy Materials," *SAMPE National Symposium,* Vol. 24, page 1039, San Francisco, California, May 1979.
30. Edelman, R. and McMahon, P., "New Polyester Systems for Automotive Composite Applications," *Thirty-fourth Annual Technical Conference,* SPI RP/C Institute, Section 2A, New Orleans, Louisiana, February 1979.
31. Delmonte, J., "Urethane or Epoxy–Which Encapsulation to Use," *Insulation/Circuits,* page 37, October 1977.
32. Mikulec, M., "RRIM–A New Process for Automotive Industry," *Thirty-fourth Annual Technical Conference,* SPI RP/C Institute, Section IIB, New Orleans, Louisiana, February 1979.
33. Tarin, P.M., Prokai, B., and Schwarz, E.G., "Effect of Particle Size Distribution on Reinforced High Modulus RIM Urethane Composites," *SPE Antec National Technical Conference,* No. 37, page 668, New Orleans, Louisiana, May 1979.
34. Block, B. and Ropars, M., "PSP Resins, New Thermosetting Binders for Advanced Composites," *SAMPE National Symposium,* Vol. 23, page 836, Anaheim, California, May 2, 1978, Re. U.S. Patent 3,994,862 (November 30, 1976).
35. Malassine, B., "PSP-6022 Resin," *SAMPE National Symposium,* Vol. 24, Page 1, San Francisco, California, May 1979.
36. Oswitch, S., "Development of a New Class of Polyimide Laminating Resins," *Composites,* page 56, March 1974; and U.S. Patent 3,700,617 to Ferro Corp. (October 24, 1972).
37. Kourtides, D. and Parker, A., "Thermoset Resins for Fire Resistant Composites," *SPE Antec National Technical Conference,* No. 37, page 586, New Orleans, Louisiana, May 1979.
38. Delano, C.B. and Harrison, E.S., "Moisture Resistant Graphite Composites from Cyanate Resins," *SAMPE National Symposium,* Vol. 20, page 243, April 1975.

4
Synthetic Resin Matrices for Use Up to 300°C

The development of high temperature resistant polymers has been encouraged by aerospace industries for composite applications which might be subject to very high speeds during supersonic flights or for the re-entry of extra-terrestrial vehicles. Epoxy systems described in the previous chapter have found their temperature limits near 200°C. High temperature resins, as typified by the polyimides, have demonstrated capabilities up to 300°C. In order to understand some of their advantages and disadvantages, the considerable development efforts during the 1970's will be reviewed. Increasing attention has been given not only to the inherent properties of the materials, but also to the availability of raw materials and the employment of processing techniques with which the aerospace industry has some degree of familiarity. As a product wins acceptance, an increasing amount of attention has been given to quality control measures to achieve uniformity of performance. At the time of writing, this phase has been reported in recent papers on the polyimides.

From a chronological viewpoint, Hergenrother and Johnston[1] have summarized the historical development of high temperature structural resins. E.I. du Pont de Nemours and Co., Inc. was awarded patents on the aromatic polyimides in the middle 1950's. Commercial polyimides in the form of high temperature insulation were among the first products made available from du Pont, though by mid-1960's, Narmco (then a division of Whittaker Corp., and now of the Celanese Corp.) made available polyimides for high temperature laminates. Polybenzimidazoles (PBI) and polyphenylquinoxalines (PPQ) also became available in limited quantities. In the late 1960's, TRW developed nadic anhydride end-capped imide-oligomers (for example, P-13N-polyimide, represented by P, the 13 representing average oligomer molecular weight of 1300, and the N the nadic end cap; see discussion further on). These products as prepolymers were not stable, and further work at NASA Lewis Research Center led to the family of solutions of monomeric polyimide precursors identified as the PMR resins.

Further developments at NASA Langley Research Center provided polymeric amines which enhanced processability (known as the LARC resins). In the meantime, du Pont offers high temperature thermoplastic polyimides of the same chemical types, without cross-linkable end caps, for use in application to C/G fibers (these are called the NR-150 B resins).

POLYIMIDES

Polyimide resins are among the candidate materials for service at high temperature up to 600°F (316°C) when used in conjunction with C/G fibers in the preparation of advanced composites. In general, the polyimide resins are higher in cost than the epoxy resin systems, and, in fact, some of the more advanced polyimides cost more per pound than graphite fibers. The economics of advanced aerospace applications will accept these cost differentials, though for the greater volume of automotive applications, recent activity in lower cost graphite fiber/polyester resins, as well as other thermosetting matrices such as the polyurethanes and the phenolics, have been examined in the previous chapter.

A valuable historical summary of the chemistry and the technology of high temperature resistant polymers was prepared by Markle of Batelle Laboratories[2] on the occasion of the fiftieth anniversary of the Organic Coatings and Plastics Chemistry Division of the American Chemical Society held in September 1974. Attention is directed to this reference because it contains a good summary of the history of the development of the aromatic polyamides and the polyimides.

Polyimide resins have been the central focus of development activity. The necessity for aromatic heterocyclic fused rings in the polymer chain to achieve heat resistance has been apparent to the polymer chemist, who has used these guidelines for research efforts. From a large scale utilization in composites, the processor thinks in terms of economical cost effective materials and processing characteristics, and the realization of a good thermal-physical profile.

Although the polyimides have been the focus of considerable attention in the past few years, and we will examine some of the earlier reports on composites which have drawn attention to them, polyimides and other high temperature polymers have been under United States government sponsored programs since the middle 1960's because of their potential for high temperature service, far beyond the capabilities of the epoxies and phenolics. In examining the earlier development in polyimides, an extra effort was needed to remove special high boiling solvents required to solubilize the resin polymer for impregnation. Furthermore, reaction by-products formed during the cure of impregnated composites required special processing to eliminate these volatiles. The cure times and temperatures have entailed gradual rises in temperature and pressure to permit volatiles to escape, and relatively long periods of cure, sometimes followed by cooling down under pressure. This has necessitated long processing times in an effort to attain void-free composites which would attain the best properties. The C/G fibers contribute to the desired strength and stiffness, and accommodations must be made for the resin matrices to attain a satisfactory performance at high temperatures up to 300°C.

It is appropriate to examine some of the earlier comprehensive studies on high temperature service laminating resins for advanced composites. It was realized that improvements in elevated temperature properties of reinforced plastics composites would be obtained through the synthesis of new, more oxidatively stable resins. Rapid gain in thermal resistive properties are achieved through modifications of the polymerization reaction so that the polymer chains were linked with oxidatively stable heterocyclic groups which replaced less thermally stable alkyl and ester linkages. The Air Force Materials Laboratory at Wright Field sponsored active research programs aimed at the synthesis of high temperature service resins (objective up to 1000°F) Aponyi[3] reviewed the state of the art in the mid-1960's. While at that time the principal reinforcements were glass fibers, the experiences gained with new high temperature resin matrices were subsequently applied to the C/G fibers. The high temperature resins reviewed by Aponyi included polyimides, polybenzimidazoles, and polybenzothiazoles.

The typical synthesis of a polyimide is the reaction between an aromatic tetracarboxylic dianhydride such as PMDA and BTDA and an aromatic diamine. Initial reaction forms a high molecular weight polyamic acid, which is soluble in solvents such as dimethylacetamide. As the prepolymer cures at elevated temperature, an aromatic linear resin is formed with evolution of water and solvent. The high glass transition temperatures of the polyimides permit good strength retention at 600°F. As will be observed from further reports arising at later times, the polyimides are currently the most promising high temperature resins.

Formation of Polyimide

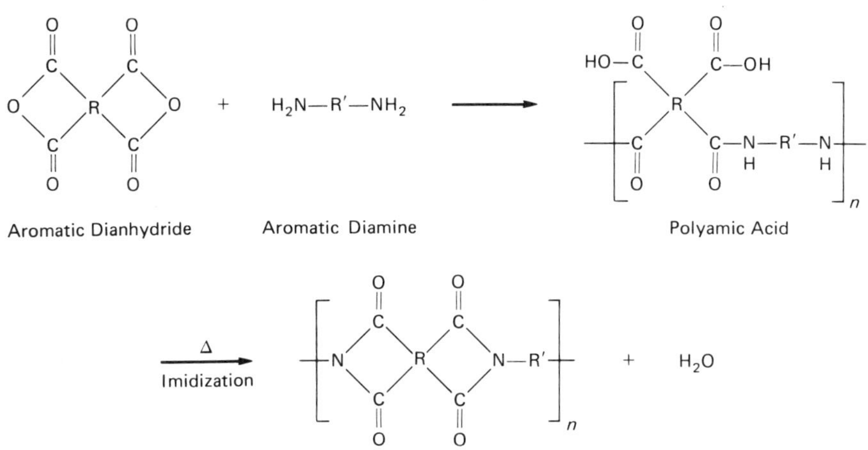

Polyimide

Developments in polyimides have followed several paths. In earlier work, condensation type polyimides have been produced by reacting an aryl dianhydride with an aryl diamine. This produced a polyamide acid through chain extension, the polymerization requiring a relatively long time. If polymerization is performed at too high a temperature, the temperature at which by-products cease to be evolved due to the chemical reaction, or because of difficulty in removing diluents, may be obscured by thermal degradation processes. These polyimide precursors were unstable and required refrigerated storage, even in solution form.

More recent work on polyimides has used partial esters of some of the polyanhydrides and have eliminated the evolution of water and other volatiles, furnishing high curing temperature, reactive end groups, which are capable of cross-linking to achieve high thermal stability. There are, in fact, thermoplastic as well as thermosetting polyimides, and while the next chapter is directed towards the thermoplastic resin matrices, it appears more suitable to encompass both types of the polyimides in this chapter. The thermoplastic polyimides have very high melting points and are classified as linear polymers with thermoplastic processing qualities.

A useful contribution by Freeman *et al.*[4] described a series of aromatic polyimides and glass cloth composites for a proposed high speed aircraft. While the variables in glass fibers and their sizing agents were examined principally, there is an organized presentation on variables in the chemistry of the heat resistant polyimide resins prepared for this study. Particular emphasis was placed upon 3, 4, 3′, 4′,-benzophenone tetracarboxylic dianhydride reactant (BTDA) and the aromatic diamines in preparing the polyimides. Aging data on E-glass laminates showed good strength retention for 250 hours at 650°F, and for 1000 hours at 600°F. It should be noted that some of the commercial polyimide resins developed in the early 1960's were based on BTDA and aromatic diamines (re Monsanto's "Skybond" resins).

The effects of molecular structure on the performance of various polyimides synthesized from pyromellitic dianhydride (PMDA) and BTDA were reported by Gillham.[5] It was pointed out that in ascertaining structure-property relationships in high temperature polymers, chemical reactions often occur in the vicinity of the high glass transitions, making it difficult to assign a value of T_g for a definite structure. Equal moles of dianhydride and diamine were reacted in dimethylacetamide at room temperature. Polyimides were prepared from the solutions of the prepolymers by thermally removing solvent and thermally effecting ring closing reactions. The following Figure 4-1 is reproduced in part from Gillham's paper.[5] The thermomechanical study by Gillham entailed a torsional braid analysis at about 1 cps. A glass cloth braid was impregnated with the sample prepolymer, and the solvent was removed during heating in a dry nitrogen stream. For isotropic homogeneous specimens subject to small strains,

Structure: Dianhydride		Structure: Diamine	Polyamic acid-viscosity	T_g
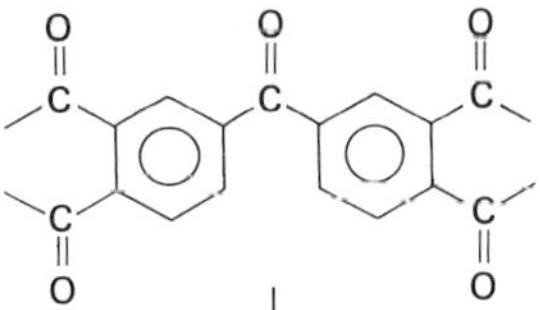	(1)	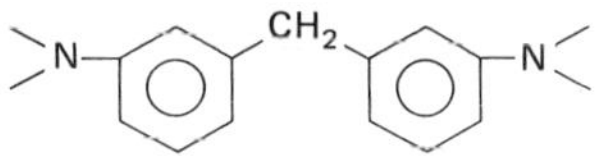	0.41	232°C
↓	(2)	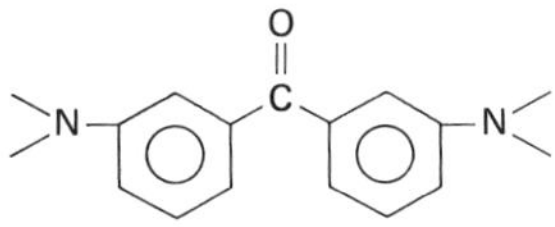	0.43	257
↓	(3)	Random copolymer of (2) and (5) (1/1)	0.55	271
↓	(4)	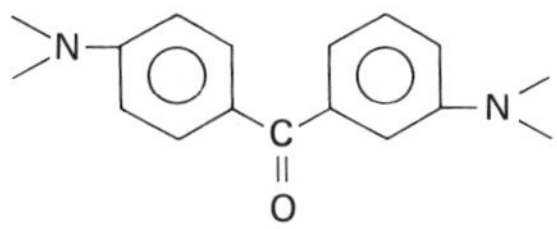	0.51	277
↓	(5)	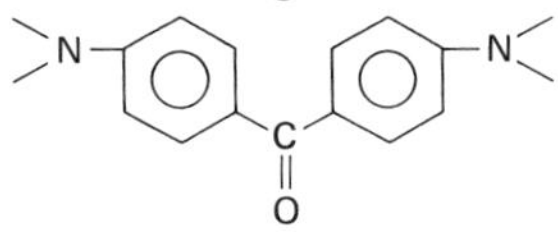	0.70	285
↓	(6)	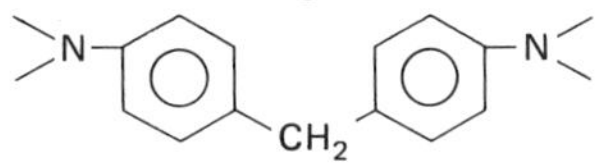	1.21	284
↓	(7)	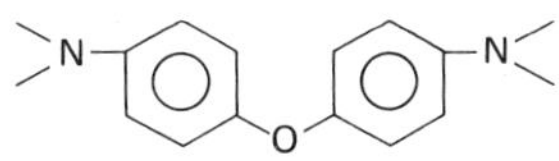	1.01	283
↓	(8)	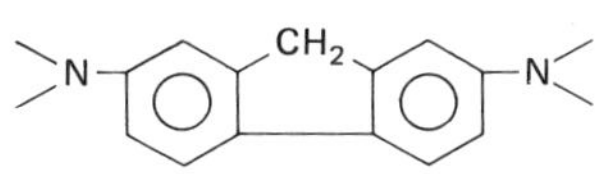	1.18	>320
↓	(9)		0.80	>320
↓	(10)	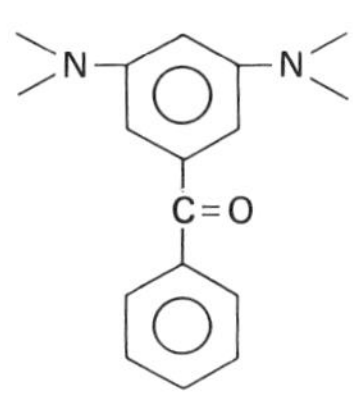	0.74	278
↓	(11)	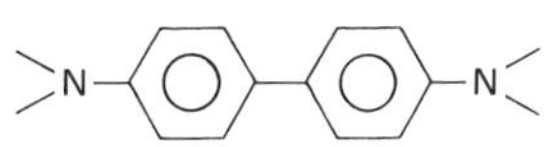	1.3	>300
↓	(12)	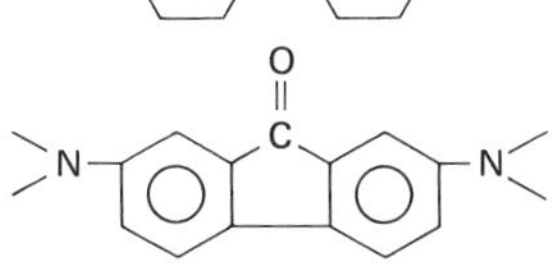	0.93	>300

(Inherent viscosity of polyamic acid (0.5% in DMAC at 25° C).[5]

Fig. 4-1. Inherent viscosity of precursor polyamic acids and T_g of cured polyimides, BTDA, and various aromatic amines.[5]

$G'=K(1/p^2)$, where G' is the in-phase elastic modulus, p is the period of the oscillation, and K is a constant depending upon geometry. Numerous curves and data on the rigidity were obtained for different polyimide resins at different temperatures, aiding in identifying the better resin systems. While PMDA polyimides yielded high glass transition temperatures (T_g) they were ill-defined because of thermally induced chemical reactions. Lower T_g and better solubility characterized the BTDA polyimide polymer resins.

Scheck and Stuckey[6] conducted a program under the sponsorship of NASA-MSFC to develop a graphite/polyimide composite, which would withstand at least 400 hours at 600°F with less than 50% loss of initial strength at both room temperature and at 600°F. Based upon the availability of the graphite fiber and the compatibility (ease of wetting by the candidate polyimides), several high modulus graphite fibers were selected for evaluation for these tests of 1971–1972 (Fig. 4-2 illustrates T_g data for aromatic polyimide). Properties of the selected graphite fibers were reported as follows.

	MODULUS (10^6 psi)	TENSILE STRENGTH (10^3 psi)	DENSITY (g/cc)
Modmor Type 1	60	250	1.94
Modmor Type 2	40	350	1.75
Type AS (Hercules)	28	350	1.75
Type HM-S (Hercules)	50	250	1.90
Type HT-S (Hercules)	32	350	1.75
GY-70 Celanese	75	325	2.03

The polyimide resin matrices selected by Scheck and Stuckey were du Pont's 4707 and 9622, Monsanto's 710, and Ciba's P-105A. The prepreg as received had a nominal 43% resin, 15% volatiles, and 20% flow, yielding the necessary tack, drape, and processing requirements, such as time, temperature, and pressure. Post-cure of panels covered graduated temperature rise from 350° to 700°F over a 20 hour period and a cool-down under pressure to 150°F. (See Table 4-1).

Table 4-1. Typical Tensile Strength of HT-S/Polyimide Composites.[6]

At 75°F	173 ksi
At 500°F	161 ksi
At 600°F (0 hours)	162 ksi
At 600°F (200 hours)	153 ksi
At 600°F (400 hours)	150 ksi
At 700°F (0 hours)	145 ksi

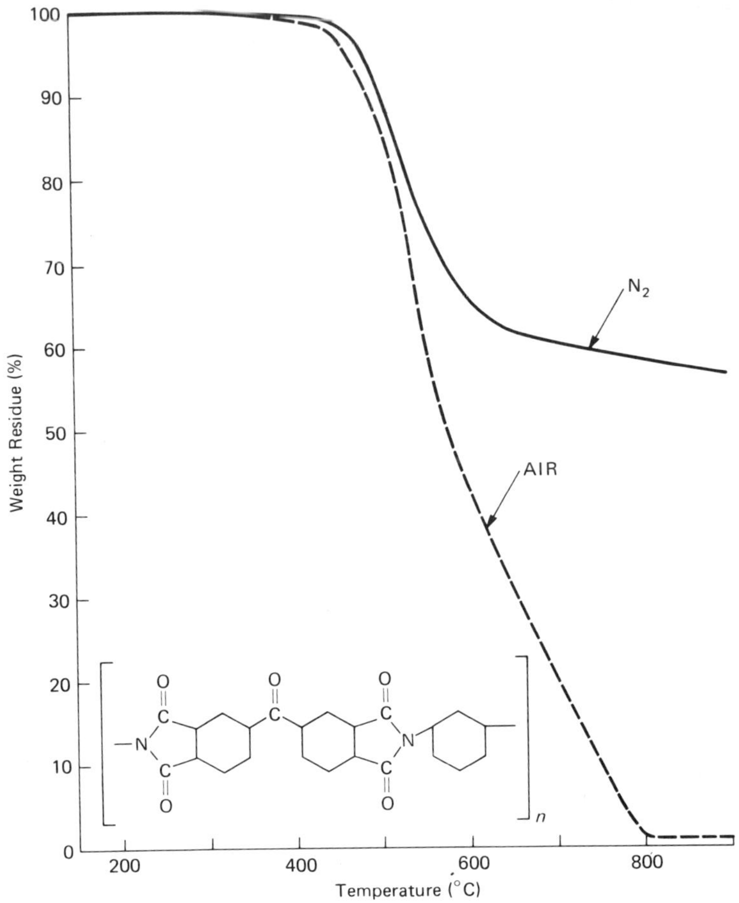

Fig. 4-2. TGA for an aromatic polyimide. $\Delta T = 150°C/hr.$[6]

The authors commented that the development of a laminating polyimide resin for the boron fibers provided some difficulty. Whereas the graphite fibers required a good flow resin to impregnate and wet the fibers, the resin in the boron composite tended to flow out during cure to leave a resin-starved, weakened system.

The processing of polyimides requires more care than the usual epoxy resin system because of steps required to fully cure composite structure at higher temperatures. These steps are designed to minimize void content in the composite through careful elimination of volatiles from the resin system, avoidance of excessive exotherm and loss of resin, and gradual build-up of pressure as polymerization of the resin system takes place.

The molding process for graphite fiber/polyimide laminates requires a staged program for raising temperatures to attain cure. While more complete lay-up procedures for molding and vacuum bag curing are detailed in the chapter on processing of advanced composites, there is tabulated below a typical commercial molding procedure[7] for graphite fiber/polyimide compounds.

1. Preheat mold and press to 400° ± 5°F.
2. Charge mold with preassembled, uncured prepreg.
3. Apply contact pressure only.
4. Raise temperature from 400° to 525°F at 5°F/minute.
5. At 525°F add 200 psi over a one minute period.
6. Hold at 525°F and 200 psi for one minute.
7. Remove pressure to 0 psi for 15 seconds, then reapply 200 psi.
8. Continue raising temperature at 5°F/minute to 600° ± 5°F.
9. Hold under pressure at 600° F for 2 hours.
10. Cool to room temperature under pressure.

To attain the optimum high temperature properties, the cured polyimide composite is further post-cured at even higher temperatures in some circumstances. Processing techniques also influence the selection of polyimide systems for manufacture of advanced composites. Hot melt addition type polyimides are used in solvent-free, wet filament winding at 125°C ("Keramid" 353, for example). Filament wound and pultruded shapes are prepared from such materials for high performance pressure vessels, tubes, fan blades, and radomes.[8] This polyimide has a high degree of flow, and the manufacturer recommends that optimum laminates are obtained with 10 to 15% resin flows. "Keramid" 353 requires post-cure at 240°C. Principal physical properties are shown in Table 4-2. Good creep and fatigue resistance are also reported.

Table 4-2. Properties of "Keramid" 353 Composites.[8]

	UNIDIRECTIONAL FIBERS					
	"MODMOR" II GRAPHITE FIBER		E. GLASS FIBERS		ARAMID 49 FIBERS	
PROPERTY	R.T.	250°C	R.T.	250°C	R.T.	250°C
Fiber content (volume, %)	60	–	60	–	60	–
Tensile strength (ksi)	161	154	153	149	175	110
Tensile modulus (msi)	25	24	6.0	5.6	9.1	7.3
Flexural strength (ksi)	212	168	176	121	94	45
Flexural modulus (msi)	24	24.7	6.1	5.8	9.5	7.2
Short beam shear (ksi)	11	7.2	11	7.2	4.0	3.0
Density (g/cc)	1.56	–	2.1	–	1.39	–

The relative thermophysical properties of some polyimides were reported by Jewell.[9] His thermogravimetric studies on ten polyimide films of varying composition showed that the thermal and thermal oxidative stability of the polymers increased directly with increasing imide content. As expected, thermal degradation processes were more rapid in air than in vacuo. The infra-red spectra showed little loss of imide bonds with progressive weight loss in air, suggesting a unit by unit degradation.

While much of the recent development effort in the United States was directed towards the solution application of polyimides, Cray[10] of the United Kingdom described a technique of processing the polyimides through a hot melt technique. His work was based upon some British polyimide developments.[11] With the objective of eliminating the costly solvent-based preimpregnation stage, this work was directed towards placing the fibers into a mold cavity together with a charge of powdered polyimide resin. Impregnation took place at 480°F, as the polymer was melted under pressure, followed by a cure time of 45 minutes at 570°F. Subsequent post-cure was carried out after removing the composite from the mold. Precision metal molds were necessary to attain the desired distribution of polyimide resin matrix. The work was performed upon the preparation of high temperature radomes from advanced composites.

Using condensation type polyimides dissolved in n-methyl pyrrolidone solvent and applied to glass fabrics, functioning as prepregs, over 2000 polyimide composites were produced for the Apollo program according to Clancy *et al.*[12] Difficulties with removing the water of condensation and the high boiling solvent in the polyimide solution were reported. Excellent, nonflammable characteristics were reported for the polyimides over other resin matrices used in fabrication of glass fiber reinforced plastics. In another paper,[13] graphite fiber reinforced polyimides were reported as not losing strength as rapidly as glass fiber reinforced polyimides at 600°F.

The adaptation of polyimides to advanced composites appeared in the development efforts of TRW, Inc. in the late 1960's.[14,15,16] Lubowitz *et al.* disclosed thermally stable polyimides prepared by condensing an aromatic polyamine (4, 4′-methylene dianiline), an aromatic dianhydride (3, 3′, 4, 4′-benzophenone tetracarboxylic acid dianhydride), and a specific monoanhydride for capping the polymer chain (3, 6-endomethylene-1, 2, 3, 6-tetrahydrophthalic anhydride or 3, 6-endomethylene-5-methyl-1, 2, 3, 6-tetrahydrophthalic anhydride (also known as methyl nadic anhydride). The monoanhydride is characterized by the formula given below.

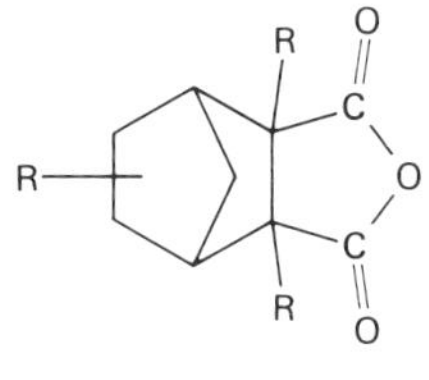

R represents a member selected from the group consisting of hydrogen and a lower alkyl radical such as methyl, ethyl, or butyl. The reaction was carried out in a solution of dimethyl formamide and toluene. High molecular weight solid prepolymers were dispersed in low molecular weight prepolymers dissolved in solvent. Applied to carbon cloth and cured in the range of 450° to 600°F, by means of increasing pressure (to permit volatile elimination), products of high heat resistance were obtained. This development was an advancement over earlier polyimide acid formation (from anhydrides and polyfunctional amines) in that the end-capping monoanhydride permitted a rapid cure as high temperatures were applied. The work culminated in the development of the P13N polymer, which was sold to Ciba-Geigy Corp. in the 1970's.

Brittle polymers from the polyimides described above and problems in processing led to the further work at NASA Lewis Research Center. This effort culminated in the patent awarded to Serafini *et al.*[17,18] The concept of using polymerizable monomer reactants (PMR resins) dissolved in easily evaporable solvents, such as methyl alcohol, led to better controls and processing. At the present time, aerospace and aircraft components are being produced from graphite fibers in a PMR resin matrix. The resin is being applied to C/G fiber reinforcements by prepreggers licensed by NASA. These polyimide impregnated graphite fibers are sold to aircraft component fabricators for the manufacture of advanced composites. The composites are characterized by low void content and interlaminar shear strengths up to 13,000 psi.

The preceding patent reference describes mixtures of monomers comprising: 1) dialkyl or tetraalkyl esters of an aromatic tetracarboxylic acid (3, 3′ 4, 4′-benzophenone tetracarboxylic acid is identified most frequently); 2) an aromatic diamine (4, 4′ methylenedianiline); and 3) a monoalkyl or dialkyl ester of dicarboxylic acid (5-norbornene-2, 3-dicarboxylic acid monomethyl ester). The necessity for forming the polyimide prepolymer prior to application to the C/G fibers is obviated. The PMR resin solutions are low in viscosity; the solvent is removed at 50° to 120°C after application to the graphite fibers, after which the prepregs are stored until ready for use. During cure, imidization takes place at about 205°C, and the high molecular weight, thermally stable polyimide forms at temperatures of 275° to 350°C. The end-capped portion of the polymer becomes reactive and cross-linking occurs to form the thermosetting structure.

The initial PMR polyimides were identified as PMR-15 (1500 molecular weight) and the components used, as well as their proportions, are shown in Fig. 4-3. A second generation of PMR resin monomeric solutions was identified as PMR-11. It is distinguished by higher temperature capabilities than the PMR-15, though the useful working life of PMR-11 solutions is about two days at ambient temperatures, as compared to the longer working life of two weeks of

PMR-15

Structure	Name	Abbreviation
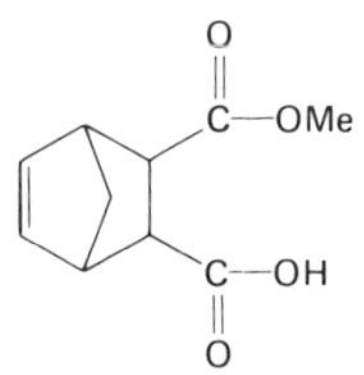	Monomethyl ester of 5-norbornene-2,3–dicarboxylic acid	NE
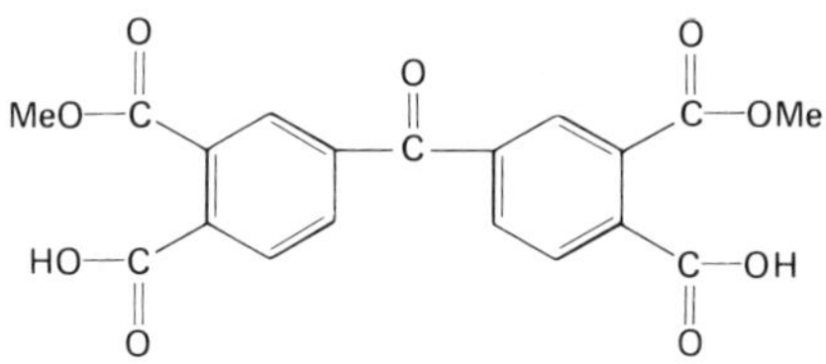	Dimethyl ester of 3, 3′, 4, 4′–benzophenonetetracarboxylic acid	BTDE
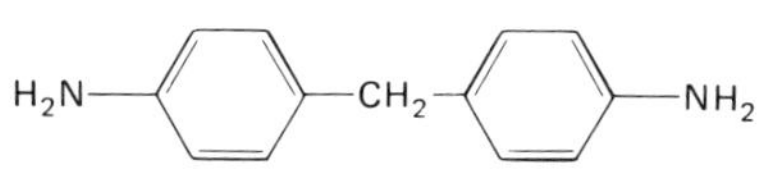	4, 4′–methylenedianiline	MDA

(a)

Fig. 4–3a. Monomers used for PMR 15 polyimide.[25]

PMR-11

Structure	Name	Abbreviation
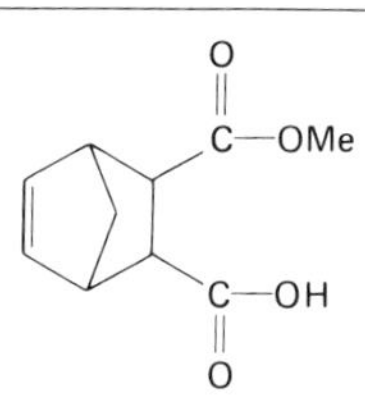	Monomethyl ester of 5-norbornene-2, 3-dicarboxylic acid	NE
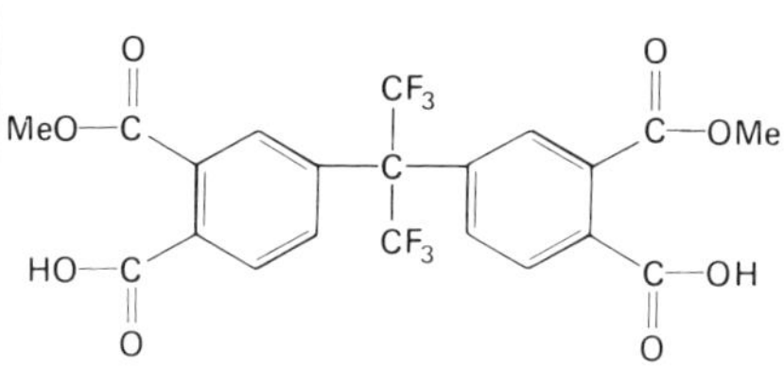	Dimethyl ester of 4, 4′–(hexafluoroisopropylidene)- bis (phthalic acid)	HFDE
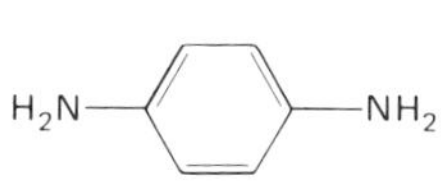	Para-phenylenediamine	PPDA

(b)

Fig. 4–3b. Monomers used for *second* generation PMR polyimides.[25]

its predecessor. The PMR-11 components are also identified in Fig. 4-3, and once again the low molecular weight amide-acid prepolymers formed are end-capped with norbornene rings, which polymerize without the evolution of volatile material. The principal components are the same reactive end cap (NE), para-phenylenediamine (PPDA), and the diacid-diester-dimethyl ester of 4, 4′-hexafluoroisopropylidene-bis-phthalic acid (HFDE).

Stoichiometry of Systems

PMR-15–2NE/3.087 MDA/2.087 BTDE

PMR-II–2NE/2.67 PPDA/1.67 HFDE

Processing of the PMR polyimides and the basic chemistry of the reaction were explained in a paper by Winters and Serafini.[19] The chemistry of the reaction is reproduced in the chart taken from their paper, as well as some of the data on the influence of formulated molecular weight on properties at elevated temperature (see Figs. 4-4 and 4-5). These data were developed upon PMR-15 formula and indicated the need for even better performance, as was obtained for PMR-11, which approached 600°F–an improvement over the earlier product.

The second generation of PMR type polyimides has demonstrated significantly improved thermo-oxidative stability[20] when tests are conducted after exposure to 600°F. Because of the faster reactivity of the PMR-11 resin system, test data at 600°F showed higher values for this resin system over the PMR-15, though the values equalized after post-cure at 600°F. PMR-15 monomeric reactants were polymerized in a methyl alcohol solution to a molecular weight of 1500. The solution was maintained at 50% solids. After impregnation of the graphite fibers on a drum winding machine, during which the volatiles were driven off by a heat lamp, the prepreg was covered with a film of "Teflon" and stored at 0°F (-17.8°C) until ready for use in fabricating the composite. T-300 graphite fibers were used in the composites, and data are shown in Table 4-3

Table 4-3. Comparison of Properties of PMR/T-300 Composites Under Autoclave Cure.[20]

	PMR-15		PMR-II	
Flexural strength (ksi)	As Cured	Post-cured	As Cured	Post-cured
R.T.	110.7	93.6		
600°F	62.3	86.2	82.6	83.4
Flexural modulus (msi)				
R.T.	8.0	7.7		
600°F	6.2	7.7	7.7	7.6
Short beam shear (ksi)				
R.T.	8.6	6.7		
600°F	4.7	6.1	5.9	6.1
Fiber volume fraction	0.53		0.55	

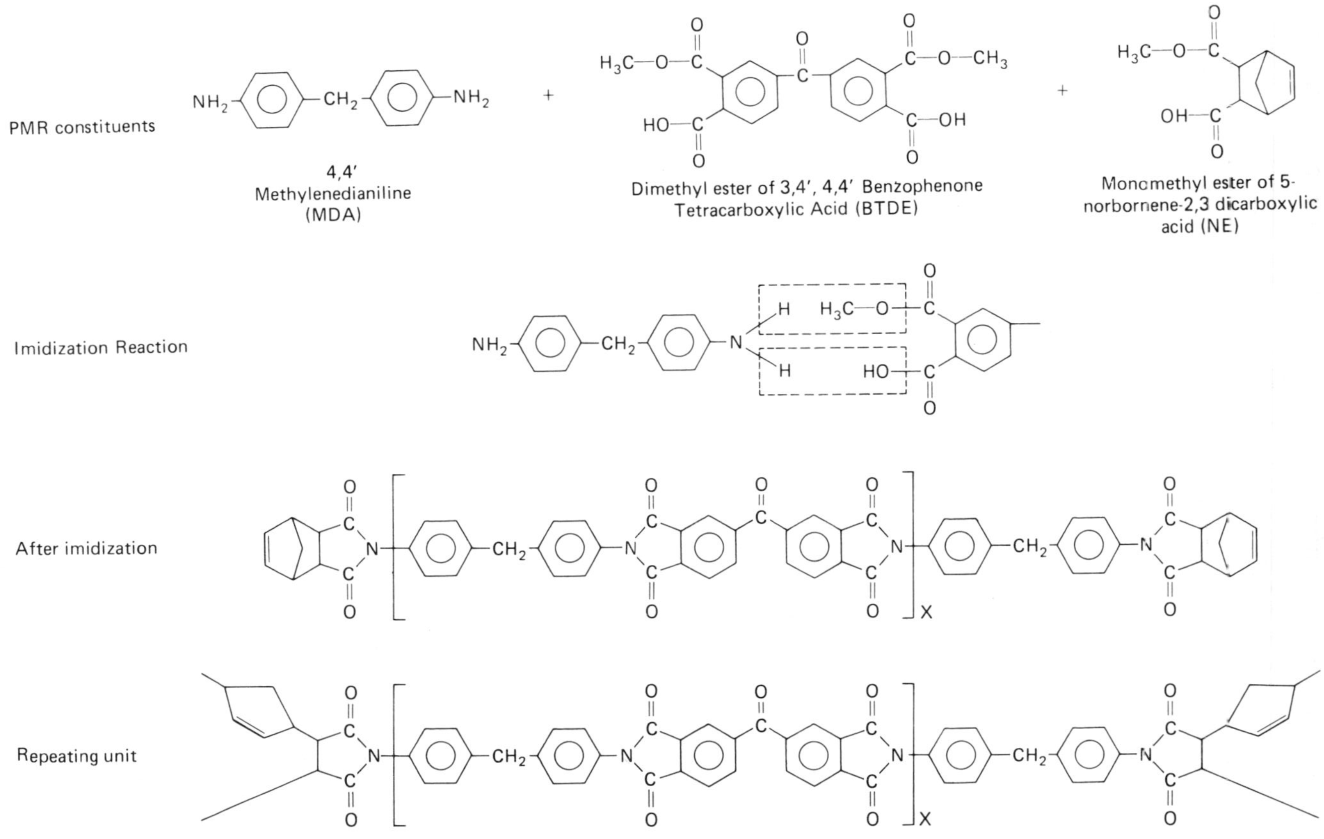

Fig. 4-4. Processing of PMR polyimides.[19]

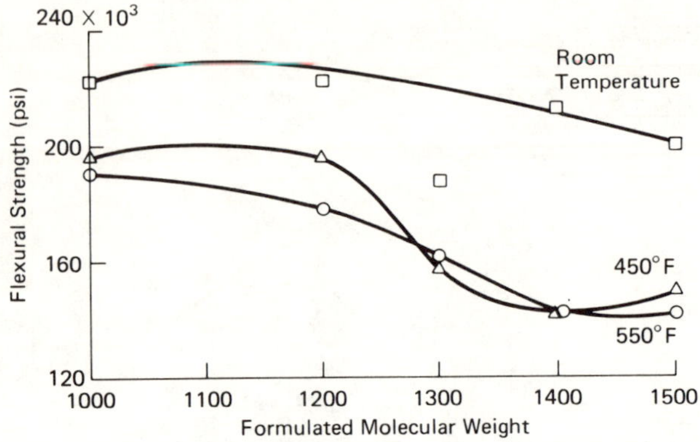

Fig. 4-5. Influence of formulated weight on polyimide/graphite composite properties.[19]

(both glass and graphite composites were produced with autoclave curing at 200 psi and die molding at higher pressure; die molding in an integrally heated tool used the following sequence).

1. Oven imidize for one hour at 400°F.
2. Insert preform in die at 450°F and hold 10 minutes at contact pressure.
3. Apply 500 to 1000 psi and raise at 8°F/minute to 600°F.
4. Hold for one hour at 600°F.
5. Eject part hot or cold.
6. Post-cure, if desired, for 16 hours at 600°F in air circulation oven.

Vannucci and Alston[21] discussed the PMR-11 polyimides and their improved high temperature performance. With resins formulated to different molecular weights in the range of 1100 to 3000, they observed significantly lower weight loss of PMR-11/HTS graphite composites after 600 hours at 600°F for the higher formulated weights.

Long term exposure to 600°F in air was reported on graphite/polyimide composites in a later reference.[22] The graphite fibers used in the test were HTS-2 and "Celion" 6000 fibers. Composites were prepared after impregnation with PMR-15 and PMR-11 monomer solutions. It was noted that data on the HTS fiber were improved considerably over earlier graphitic materials, yielding a significant increase in composite lifetime on exposure to air at elevated temperatures. Comparison data on PMR-15 graphite composites versus PMR-11 composites indicated higher values in flexural strength at 600°F and interlaminar shear at 600°F for the PMR-15 (see Figs. 4-6 and 4-7.[23])

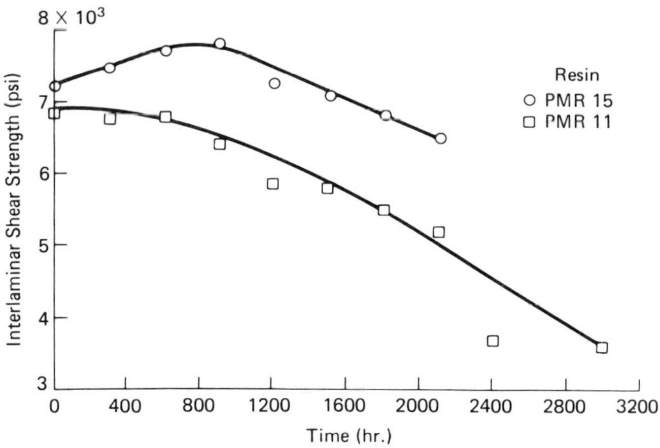

Fig. 4-6. I.L.S. stability of polyimide composites at 600°F.[23]

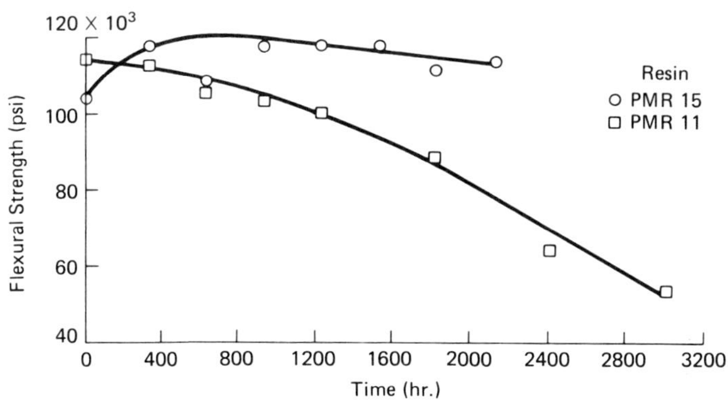

Fig. 4-7. Flexural strength stability of polyimide composites at 600°F.[23]

In some recent work, a variety of graphite fibers were fabricated to void-free composites, using PMR-15 solutions for impregnation. They were post-cured at 600°F and thermally treated to 700°F.[22] While there was a marked improvement in room temperature modulus and strength properties for short exposure, there was a marked fall-off in the interlaminar shear strength for the graphite fibers. Levels of shear strength and gradations of bonding appear to correlate. Considerable micro-cracking in the resin was evident after 700°F exposure.

During 600°F exposure, comparable fiber weight loss resulted from the use of either the first or second generation PMR polyimide matrix resins. PMR-15 matrix resins were subject to increased weight loss caused by thermo-oxidative interaction with the graphite fibers, related to the level of impurities on the fiber. PMR-11 was subject to less weight loss due to deleterious resin/fiber thermo-oxidative interactions. PMR-15 type matrix resins allowed the formation of significant amounts of loose surface fiber during prolonged 600°F exposure, as compared to PMR-11 matrix resin. Both PMR matrix resins formed a hydrazine indigestible residual resin on the graphite fiber surface. Data from the subject reference are shown in Fig. 4-8 comparing the resin weight loss of composites of PMR-11 and PMR-15 (and an earlier version of PM-13) on HT-S-1 and "Celion" 6000 graphite fibers at high temperature in air at 600°F.[24] Alston also shows weight loss of the C/G fibers over the same exposure period (Fig. 4-9), and the better behavior of "Celion" 6000 over HT-S-1, presumably related to the level of impurities on the fiber. In his conclusions, he drew attention to the fact that the PMR-15 type of matrix resins allowed significant amounts of loose carbon fiber on the surface, during prolonged exposure to 600°F, as compared to PMR-11 resin.

Serafini[25] prepared a status review of the polyimides, indicating the commercial availability of prepregs with PMR-15 resin systems. Because of their excellent processing characteristics, they are being used in a number of diverse structural components. The inner cowl for an experimental turbofan engine, oil tank brackets, and an aft body flap for a shuttle orbiter were cited as application examples.

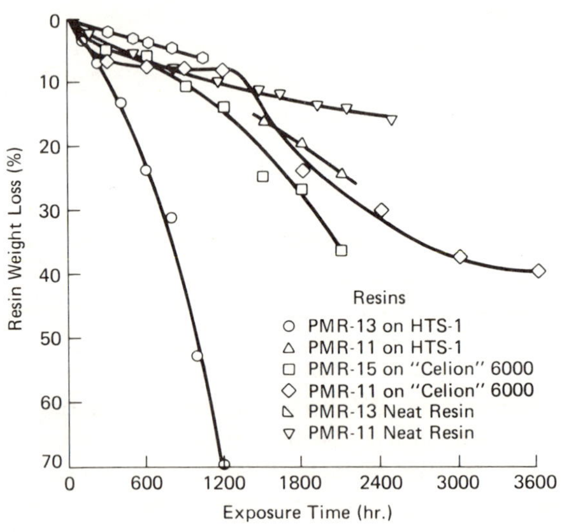

Fig. 4-8. Comparison of resin weight losses at 600°F.

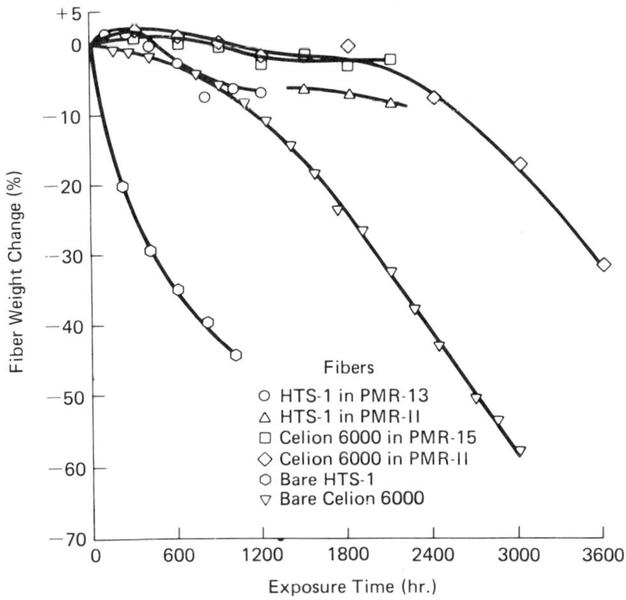

Fig. 4-9. Comparison of fiber weight changes at 600°F.[24]

Further developments in polyimide prepregs similar to the PMR resins were introduced as NASA Langley Research Center.[26] A solventless prepreg which retains good drape and tack was prepared with Hercules HT-S graphite fiber. Diethyl esters of benzophenone tetracarboxylic acid were used in conjunction with the norbornene end-capping agent (NE) described earlier, and a new liquid polyamine identified as "Jeffamine" AP-22 (experimental and limited availability). Ethanol solvents were also employed to aid some of the processing and application of polyimide resin to the fibers. Good retention of physical properties was reported after 2000 hours at 525°F. Interlaminar shear strengths were in the range of 50 to 60 MPa after aging and testing at 525°F. Fabrication development activities are reported at Boeing Aerospace Company and at Rockwell International.

The stability of the PMR polyimide solutions has been of concern. The influence of processing and storage variables on the final composite properties have been examined upon PMR-15 polyimide HT-S graphite prepreg. The prepreg quality was determined by resin content, solvent content, tack, flow, fiber collimation, yarn count, uniformity and thickness. Chemical tests involved gas chromatography, gel permeation chromatography, Fourier transform spectrophotometry, and T_g thermal tests. Basic properties determined on a commercial prepreg[27] are given below. The PMR-15/graphite prepreg contained 38.4% resin and volatiles of 11.0%.

PMR-15/HT-S COMPOSITE[27]

Flexural strength, MN/m^2 (ksi)	R.T.	1477	(214)
	600°F	890	(129)
Flexural modulus, GN/m^2 (msi)	R.T.	115	(16.7)
	600°F	112.6	(16.3)
Short beam shear, MN/m^2 (ksi)	R.T.	108	(15.6)
	600°F	59	(8.6)

In a further report, the PMR polyimide resin system studied was an alcoholic solution of three monomer reactants: an ester of 5-norbornene-2,3-dicarboxylic acid (NE), MDA, and a diester of 3, 3′, 4, 4′-benzophenone tetracarboxylic acid (BTDE). The ester studies included the tri- and tetraesters formed by further esterification of BTDE in the alcohol solvent. While storage at -18°C is sufficient to retard formation of higher ester impurities, the concentration of impurities was marked when stored at 5°C and higher. Correlation of decreased interlaminar shear strength and flexural strength in graphite composites was determined with increasing ester content—and a limit of 10% higher ester content was established. Data showed a 20% fall in flexural strength with this limit. Presumably, the more reactive PMR-11 would have a shorter storage time.[28,29]

Because of the reactivity at room temperature of the basic components forming the PMR polyimide resins, it has been desirable to premix the materials and apply promptly to C/G fibers as prepregs, which are refrigerated before application. The problem is quite basic, in that those polyimide systems which are low enough in viscosity for impregnation purposes do not lend themselves to convenient storage and are better processed early into finished components, before polymerization advances too far either in the solutions or on the prepregs for good processing. The situation is further complicated by the fact that reactions occurring during storage are not necessarily the same reactions that are anticipated during the processing of the polymer.

As graphite/resin matrices encounter higher operating temperatures, above 500°F (292.5°C), the influence of contaminants as contributing factors to more rapid thermal degradation becomes more apparent. Gibbs *et al.*[30] reported that traces of residual sodium sulfate on the carbon fibers was one of the principal destabilizing factors, and fibers with higher traces of sodium ions on their surface yielded composites which were less stable. These observations have resulted in graphite fibers with less contaminants on their surface, and evidence of greater selectivity in applying and selecting sizing agents has been apparent as fabricators report more recent "batches" of graphite yielding better results, to enhance bond to resin matrices. Much of this work has been reported for high temperature epoxy/graphite composites, though undoubtedly studies are in progress upon polyimide binders with C/G fibers. The exposure of hybrid structures of glass and graphite fibers may cause instability factors due to the chemicals of one system affecting the other at high operating temperatures. On the other hand, it is reported that the presence of glass fibers, which melted at low temperatures, aided the retention of graphite fibers during a fire and simulated explosion.[31] Under thermal decomposition of epoxy/graphite composites, it has been reported that graphite fiber release may represent a hazard to electrical equipment. The melted glass fibers of a hybrid construction clumped and retained the graphite fiber. See chapter on electrical properties of C/G composites.

THERMOPLASTIC POLYIMIDES

Returning to the subject of polyimide/graphite composites, it is also noteworthy that higher and higher temperature performing materials are not always the objective. By concentrating upon the more readily obtainable lower temperature performance, Gibbs[32] has emphasized the processing and performance advantage of some thermoplastic type polyimides. While the monomeric precursor solutions for the polyimides described in this paper appear similar to the previously referenced thermoset type polyimides, the absence of potentially reactive end-caps and drastic processing measures obviate the cross-linking, and hence the thermoplastic, linear polymer structure is formed. The prepreg based on "Thornel" 300 graphite fibers and NR-150A2S5X resin yields high quality composites under a wide range of conditions. Low void laminates were produced under the processing measures that were outlined. Properties reported in this paper by Gibbs are for T-300 graphite/NR-150A2S5X polyimide composites (see Table 4-4).

It should be noted that the laminates prepared and post-cured in this manner did not have the micro-cracks in the matrix usually associated with the more brittle resin. Furthermore, the improvement in ILSS over other polyimides is noteworthy. The NR-150 solutions are expensive–about $75 /lb (1979). Products formed and molded require high pressures above softening temperature of the resin binder, and further heating may cause release of traces of volatiles which would limit the thickness that could be processed. End products are not brittle and show good fatigue life and thermal oxidative stability at high temperatures equal to, if not better than, the PMR polyimides, which are thermosetting by reason of their cross-linking of polymer chains.

A second generation of NR-150 polyimide binders has been developed to

Table 4-4. Physical Properties of NR150 Polyimide/Graphite Composite.[32]

Percent volume of fibers	57%
Volume of voids	Less than 1%
Weight of resin	36%
Density (g/cc)	1.59
Flexural strength (psi), 73°F	238,000
500°F	122,000
Flexural modulus (psi), 73°F	17×10^6
500°F	16×10^6
Short beam shear strength (psi), 73°F	16,800
500°F	6400
Molding	Vacuum bag/caul plate consolidated at 428°F and molded at 200 psi for 4 hours at 600°F Post-Cure–270°-700°F/33 hours (8 hours/700°F); weight loss during post-cure = 1.3%; laminate = 90 mils thick

replace the first generation products. The new binders are designated NR-150A2 and NR-150B2 and replace the old NR-150A and NR-150B.

Glass Transition Temperatures

NR-150 A2	280°-300°C (536°-572°F)
NR-150 B2	350°-371°C (662°-700°F)

The second generation polymers are claimed to be tougher and more oxidatively stable at high temperatures. Excellent solution stability at room temperature is reported.[33]

Gibbs and Ness[34] presented an outstanding paper on the quality control techniques for NR-150 polyimide binders. In studies involving NR-150 polyimide precursor solutions, both amide acid oligomer formation and esterification were shown to be possible side reactions. Quantitative titration techniques were developed to analyze for total acid, ester, and amide groups. In view of the effect of temperature on the amide acid formation and esterification reactions, storage at 39°F or below is recommended for all NR-150 polyimide precursor solutions and prepregs to minimize further side reactions. Authors also used Karl Fischer titration of by-product water as a useful overall measure of side reactions.

Data has been reported by du Pont on the aging characteristics of NR-150B2 polyimides in combination with carbon fibers from polyacrylonitrile ("Celion" 6000). Tested at 600°F after aging at 600°F, the following values were reported for composites of C/G fibers and NR150 B2 polyimide binders:

PHYSICAL TESTS AT 600°F (du PONT)

	ORIGINAL	AFTER 200 HOURS	AFTER 500 HOURS
Flexural strength (ksi)	104.0	119.0	132.0
Modulus of elasticity (10° psi)	15.6	15.3	15.1
Interlaminar shear strength (ksi)	6.0	6.5	6.3
Weight loss (%)	–	1.3	2.2

Further chemical quality control measures are described by Hoeger[35] for the NR-150 polyimide precursor solutions. He also presented a chart on the basic chemistry of the NR-150 binder solutions which are used with advanced composites. This chart is reproduced in Fig. 4-10. Best resolution was obtained in the analysis with reverse phase ion pair chromatography on a C_{18} bonded column using an aqueous acetonitrile gradient. The NR-150 solutions were stable when stored below 4°C.

"Torlon," another thermoplastic type of product, is a poly (amide-imide), a reaction product of Amoco's trimellitic anhydride and aromatic diamines. It is called amide-imide because the polymer chain consists of aromatic nuclei joined by amide linkages alternating with imide. Although the operating temperatures (up to 500°F) which characterize this material are lower than the high heat resistant polyimides, the presence of amide groups impart flexibility and elongation.

The properties of graphite fiber reinforced "Torlon" have been reported by Chen.[36] Because of the high melt viscosity of the resin and the additional pressure required for high filler loadings, molding takes place at high process temperatures. Heat treatment of molded parts to permit the escape of traces of water of condensation is recommended. In this paper there is an orderly report on 25 to 50% by weight of chopped graphite fiber in the "Torlon" molding composition. The table of data is reproduced here, as Table 4-5 and should be compared with graphite fiber filled thermoplastics reported in the next chapter. Noteworthy is the large increase in modulus of elasticity with higher graphite fiber loading, as well as the marked decrease of thermal expansion coefficients. In further papers, additional engineering properties were reported of injection molded polyamide-imide/graphite parts.[37] It was observed that the fatigue strength was substantially improved by the graphite fiber reinforcement, and that the stress cracking resistance of molded parts in contact with aircraft engine turbine oil was improved by the graphite fibers. Continuous use temperature by UL Method 746 has been established at 210°C.

Table 4-5. Engineering Properties of Graphite-Fiber Reinforced "Torlon."[36]

GRADE	XG25	XG30	XG40	XG50
Graphite fiber content (%)	25	30	40	50
Molding shrinkage (%)	-0.07	-0.10	-0.11	-0.12
Heat treatment shrinkage (%)	0.27	0.26	0.25	0.25
Total shrinkage (%)	0.20	0.16	0.14	0.13
Specific gravity	1.41	1.41	1.42	1.42
Deflection temperature (°F)	517	516	519	522
Water absorption (%)	0.23	0.24	0.22	0.21
Coefficient of linear thermal expansion (in./in./°F)	6.6×10^{-6}	5.6×10^{-6}	5.5×10^{-6}	4.1×10^{-6}
Spiral flow (in.)	4.75	4.25	3.75	3.25
Cavity pressure (psi)	16,000	14,400	12,600	10,600
At room temperature:				
Tensile strength (psi)	25,000	24,000	24,500	24,000
Tensile elongation (%)	2.9	3.3	3.7	3.0
Tensile modulus (psi)	2,790,000	2,870,000	4,250,000	4,430,000
Flexural strength (psi)	39,400	40,000	45,100	43,300
Flexural modulus (psi)	2,450,000	2,610,000	3,410,000	3,480,000
Notched izod (ft-lb/in.)	1.31	1.41	0.93	0.77
Hardness, Rockwell Scale	E94	E98	E88	E87
Shear strength (psi)	15,800	15,500	15,200	15,000
At 300°F:				
Tensile strength (psi)	18,000	17,500	17,200	15,000
Tensile elongation (%)	5.2	4.8	4.7	5.5
Tensile modulus (psi)	1,480,000	1,880,000	3,170,000	3,720,000
Flexural strength (psi)	21,700	30,100	33,400	34,400
Flexural modulus (psi)	1,661,000	2,550,000	3,179,000	3,288,000

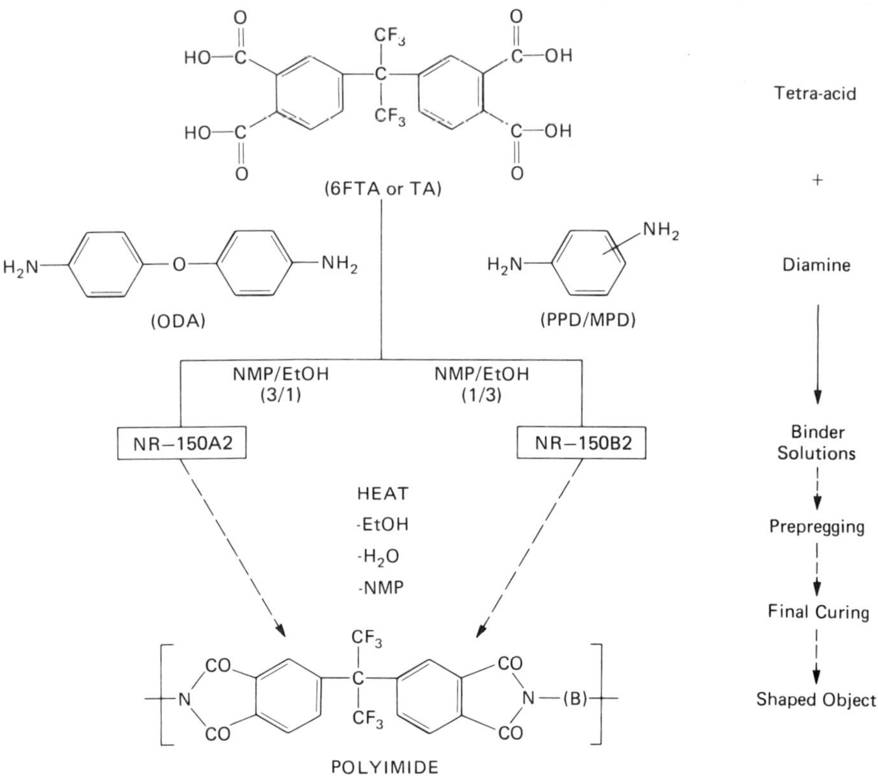

6FTA = hexafluoroisopropylidene-bis-(4-phthalic acid)
NMP = N-methyl pyrrolidone
MPD = m-phenylenediamine
(B) = bivalent radical derived from base (diamine)
PPD = p-phenylenediamine
TA = tetra-acid
ODA = oxy-dianiline

Fig. 4-10. Basic chemistry of NR-150 binder solutions and cured products made therefrom.[35]

Comparisons of graphite filled polyamide-imides have been reported by Amoco Chemicals Corp. on the following materials: "Torlon" 7030–30% graphite fiber; "Torlon" 5030–30% glass fiber; "Torlon" 4203–unfilled poly (amide-imide); "Ryton" R-4-Phillips poly-phenylene sulfide–40% glass. The relationships between tensile thrength and temperature are shown in Fig. 4-11.[38] Data were obtained on injection molded specimens and show good comparative performance for the graphite filled compounds over a broad spectrum of temperatures. Molded "Torlon" parts are shown in Fig. 4-12.

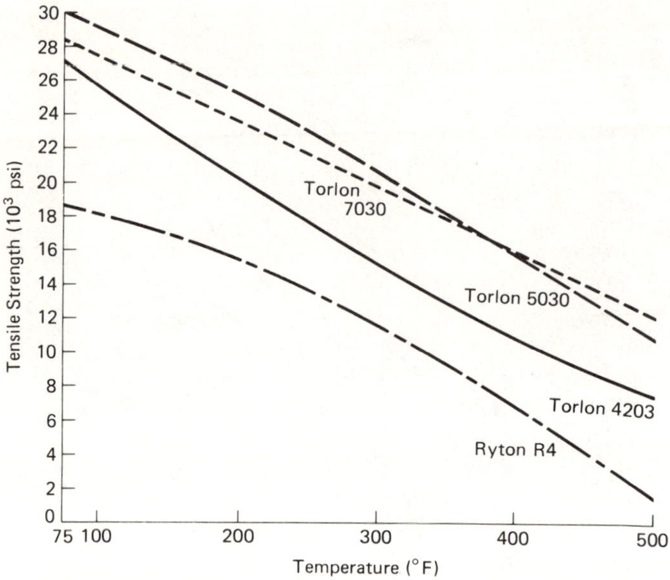

Fig. 4-11. Tensile strength versus temperature for "Torlon" molded materials.[38]

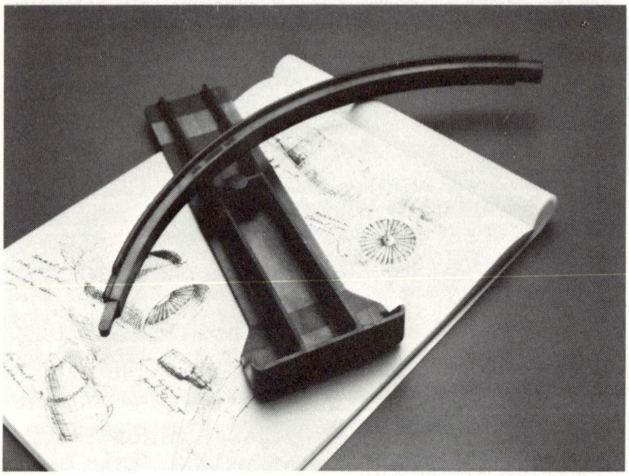

Fig. 4-12. Molded instrument bracket and shroud segment for jet engine fan molded of graphite fiber reinforced "Torlon." (*Courtesy Amoco.*)

Jones reported on a program to select graphite/polyimide composite materials in a very thin tape form for use in structural applications on the space shuttle orbiter. The materials were potentially capable of service to 500°F. The program concentrated upon a new generation of bis-maleimide addition reaction types of polyimide binders in composites with graphite fibers.[39] Bis-maleimides such as "Hexcel" HX-560, modified "Kerimid" 353, and Narmco 5230 were found as acceptable matrices for use in composites with "Thornel" 300 graphite fibers. All composites were coated with an aluminum filled polyimide resin to prevent degradation of the surface resin to fiber bond during aging at 500°F.

Polybutadiene bis-imide resins offer advantages of hot melt processing on graphite fiber matrices. Without solvent additions, void-free laminates can be obtained. Blends of hydroxy and carboxy terminated 1, 2-polybutadiene chains extended with succinic anhydride provide excellent resistance to moisture degradation and high strength retention at 350°F.[40]

ACETYLENE TERMINATED POLYIMIDES

A new class of thermosetting polyimides which cure by an addition polymerization process was developed at Hughes Aircraft Company.[41] Structural composites were fabricated at about 251°C (485°F) at pressures of 200 psi. Prolonged post-cure at 371°C (700°F) yielded the best physical properties. Molding compounds containing chopped Hercules AS graphite fiber showed 85% strength retention after 250 hours in air at 316°C (600°F) and 45% strength retention after 50 hours in air at 371°C (700°F). Differential thermal analysis of the resin showed the onset of exothermic cure commencing at 220°C.

The acetylenic substituted polyimides offered even higher heat resistance than the previously discussed polyimides. They have been used as resin matrices for graphite composites. Formerly identified as HR-600 by Hughes Aircraft Company, they have been offered commercially more recently by Gulf Oil Chemical as "Thermid" 600. These high temperature resin matrices are hard, brittle resins. There are a number of new acetylenic polymers which have been offered recently, and the materials are underoing further evaluation and development. Like the PMR polyimide resins discussed earlier, the acetylenic terminated polyimides do not liberate significant quantities of volatile by-products, which contribute to undesirable voids in the composites. They differ from previous polyimides in the utilization of acetylene group on the ethynylated amines. The acetylenic end cap provides the means for chain extension as well as cross-linking to a thermosetting structure during thermal polymerization. Cure or chain extension takes place at temperatures above 200°C. Mechanical properties of the neat resin, HR-600 ("Thermid"600), were reported by Bilow *et al.*[42] as follows:

Tensile strength	14,000 psi
Tensile modulus	550,000 psi
Flexural strength	18,000–21,000 psi
Flexural modulus	650,000 psi
Compressive strength	up to 66,000 psi

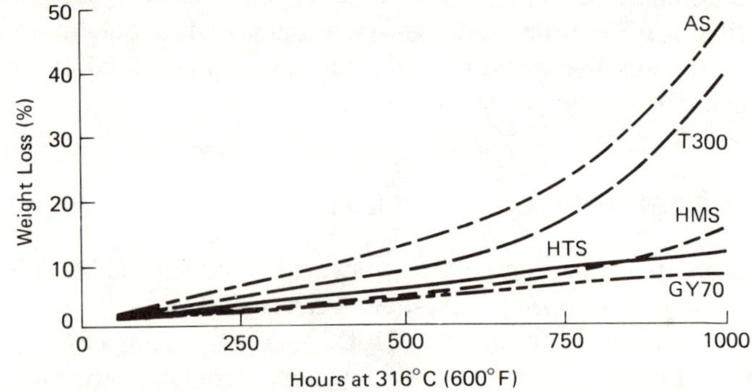

Fig. 4-13. Weight loss of graphite laminates at 316°C.[43]

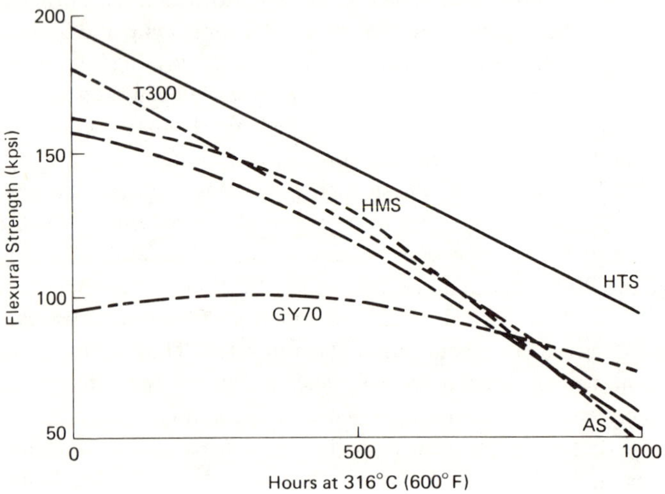

Fig. 4-14. Flexural strength of thermally aged graphite composites.[43]

The cure was 2 hours at 250°C, 2000 psi, and the post-cure in air was 16 hours at 316°C. The resin precursors were identified as 1, 3-bis-(3-aminophenoxy) benzene, benzophenone tetracarboxylic acid dianhydride (BTDA), and meta-aminophenylacetylene, which are initially reacted in solution. Both the imidized prepolymers and their polyamic precursors are soluble in dimethylformamide or n-methylpyrollidone, and the solutions are used as varnishes for the impregnation of graphite fabrics. Some additional data on the weight loss of graphite composites at 316°C and the ambient flexural strength of thermally aged graphite composites are shown for several graphite fibers in Figs. 4-13 and 4-14, (taken from Bilow and Landis).[43] GY-70 and HT-S graphite fibers showed the best percentage strength retention after 1000 hours at 316°C. Additional studies showed a 25% loss of interlaminar shear strength of HT-S graphite/"Thermid" 600 after 500 hours at 316°C.[43] Thirteen new acetylenic polyimides were reported.

POLYBENZIMIDAZOLES

Although the polybenzimidazoles (PBI) have not appeared as prominently in the technical press as the polyimides, it is proper that we examine some of the reports that appeared, particularly as they might have a bearing upon our recognition of the basic requirements of advanced composites. PBI resins were first synthesized in the late 1950's, exhibiting excellent thermal stability. Oxidative resistance is about 300°C in air. Basically, the polymer is produced by reacting aromatic diamines with aromatic diacids to form relatively low molecular weight prepolymers. Polymers prepared from 3, 3′-diaminobenzidine and diphenyl isophthalate have shown good stability. By replacing the diphenyl ester with isophthalamide in the reaction, problems relating to evolution of phenol and water during cure were reduced. One must recognize the carcinogenicity potential of some of the benzidine compounds, which limit their utilization in aircraft fabricating processes. For a detailed summary of work in this field, including the extensive amount of work on preparation of PBI polymers by Levine at Whittaker Corporation, consult the references given at the end of this chapter. The published results on PBI systems have been primarily associated with glass cloth laminates. Marvel's classic work on PBI was an earlier milestone which stimulated work in this area. Mayor felt that flexural strength properties of graphite/PBI composites did not greatly differ from other resin composites of equivalent reinforcement.[44,45,46]

MISCELLANEOUS HIGH TEMPERATURE MATRICES

There are numerous miscellaneous thermosetting resin matrices which have appeared for potential use in high temperature composites with glass and with graphite. In recent years, several important programs have been sponsored on the polyquinoxalines. These polymers were demonstrated to have good processability in the preparation of low void laminates.

Polyphenylquinoxalines

Hergenrother and Levine[47] covered several types of polyquinoxalines, and their uses in laminates, as adhesives, and in conjunction with explosives. Further efforts at Boeing on the polyphenylquinoxalines led to reports on laminated structures. Their assessment of the potential of polyphenylquinoxaline/graphite was a short term structural material for use up to 700°F.[48] In processing these polymers, amino-benzidine derivatives were used. These polymers have had relatively low softening points and a tendency to volatilize (i.e., not yield high char residue) at very high temperatures. Developmental resins have been available from Whittaker Corp.

Peritt[49] discussed details on the molding of the high temperature linear polymeric resins developed at the Naval Ordnance Laboratory. They addressed themselves to the problems of minimizing the voids through mold venting. Peritt reported that polyphenylquinoxaline has a glass transition temperature of 365°C and requires molding temperatures of about 400°C—a very high range for most manufacturing establishments. With Courtauld's HM-S graphite fiber, void contents were in the 6 to 9% range. For best high temperature performance, high temperature post-cure which presumably induces cross-linking through thermal and/or thermal-oxidative degradation. Properties of the polymer are controllable through synthetic modifications on the polymer backbone chain, and the pendant phenyl group.

Alston[50] reduced the thermoplasticity of polyphenylquinoxaline matrix resins, introducing cross-linking by reaction of terminal nitrile groups.[50] PMR technology, and use, of high molecular weight resin solutions considerably improved the ease of composite fabrication of linear polyphenylquinoxaline matrix resins. Acceptable non-thermoplastic mechanical properties were obtained at 600°F for polyphenylquinoxaline/HM-S graphite fiber composites.

Pyrolyzed Structures

Some of the highest heat resistant resin systems have been based upon the fabrication of a resin/graphite composite and subsequent copyrolysis at temperatures up to 1000°C. The organic resin binder decomposes in most cases, leaving behind a charred carbonaceous structure supporting the graphite fibers. Further impregnation, cure, and further pyrolysis again result in a charred structure, but with a higher amount of char. The carbon/carbon composites which have been prepared are discussed in greater detail in a later chapter on matrices for use at temperatures above 300°C. High temperature nose cones for rockets were among the first applications of carbon fiber/carbon matrices (carbon/carbon composites), while aircraft brakes are finding that these systems are very useful, and that they out-perform metals.

REFERENCES

1. Hergenrother, P.M. and Johnston, N.J., "Status of High Temperature Laminating Resins and Adhesives," *ACS Organic Coatings and Plastics Chemistry,* Vol. 40, page 460, April 1979.
2. Markle, R., "Applied Polymer Science" Chapter 39, p. 560, in *ACS Organic Coatings and Plastics Chemistry,* Craver, J.K. and Tess, R.W. (Editors), 1974.
3. Aponyi, T.J., "Progress in High Temperature Resins," *Twenty-third Annual SPI Conference,* page 27, Palm Springs, California, April 1966.
4. Freeman, J.H., Frost, L.W., Bower, G.M., Traynor, E.J., Burgman, H.A., and Ruffing, C.R., "Resins and Reinforced Plastic Laminates for Continuous Use at 650°F," *Polymer Engineering and Science,* Vol. 9, No. 1, page 56, January 1969.
5. Gillham, J.K. and Gillham, H.C., "Polyimides–Effect of Molecular Structure and Cure on Thermomechanical Behavior," *Polymer Engineering and Science,* Vol. 13, No. 6, page 447, November 1973.
6. Scheck, W.G. and Stuckey, J.M., "Development and Evaluation of Graphite and Boron Polyimide Composites," *SAMPE Symposium on Nonmetallic Materials,* Vol. 4, page 9, Palo Alto, California, October 1972.
7. Bulletin:Fiberite Corp., Procedure M–8, June 1978.
8. Bulletin: "Keramid" 353, Rhodia, Inc., February 1975.
9. Jewell, R.A., "Relative Thermophysical Properties of Some Polyimides," *Applied Polymer Science,* Vol. 15, 7, pages 1717–1727, 1971.
10. Cray, M.C., "Recent Advances in Processing of Polyimide Resins," *SAMPE Nineteenth National Symposium,* Vol. 19, pagé 663, Buena Park, California, April 23, 1974.
11. Dine-hart, R.A. and Wright, W., "Development of a Melt Processable Polyimide Resin System," *Plastics Institute RPG, Conference Paper 7,* London, 1973.
12. Clancy, H.M., Kaumeyer, R.A., and Yoshino, S.Y. (North American Rockwell Corp., Space Divisions), "Development of Polyimide-glass Fabric Laminates with Nonflammable Characteristics in High Oxygen Spacecraft Environments," Company Funded Research (1968), Report 11–4B–4.
13. CA–74–23309Z, February 1971; and *Composite Materials,* Vol. 4, page 390, 1970.
14. Lubowitz, H.R., U.S. Patent 3,528,950, "Rapid Cure Polyimide Prepolymers," filed July 3, 1967, granted September 1970.
15. Lubowitz, H.R., Wilson, E.R., Jones, J.F., Burns, E.A., and Braswell, T.F. (TRW, Inc.), U.S. Patent 3,632,428, "Thermally Stable Coatings," filed November 30, 1967, granted January 4, 1972.
16. Lubowitz, H.R., Kendrick, W.P., Jones, J.F., Thorpe, R.S., and Burns, E.A. (TRW, Inc.), U.S. Patent 3,647,529, "Reinforced Structural Plastic, filed July 10, 1967, granted March 7, 1972.

17. Serafini, T.T., Delvigs, P., and Lightsey, G., U.S. Patent 3,745,149, "Preparation of Polyimides from Mixtures of Monomeric Diamines and Esters of Polycarboxylic Acids," filed September 29, 1971, granted July 10, 1973.
18. Serafini, T.T., Delvigs, P., and Lightsey, G.R., "Thermally Stable Polyimides from Solutions of Monomeric Reactants," *Journal of Applied Polymer Science,* Vol. 64, pages 905–915, April 1972.
19. Winters, W.E. and Serafini, T.T., "PMR Polyimides–Processable High Temperature Composite Matrix Resins," *SAMPE Symposium,* Vol. 20, page 629, April 1975.
20. Winters, W.E. and Cavano, P.J., "PMR-11 Composites," *Tenth National SAMPE Meeting,* Vol. 10, page 661, New York, October 1978.
21. Vannucci, R.D. and Alston, W., "PMR–Polyimides with Improved H.T. Performance," *Thirty-first Annual Technical Conference,* SPI RP/C Institute, Section 20A, Washington, D.C., February 1976.
22. Potker, I. and Stern, B.A., "Structural Performance to 700°F of Graphite/Polyimide Composites," *SAMPE National Symposium,* Vol. 23, page 775, Anaheim, California, May 1978.
23. Delvigs, P., Alston, W.B., and Vannucci, R.D., "Effect of Graphite Fiber Stability on the Properties of PMR Polyimide Composites," *SAMPE Twenty-fourth Symposium,* Vol. 24, page 1053, San Francisco, California, May 1979.
24. Alston, W.B., "Resin/Fiber Thermo-oxidative Interactions in PMR Polyimide/ Graphite Composites," *SAMPE Twenty-fourth National Symposium,* Vol. 24, page 72, San Francisco, California, May 1979.
25. Serafini, T.T., "Status Review of PMR Polyimides," *ACS Organic Coatings and Plastics Chemistry,* Vol. 40, page 469, April 1979.
26. St. Clair, T.L. and Jewell, R.A., "LaRC-160–A New 550°F Polyimide Laminating Resin," *SAMPE Meeting,* Vol. 8, page 82, Seattle, Washington, October 1976.
27. Hoggatt, J.T., Hunter, A.B., and Sheppard, C.H., "Reproducibility of PMR-15 Graphite Polyimide Material," *Thirty-third Annual Technical Conference,* SPI RP/C Institute, Section 15E, Washington, D.C., February 1978.
28. Lauver, R.W. and Vannucci, R.D., "Characterization of PMR Polyimides Correlation of Ester Impurities with Composite Properties," *SAMPE Twenty-fourth Annual Symposium,* Vol. 24, page 522, San Francisco, California, May 1979.
29. Lauver, R.W., Alston, W.B., and Vannucci, R.D., "Stability of PMR Polyimide Monomer Solutions," *Thirty-fourth Annual Technical Conference,* SPI RP/C Institute, Section 23A, New Orleans, Louisiana, February 1979.
30. Gibbs, H.H., Wendt, R.C., and Wilson, F.C., "Carbon Fiber Structure and Stability Studies," *Thirty-third Annual Technical Conference,* SPI RP/C Institute, *Section* 24F, Washington, D.C., February 1978.
31. Tompkins, S. and Brewer, W.D., "Effect of Boron and Glass Hybrid Epoxy Composites on Graphite Fiber Release in an Aircraft Fire," *SAMPE Thirty-fourth National Symposium,* Vol. 24, pages 620–630, San Francisco, California, May 1979.
32. Gibbs, H.H., "NR-150A2–The 500°F Binder with Epoxy-Type Moldability," *SAMPE Twenty-fourth National Symposium,* Vol. 24, pages 11–20, San Francisco, California, May, 1979.
33. Gibbs, H.H., "Second Generation NR-150 Polyimide Binder Systems," *SAMPE National Symposium,* Vol. 20, page 212, April 1975.
34. Gibbs, H.H. and Ness, J.R., "Development of Quality Control Techniques for NR-150 Polyimide Adhesive and Binder Materials," *SAMPE National Symposium,* Vol.

23, page 806, Anaheim, California, May 1978; and *SAMPE Journal,* page 11, January/February 1979.
35. Hoeger, E.F., "Chemical Quality Control of NR-150 Polyimide Precursor Solutions," *SAMPE Thirty-fourth National Symposium,* Vol. 24, page 533, San Francisco, California, May 1979.
36. Chen, Y.T., "Graphite Fiber Reinforced Torlon," *SAMPE National Symposium,* Vol. 23, page 826, Anaheim, California, May 1978.
37. Chen, Y.T. and Watson, R.W., "Engineering Properties of Injection Molded Poly(amide-imide) Thermoplastic Parts," *Thirty-seventh SPE ANTEC Conference,* p. 623, New Orleans, Louisiana, May 1979.
38. Amoco Bulletin: TAT-18a, "Elevated Temperature Properties," April 1979.
39. Jones, J.S., Rockwell, "Graphite/Polyimide Composite Material Systems for Potential Space Shuttle Use," *SAMPE National Symposium,* Vol. 21, page 438, April 1976, Los Angeles.
40. Vaughn, R.W., O'Rell, M.K., and Buyny, R.A., "TRW Polybutadiene Bis Imide Matrix Resins," *SAMPE National Symposium,* Vol. 23, page 528, Anaheim, California, May 1978.
41. Bilow, N., Landis, A.L., and Aponyl, T.J., "Thermosetting Acetylene Substituted Polyimides," *SAMPE National Symposium,* Vol. 20, page 618, San Diego, California, April 1975.
42. Bilow, N., Keller, L.B., Landis, A.L., Boschan, R.H., and Castillo, A.A., "New Developments in Acetylene Substituted Polyimides," *SAMPE National Symposium,* Vol. 23, page 791, Anaheim, California, May 1978.
43. Bilow, N. and Landis, A.L., "Acetylene Substituted Polyimides," *ACS Division of Polymer Chemistry,* Vol. 19, No. 2, page 23, September 1978.
44. Hall, R.J. and Levi, D.W., *Polybenzimidazoles–A Review,* Plastics Tech. Evaluation Center, Picatinny Arsenal, Dover, New Jersey, 44 pages, July 1966.
45. Marvel, C.S. and Vogel, H., *Journal of Polymer Science,* Vol. 50, page 511, 1961.
46. Mayor, R.A., CA-88-7764C (1978), *SAMPE National Conference,* Vol. 8, page 478, 1977.
47. Hergenrother, P. and Levine, H.M., "High Temperature Structural Resins," NRD-MJO-325A, 76 pages, May 1978.
48. Hoggatt, J.T. and Hergenrother, P.M., "Development of Polyphenyl Quinoxaline Graphite Composites," NASA-CR-121109, BAC-2-5540, 140 pages, January 1973.
49. Peritt, H., "Fabrication of Polyphenylquinoxalines (PPQ) Shingled Carbon Composites Using Vented Compression Mold," *Thirtieth Annual Technical Conference,* SPI RP/C Institute, Section 14G, Washington, D.C., February 1975.
50. Alston, W.B., "Nitrile Crosslinked Polyphenylquinoxaline/Graphite Fiber Composites," *SAMPE National Symposium,* Vol. 2, page 114, Seattle, Washington, 1976.

5
Thermoplastic Matrices

In previous chapters, emphasis was placed upon thermosetting resins and high temperature resins (including thermoplastics) that have been used for the preparation of advanced structural composites from carbon and graphite fibers. Epoxy resin systems and polyimides were among the more prominent materials used as resin matrices, the epoxies for service to 200°C and the polyimides for service to 300°C. In this chapter, a broader spectrum of thermoplastic resins will appear because of the large number of commercially available materials. In many instances, the same compelling reasons for their selection will be apparent—namely, good heat resistance, good resistance to adverse environmental influences, and the ability to replace metal components in many circumstances. The fact that carbon and graphite fibers will be singled out for attention attests to the good physical properties that are achieved in combination with them—notably improved strength and, even more important, improved stiffness as measured by the modulus of elasticity of the composites and good fatigue strength and dimensional stability. Glass fibers have always had much to offer, including strength, heat resistance, and cost advantages, and will be cited in several instances for comparison. The quality of stiffness and good dimensional stability, so important to the design of products and machines, arises in many examples. This explains the growing attention to C/G fibers in conjunction with thermoplastic resin matrices.

Advanced composites prepared from C/G fibers and thermoplastic resins have some advantages and disadvantages as compared to the thermosetting resins discussed in earlier chapters. The thermoplastic resins, in most instances, are offered to the fabricator as completely cured or polymerized materials requiring heat and pressure to form and shape them. This is a distinct advantage over the complex cure and post-cure requirements of the thermosetting resins. The latter must penetrate the C/G fibers during processing within a limited period of time when they possess high fluidity (low viscosity), before conversion to a relatively infusible state. With the availability of long processing times, large complex structures of varying thicknesses are fabricated into complete products from C/G fiber composites. The strength, stiffness, and low weight achieved with C/G reinforcements out-perform most metals. Thermosetting resin matrices, in combination with C/G fibers and other fibers (hybrid constructions) have optimized strength through utilization of long continuous lengths of unidirectional fibers or woven products, frequently with prior impregnation of the fibers before the laminating operation into a composite structure. Hence, C/G fabrics and tapes impregnated with thermosetting resins are offered as "prepregs" and are usually stored under refrigerated conditions until ready for fabrication.

On the other hand, the thermoplastic resins are more usually combined with short, chopped C/G fibers which may be partially replaced with lower cost fibers. In consequence, C/G filled thermoplastic molding materials have appeared and are finding their way into many industrial applications—principally outside of aerospace. There will be observed significant applications to aircraft components, though greater promise exists for these materials in automotive applications. Another advantage for the thermoplastic/graphite composite materials lies in the relative ease with which "hybrids" may be developed. Hybrids may involve multiple types of fibers with the same resin binder, arriving at some compromise in properties. These situations more readily appear outside aerospace, where physical property requirements are not as stringent. To cite well known examples: improvements in impact strength may be obtained with glass fibers or aramid fibers used in conjunction with graphite fibers, and, for electrical devices, there is evidence that in the event of external fires, the glass fibers on melting will restrain the graphite fibers. The other side of the picture is, of course, sacrifice of some of the stiffness and good fatigue resistance associated with graphite fiber reinforcements.

Thermoplastic matrices may also be used with woven graphite fabrics, in a laminating operation rather than as molding materials. Processing costs favor the use of thin films of thermoplastics, between which are sandwiched the layers of C/G woven goods. The other alternative is to dissolve the thermoplastic in a solvent (difficult for some of the heat resistant thermoplastics), apply to the woven goods, and remove the solvent. A recent paper[1] cites the advantages of using dry films of polyphenyl sulfone as a matrix in laminated C/G products. Much of the work and development of thermoplastic composites has been due to the efforts of groups such as LNP Corp. and Fiberite, Inc., who have actively engaged in the preparation of thermoplastic molding compounds with C/G fibers.

It should also be mentioned that included among the high heat resistant thermosetting resins described in the previous chapter, there are certain grades which are distinctly thermoplastic and in which no attempt is made to cross-link the polymer chains during polymerization. Notable examples are some of the thermoplastic polyimides (such as du Pont's NR-150B resin series of precursor polymer solutions). The thermoplastic qualities are emphasized because such composites will offer improved resistance to micro-cracking (see the chapter on mechanical properties) under thermal cycling.

BLENDING OF FIBROUS FILLERS AND RESINS

In examining the many available thermoplastic polymers, it seems evident that the great majority of them may be blended with C/G fibers by various processing techniques. To summarize key methods available, the following are noted.

Intensive Internal Mixing

Under heat, thermoplastics are softened, and under the high shear action of the mixing blades and the inclusion of fillers, the temperatures will rise further. As the volume load increases, power demands are greater and there are practical limits of how much filler and fiber addition can be made. While the thermoplastics will not advance in cure, as do the thermosets, the shearing action will break down fibers into smaller fragments so that their value as additives will be downgraded.

Mixing Rolls

Historically, these have been used for plastics and for rubber compounding, producing an intimate mixing of fillers with the resins. The process is slower than internal mixers, and tends to break down fibrous fillers.

Extrusion Mixers

Several commercial extrusion mixers are designed for blending of resins and fillers on a continuous basis. While fillers, additives, and their resin binders may be automatically proportioned into a mixing hopper, partial blends of material components are usually effected beforehand. Rapid, intensive mixing of the screw-type extruders yields strips or ribbons of products, which are chopped to the size required for molding. Sufficient heat is maintained on the extrusion barrel to obtain thorough blending and welding together of material components.

Solvent Blenders

These units are generally enclosed to maintain volatiles in a confined space. While there is no limit on the proportioning of thermoplastic resins (providing they are readily soluble) and fibrous fillers, the added costs of solvent removal must be faced not only in preparing the compound, but in molding composites under special pressure and temperature cycles to remove traces of volatiles.

Inert Carrier Process

There is no limit to the proportioning of graphite fillers and resins and no solvent action is required. Resin systems are dispersed into inert carriers with high speed rotating blades to form a latex-like emulsion. Thermoplastics would have to be liquefied by hot melting or partial solubilizing as they are fed into the inert carrier. The inert carrier is usually water. The problems of introducing additives, such as sizing agents for the fillers and other modifiers, will require special time sequencing of material additions into the emulsion. When desired proportions and mixing are achieved, the water is filtered off and the thermoplastic/graphite fillers are oven dried—a problem of timing for thermosets, but not for thermoplastics. Very high ratios of fillers to binders can be obtained in this process. The alternative of freeze-drying can also be considered in processing these materials.

There are other processes for producing the blends, though the above are typical of commercial practices available. While it may be questioned why these details are brought into this text, the reader is encouraged to be familiar with some of the basic processing methods, in order to evaluate reasons for the presence of impurities of unexpected by-products. An intimate blend of resin and fibrous C/G is essential for good composites, a condition aided by "sizing" agents or surface treatments on the fiber surface. What is available in industry today as coupling or sizing agents or special surface treatments applied to the C/G fibers in anticipation of blending with the resin is a complex subject in itself, and will be discussed in a separate chapter.

CHARACTERISTICS OF THERMOPLASTICS

As noted earlier, the subject of thermoplastics is a large one, and to present material of value to the designer and fabricator of advanced composites of C/G, this text will be selective in presenting data on those materials which have been introduced to composite technology. There will also be references to earlier work on the use of glass fibers and thermoplastics in industrial molding compounds—a very viable alternative to consider. The use of C/G reinforcements can only be considered if the presence of these fillers will offer obvious property and economic advantages over the other fibrous fillers. As will be seen from data and charts presented, these advantages appear in improved strength, dimensional stability, and stiffness of the composites. Whereas laminated composites are prepared principally with unidirectional fibers and woven fabrics, the composites prepared for molding compounds utilize finely divided and randomly distributed fibrous fillers, and hence do not have the structural capabilities of the laminated products.

A particularly useful reference with a good overview of thermoplastics was prepared by N. Platzer.[2] In examining the rigidity and maximum use temperatures of thermoplastics, Fig. 5-1 (from Platzer's paper) illustrates the temperature dependence of the short-term modulus of elasticity in flexure of several thermoplastics. Thermoplastic polyimides and polysulfones retain their moduli at higher temperatures than do the other thermoplastics illustrated. Note that at 20°C (R.T.), most rigid thermoplastics have flexural moduli of elasticity in the range of 350 to 550 ksi.

Long-term rigidity of thermoplastics is reflected in their ability to withstand creep and cold flow, and to maintain their initial modulus of elasticity. Platzer compares the creep of different thermoplastics under sustained stress by plotting isochronos creep (percent strain after 1000 hours) as a function of different loadings at room temperature. The long-term rigidity of several thermoplastics are compared in Fig. 5-2. Some materials, such as high impact polystyrene, ABS, polypropylene, and rigid PVC, are more load sensitive than others. Modifica-

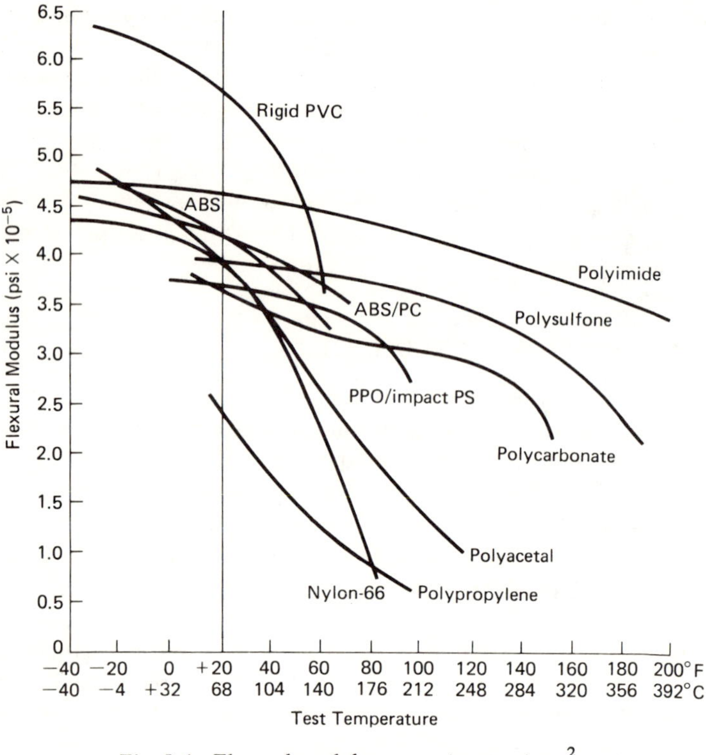

Fig. 5-1. Flexural modulus versus temperature.[2]

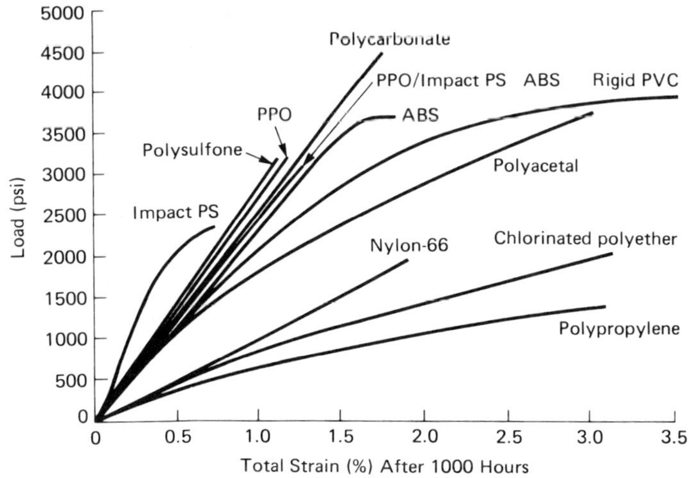

Fig. 5-2. Creep of thermoplastics under tensile loads at room temperature.[2]

tion with plasticizers will greatly increase creep rates. Creep, like the modulus of elasticity, is also heat sensitive, especially near the glass transition temperature. The isochronos creep at the same loading over a broad temperature range is the least for polyphenylene oxide, polysulfones, and polycarbonates. It indicates that polysulfones, polyphenylene oxide, and polycarbonates may be used more readily under permanent load at higher temperatures, whereas many other thermoplastics will undergo viscoelastic flow. The inclusion of the high modulus, low creep graphite fibers into the compositions will help to reduce creep of thermoplastics under stress. The improvement will depend upon volume percent of C/G fibers.

One of the important disadvantages of thermoplastics is their significant thermal expansion and contraction with change in temperature—more than that experienced with the more common metals. In combination with sufficient C/G fibers as reinforcements, these differences are reduced considerably. Another compounding advantage that might be noted in the presence of graphite fibers is the lowering of electrical resistivity, and the attainment of good lubricity features which may influence some designs using C/G composites. Lowered electrical resistivity will be accompanied by reduction of static charges on the surfaces of components.

The list of thermoplastic resins is very extensive, many of the resins fulfilling large volume uses in industry. The polyethylenes and polypropylenes dominate plastics activity with their multibillion pound markets. Of immediate interest to us is a smaller segment of materials, sometimes referred to as "engineering" thermoplastics, characterized by the following qualities:

1. Better than average glass transition temperature–improved through the inclusion of glass or C/G fibrous fillers.
2. Good average mechanical properties which enable them to fulfill many product and component designs, with predictable fatigue and creep characteristics. High modulus graphite fibers vastly improve these properties.
3. Cost advantages qualify engineering thermoplastics over competitive thermosets used in advanced composites.
4. The processing of engineering thermoplastics with C/G fibers has distinct advantages over the sophisticated thermosets which are designed to polymerize during application and hence are subject to time and temperature variables. Thermoplastics have been manufactured to the desired degree of molecular weight and distribution and hence are ready for use. After thermoplastics have been molded and shaped under heat and pressure, cooling down under pressure will usually be necessary.
5. Thermoplastics are generally compounded with randomly distributed fibers and, as such, are intended for molding in closed molds to achieve final shape. This is in contrast to the thermoset matrices which are used principally with unidirectional fibers and woven fabrics for optimizing the strength and stiffness factors. Noteworthy, however, is some of the current work using C/G fabrics laminated with films of thermoplastics to achieve advanced composites.

C/G FIBERS WITH ENGINEERING THERMOPLASTICS

The development of engineering thermoplastics with C/G fibers has been reported. There have been good comparisons not only of the "engineering" thermoplastics, but also of thermoplastic candidates from the polyimides (as well as of thermoplastic polyesters). Several of the excellent tabular comparisons appearing on the following pages are due to the published work of LNP Corp.[3,4]

Some of the earlier work entailing a detailed comparison of the thermal properties of thermoplastic resins and their composites with glass fibers was covered by Arkles *et al.*[3] A thorough investigation into the performance of glass fiber reinforced thermoplastics included physical ASTM tests at ambient temperatures and at elevated temperatures, as well as tests of creep resistance. The suitability for high temperature service is influenced by the use to which the materials are to be subjected. Both glass and carbon fiber result in substantial improvements in most of the thermoplastics in their resistance to thermal environments. Some of the data developed from glass filled thermoplastics are shown in Table 5-1. It will be apparent that decreases in mechanical properties at elevated temperatures are normally associated with thermal transitions in the base polymer.

The stability in tensile strength after 1000 hours of exposure to 400°F is also significant.

Although equivalent conditions, such as glass fiber content, were not employed for all samples, the comparisons between the thermoplastics are interesting. While data for the thermoplastic polyimide are included from the subject reference, it should be remembered that since publication of this information there have been improvements in heat resistant polyimides both in thermoplastics and in thermosetting matrices. One further observation by the authors emphasizes the superior ability of glass reinforced nylon 6/6 and glass reinforced polyphenylene sulfide to withstand applied stress at 300°F—this was a totally unexpected result.

In one of the first detailed papers on the addition of chopped C/G fibers (0.25 to 0.5 inch typical) for the reinforcement of thermoplastics, Arkles *et al.*[4] cited the fact that carbon fiber reinforced compounds are available in nylon, polypropylene, polyphenylene sulfide, polybutylene terephthalate, polycarbonates, polysulfones, ethylene tetrafluoroethylene copolymers, and vinyl fluoride tetrafluoroethylene copolymers. Adding carbon fibers lowers the coefficient of friction, and contributes to important applications in bearings and valves. Toughness on impact of fiber filled molded parts appeared more favorable with the glass fibers over the C/G fibers. These basic observations have encouraged the selection of hybrids of aramid, glass, and carbon fibers in engineering components. The tabular comparisons taken are reproduced in Table 5-2. Different levels of carbon and glass fibers are compared with nylon 6/6 matrix. Note also the reduced mold shrinkage and lower thermal expansion with C/G fiber composites, as compared with glass fiber reinforcement.

The data in the table speaks for itself, as it is quite apparent that not only are there mechanical advantages in the case of the nylon 6/6 matrix filled with C/G fibers over glass fibers, but also that the greatly reduced coefficient of expansion and mold shrinkage makes it possible for such composites to be used in telescope mountings, as well as in optical mounting benches. The property advantages over the neat resin are apparent in the data. The fiber reinforcement of nylon 6/6 is a good example of synergism of two diverse materials comprising the composite, particularly for C/G fibers.

While in an earlier chapter some of the basic characteristics of C/G fibers have been reviewed, herewith is a comparison of the glass and carbon fiber products used in the table below. Note the low and negative coefficient of thermal expansion of the carbon fiber.[5]

Table 5-1. Tensile Strength at Elevated Temperature (ksi).[3]

BASE RESIN	GLASS CONTENT (weight %)	AT 73°F INITIAL R.T.	AT 200°F	AT 300°F	AT 350°F	AT 400°F	AFTER 1000 HOURS AT 400°F
Polyimide	30	13.0	6.2	4.8	3.1	2.3	13.4
Ethylene tetrafluoroethylene	20	11.3	6.8	4.0	2.0	–	11.2
Polyphenylene sulfide	40	23.2	11.2	8.1	4.8	1.1	18.6
Nylon 6/6	50	31.0	16.0	12.4	7.3	2.2	15.6
Polysulfone	40	17.3	14.9	2.3	1.1	–	4.3*

*After 750 hours.

Table 5-2. Nylon 6/6 Composites Data. (LNP Corp. Product Designations).[4]

PROPERTY	RC-1000 POLYMER	RC-1004 20% CARBON	RC-1006 30% CARBON	RC-1008 40% CARBON	RF-1008 40% GLASS	RF-HYBRID* 20% GLASS 20% CARBON
Specific gravity (g/cc)	1.14	1.23	1.28	1.34	1.46	1.40
24-Hr water absorption (%)	1.5	0.6	0.5	0.4	0.6	0.5
Mold shrinkage (t - 1/8 in.) (in./in.)	0.015	0.002–0.003	0.0015–0.0025	0.0015–0.0025	0.0035–0.0040	0.0025–0.0035
Tensile strength (psi)	11,800	28,000	35,000	40,000	31,000	34,000
Flexural strength (psi)	15,000	42,000	51,000	60,000	42,000	49,000
Flexural modulus (psi × 10)	0.4	2.4	2.9	3.4	1.6	2.8
Shear strength (psi)	9,600	12,000	13,000	14,000	12,700	13,000
Izod impact (ft-lb/in. notch)	0.9	1.1	1.5	1.6	2.6	1.8

Table 5-2. Continued

PROPERTY	RC-1000 POLYMER	RC-1004 20% CARBON	RC-1006 30% CARBON	RC-1008 40% CARBON	RF-1008 40% GLASS	RF-HYBRID* 20% GLASS 20% CARBON
Heat distortion temperature (264 psi-°F)	150	495	495	500	500	500
Coefficient of thermal exp. (in./in./F° × 10^{-5})	4.5	1.40	1.05	0.8	1.4	1.15
Thermal conductivity (BTU/hr/ft^2/°F/in.)	1.7	5.50	7.0	8.5	3.6	6.4
Surface electrical resistance (ohms/cm^2)	10^{15}	25–30	3–5	1–3	10^{15}	–

*Data in this column is taken from Theberge *et al.*[5]

	CARBON FIBER	"E" GLASS FIBER
Young's modulus of elasticity (psi)	30×10^6	10.5×10^6
Tensile strength (psi)	300,000	500,000
Tensile elongation (%)	1.2	4.8
Mean fiber diameter (in.)	0.0003	0.0003-0.0005
Longitudinal coefficient of thermal expansion (in./in./F° $\times 10^{-6}$)	-0.55	1.6

The properties of the 50/50 hybrid composite of glass and carbon fibers and nylon 6/6 are intermediate between the two types of fiber reinforcement. Further tests were performed comparing the physical properties of both glass and carbon filled nylon 6/6 in the dry-as-molded state with moisture conditioning of the molded parts at 50% relative humidity. Flexural and tensile strengths fell about 20%, and flexural modulus values fell about 30%, with slightly smaller losses experienced for the carbon filled materials.

Further data comparing 30% carbon fiber with 30% glass fibers for the thermoplastic matrices of polysulfones, thermoplastic polyester, and polyphenylene sulfide appeared in Theberge's paper of 1978.[6] Some of these comparative data are shown in the accompanying table for 30% by weight of fiber additions. Once again the advantage in physical and thermal properties appear to with the carbon/graphite fiber reinforcements over the glass fiber reinforcements.

A blend of 50% nylon and 50% high modulus graphite fibers ($E = 30 \times 10^6$ psi) has been compared with aluminum castings (A-356 T61).[6a] Injection molded graphite fiber/nylon 6/6 compared very favorably at temperatures up to 160°F. Physical properties reported by Valentine were as follows:

PHYSICAL PROPERTIES OF GT-50N (Reference [6a])

Tensile strength	RT.-	33,700 psi
	160°F	22,900 psi
	250°F	16,600 psi
Tensile modulus	RT	6.4×10^6 psi
	160°F	4.5×10^6 psi
	250°F	2.7×10^6 psi
Flexural strength	RT	69,100 psi
	160° F	39,800 psi
	250° F	28,400 psi

More recent products involve fiber loadings of chopped C/G fibers from pitch precursors, rather than the PAN precursors used in earlier work. Improved electrical conductivity and lubricity were reported.[7]

Table 5-3. Fiber Filled Thermoplastic Composites.[6]

PROPERTY	ASTM TEST	UNIT	POLYSULFONE NEAT RESIN	POLYSULFONE WITH 30% GLASS	POLYSULFONE WITH 30% CARBON	THERMOPLASTIC POLYESTER NEAT RESIN	THERMOPLASTIC POLYESTER WITH 30% GLASS	THERMOPLASTIC POLYESTER WITH 30% CARBON	POLYPHENYLENE SULFIDE NEAT RESIN	POLYPHENYLENE SULFIDE WITH 30% GLASS	POLYPHENYLENE SULFIDE WITH 30% CARBON	FIBERITE RTP-287
Specific gravity	D792		1.24	1.46	1.37	1.32	1.52	1.47	1.34	1.56	1.45	1.33
24-Hr water absorption	D570	%	0.20	0.20	0.15	0.08	0.06	0.04	0.20	0.04	0.04	0.4
Mold shrink (1/8 in. specimen)	D955	in./in.	0.007–0.008	0.002–0.003	0.001–0.002	0.017–0.023	0.003–0.004	0.001–0.002	0.010	0.002	0.001	0.002
Tensile strength	D638	ksi	10.2	18.0	23.0	8.0	19.5	20.0	10.8	20.0	27.0	34.0
Tensile elongation	D638	%	50–100	3–4	2–3	10	3–4	2–3	3–4	3–4	2–3	3.0
Flexural strength	D790	ksi	15.4	24.0	30.0	13.0	28.0	29.0	20.0	29.0	34.0	45.0
Flexural modulus	D790	msi	0.39	1.2	2.4	0.35	1.35	2.0	0.6	1.6	2.45	2.4
Shear strength	D732	ksi	9.0	9.5	9.5	7.0	8.0	8.0	–	–	–	–
Izod impact (notched, ¼ in.)	D256	ft-lb/in.	1.2	1.8	1.2	0.3	1.6	1.2	0.3	1.4	1.1	1.5
Heat distortion temperature at 264 psi	D648	°F	345	365	365	155	430	430	278	500	500	500
Coefficient of thermal expansion	D696	in./in./°F × 10^{-5}	3.1	1.4	0.6	5.3	1.2	0.5	3.0	1.3	0.6	[illegible].8
Thermal conductivity	C177	BTU/hr/ft^2/°F/in.	1.8	2.2	5.5	1.1	3.2	6.5	2.0	2.8	5.2	[illegible].0

Data on the shrinkage and warpage of fiber filled thermoplastics were examined, and guidelines offered on how filler and reinforcement systems can be used to reduce warpage problems in the production of plastics parts.[8] Studies included glass and carbon fiber filled polycarbonate and nylon 6/6 composites.

Improvements in the engineering thermoplastics will continue to enhance their qualities as matrices for C/G fiber fillers. A good example is more recent information on new high molecular weight polyphenylene sulfides (PPS). Reference is made to a new process for direct production of the high molecular weight thermoplastics.[9] Higher tensile strength and higher impact qualities are reported over earlier types of PPS, as well as improved moldability in thick sections (3/4 inch).

Carbon fibers are also added to thermoplastics to improve the electrical conductivity and assist in the dissipation of static electric charges.[10] A case in point is a carbon fiber reinforced polysulfone foam molding which exhibited a specific gravity of 0.7, a tensile strength of 3480 psi, and a tensile modulus of 925 ksi.

ENGINEERING THERMOPLASTICS

Reviews of some of the more important engineering thermoplastics which have appeared in programs of advanced composites, particularly in conjunction with C/G fillers, are the following materials.

Nylon was one of the first polymers to appear during Carother's work on condensation polymers initiated at du Pont in 1928. Fibers and molding compounds based upon nylon 6/6 from the condensation of six carbon chain diamine and six carbon chain dibasic acid offered interesting properties (see structure below). It possessed a sharp transition from solid to melt, had high service temperature, and had a combination of rigidity and toughness which made it suitable for gears and bearings. Property enhancement with carbon fibers has made the product even more attractive. Other products than the nylon 6/6 have fulfilled numerous industrial and commercial requirements.

$$n\mathrm{H_2N(CH_2)_6NH_2} + n\mathrm{HO\overset{O}{C}(CH_2)_4\overset{O}{C}\text{-}OH} \rightarrow [\mathrm{\overset{H}{N}\text{-}(CH_2)_6\overset{H}{N}\,\overset{O}{C}(CH_2)_4\overset{O}{C}}]\ n\text{-}\mathrm{H_2O}$$

Nylon 6/6

Because silane coupling agents for enhancing good interfacial bonds for glass fibers were deemed not satisfactory for carbon fibers blended in a nylon matrix, O'Connor and Lewis[11] presented data on the use of low viscosity monomers to wet the surface of the carbon fibers. The monomer used was the monomer salt of nylon 6/6 (hexamethylenediammonium adipate) dissolved in hot methyl alcohol. Chopped carbon fibers of Hercules AS and "Thornel" 50 were coated with the monomer, which was polymerized at 180° and 220°C. It was estimated that 2.9% polymer coating was retained on the fibers. Composites with nylon 6/6 show a modest increase in impact and no increase in tensile or flexural properties. Scanning electron microscopic photographs show good adhesion between matrix resin and carbon fibers.

Polycarbonates offer good qualities as engineering thermoplastic polymers. Fox[12] reviewed early development work upon the polycarbonates, at General Electric Company, pointing out that, like the nylons, they were directed towards metal replacement. Unlike the nylons, polycarbonates do not crystallize readily, and are essentially amorphous materials with high softening points. The glass transition temperature (T_g) near 150°C assures polycarbonates with good mechanical integrity up to this temperature. Compounded with carbon fibers, even superior results can be obtained. Composites of polycarbonates with C/G fabrics laminated with polycarbonate films have been more in evidence than molding compounds with chopped fibers. As transparent sheets and as formed structural shapes, polycarbonates are used extensively in aircraft interiors and in medical equipment. The chemical structure of the polycarbonates is depicted below.

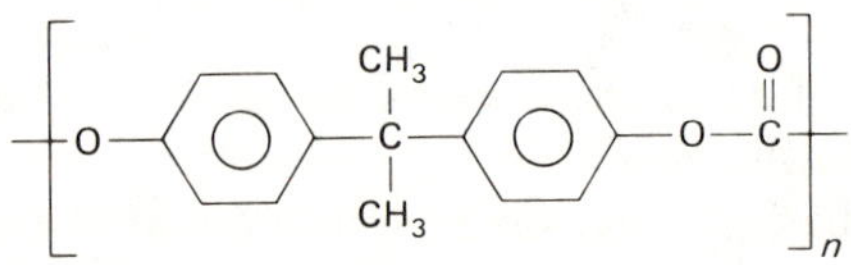

Kardos *et al.*[13] reported upon the preparation of graphite fiber filled molding compounds of polycarbonate. Blends of fiber, along with a 60-100 mesh of polycarbonate powder, were obtained in a 75% methanol, 25% water solution, and a preform mat of random fibers was produced. The mats were dried for 8 hours at 110°C under vacuum and then molded. Volume content of the fibers was 20% for both the glass samples and 20% for the HMG graphite fibers (sized with a standard epoxy resin). The mats were compression molded at 5000 psi, heated to 275°C, and cooled after 10 minutes, or subjected to a further annealing for 3 hours at 245°C, with surprising results (see data in Table 5-4). With regard to the "annealing," Kardos comments that when the system is held near the T_g for a period of time, morphological changes begin to occur which involve gradual alignment of these structures to form lamellae and, subsequently, spherulites. He made these observations for the polycarbonates, polysulfones, and thermoplastic polyethylene terephthalate.

The remarkable increase in modulus for the carbon filled polycarbonate is quite apparent from the data in Table 5-4. The authors admit that the annealing process is not fully understood. Apparently, it is not a characteristic of the

Table 5-4. Effect of Thermal Treatment on Polycarbonate Composites (Filled Systems—20% Volume Fiber).[13]

	MOLD TO TEMPERATURE (275°C)	
PROPERTY	COOLED	THREE HOURS ANNEALING AT 245°C
Glass fiber composite		
Tensile strength (psi)	11,000	11,200
Tensile modulus (× 10^5 psi)	8.20	9.28
Elongation (%)	2.0	3.0
Carbon fiber composite		
Tensile strength (psi)	13,700	16,300
Tensile modulus (× 10^5 psi)	29.0	37.8
Elongation (%)	0.8	0.8
Unfilled resin		
Tensile strength (psi)	9100	8900
Tensile modulus (× 10^5 psi)	3.30	3.11
Elongation (%)	5.3	4.9

polycarbonate alone, otherwise improvements would have been evident on the glass fiber reinforcements.

Polysulfones have been the subject of significant studies in the preparation of advanced composites with C/G fibers. Polyphenyl sulfones have received considerable attention (see the typical structure of polysulfone below).

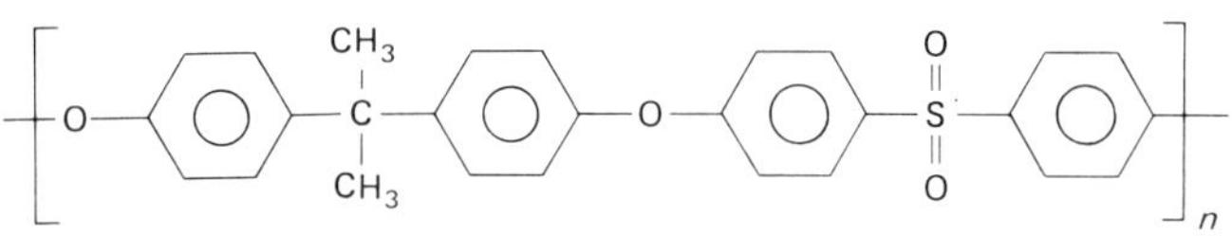

The product was originally developed by Union Carbide, by reaction of the anhydrous disodium salt of bisphenol A upon bis(para-chlorophenyl) sulfone. With good hydrolytic stability and high thermal endurance, polysulfone functions as a good base for advanced composites. It has an added advantage of low flammability and low smoke evolution—which are important requirements in aerospace and automotive applications. As with other amorphous thermoplastic polymers, susceptibility to attack by organic solvents is a drawback.

In tests conducted at Boeing Aerospace Company, materials and processes studies indicate that graphite fiber reinforced polysulfone composites have potential cost saving benefits for aricraft structures.[14] Prototype graphite/polysulfone composite sheets were developed to replace aluminum center body skins on the supersonic Firebee target drones. A 20% cost saving and 16% weight reduction were attained in the designs, as well as compliance with ultimate load conditions. The development activities in this program were limited to cylindrical fuselage sections. Fabrication details on the program have been reviewed.[15]

May *et al.*[16] at General Dynamics recognized that thermoformable composites of graphite/polysulfones provided an opportunity for major cost reductions in fabricating hardware. Because higher processing temperatures were required than for epoxies, General Dynamics decided upon heating and cooling ceramic tooling for shaping parts. Landing gear door skins were illustrated. Tests results using polysulfone binder and type AS graphite filaments (Hercules) yielded the results shown in Table 5-5, which indicate the effect of fiber orientation on the laminate properties (fiber volume: 57 to 58%).

Thermoplastic components are the subject of recent studies,[17] where the cost advantages in manufacturing are noted over graphite/epoxy composites. Reprocessability features are observed, in that if voids are visible and wrinkling occurs in the fabricated part, a thermoplastic such as a polysulfone may be reprocessed to recover the component. Disadvantages associated with thermoplastics are attributed to poor resistance to hydraulic fluids or fuels associated with aerospace operations.

Table 5-5. Graphite Unidirectional C/G Fibers and Polysulfone Laminates.[16]

PROPERTY			
Orientation of fibers	0°	60°	90°
Tensile strength (ksi)	204	68.7	6.4
Tensile modulus (msi)	16.7	6.6	1.2
Compressive strength (ksi)	158	53	–
Flexural strength (ksi)	212	–	–

POLYESTERS

Thermoplastic polymers represent a large potential group of thermoplastic resin matrices, which, in combination with graphite fibers, offer the advantage of lower material costs—a feature which has excited automotive engineers. Polyethylene glycol terephthalate (PET) and poly-1, 4-butanediol terephthalate (PBT) are among the prominent candidate materials. Glass fiber reinforced thermoplastics, introduced in the early 1970's by Celanese, Eastman Kodak, and General Electric as molding compounds, had the desired processability and suitable balance of mechanical properties to make the polyesters viable matrices for composite materials. More recent compounding with graphite fibrous fillers in conjunction with other fibers have qualified the materials for a growing market in the automotive industry. Both PBT and PET resins are crystalline materials, and processed in hot molds to achieve a high level of crystallinity.

PBT is the basis of a number of high performance products. With melting points in the neighborhood of 224°C, PBT molding compounds are processed at 250°C (482°F), where the low molecular weight resin (36,000 to 50,000) has a melt viscosity of 2000 poises. The fiber filled molding compounds have a heat distortion about 430°F (at 66 psi load). There is good creep resistance at room temperature and at temperatures of 260°F. Low frictional characteristics of carbon fiber filled polyester are surpassed only by resins from tetrafluoroethylene, and hence PBT materials are used in pump housings, impellers, appliances, and gears. In the automotive field, they will withstand the solvents and temperatures encountered in painting operations. Because PBT resins crystallize rapidly, they enjoy short molding cycles. Mold temperatures are 100° to 200°F (37° to 93°C), and the cycle time is from 5 to 50 seconds, depending on the size of the posts. The unfilled grades were priced in 1979 at $1.00/lb.[18]

The PET thermoplastics are more widely used for synthetic fibers, films, and blow molded parts. The molecules are readily oriented by the drawing process to yield high strength products. Another thermoplastic polyester of note used with fiber reinforcements for high impact and improved heat resistance is the copolyester of glycol modified poly(cyclohexanedimethanol) terephthalate. With the price advantage offered by the polyesters, one looks to such materials for the injection molding of both glass and graphite fiber filled products for industrial purposes, requiring improved stiffness, heat resistance, and resistance to creep under stress.

MISCELLANEOUS THERMOPLASTIC MATRICES

Other noteworthy applications of thermoplastic resin matrices with C/G fiber should be recognized. Both Russian and Japanese developments in graphite and carbon fiber filled polytetrafluoroethylene (PTFE) should be noted. In some cases, the blends are produced in inert carriers or in water emulsions. Molding the PTFE requires temperatures of 370°C. Applications are principally on self lubricating bearings.[19]

The potential of thermoplastic polyphenylquinoxalines and poly(amide-imides) requires particular attention. These materials, and others of a similar nature—as well as thermoplastic polyimides, have been discussed earlier.

The use of reinforced thermoplastic composites for the fabrication of space beams suggests the use of thermoplastic matrices and graphite fibers as principal structural elements. The potential forming and bonding techniques for this system have inherent advantages over the thermosetting materials. Lubin and Poveromo,[20] after studying various thermoplastics for the proposed space solar collector, have recommended the use of polymethyl methacrylate. The presence of C/G fibers, as already pointed out, will do much to enhance the stiffness and dimensional stability, hopefully under the radiation and thermal exposures of outer space. The crazing of acrylics by solvents may prove a negative factor. It is noteworthy that the proposals for installing fabricating equipment in future orbitting vehicles has reached the planning stage, and a schematic rendition of the equipment for shaping the fiber reinforced structures for the space platform has been presented by the Goldsworthy Engineering Company of Torrance, California. Raw materials supplied to the orbiter vehicle will entail C/G fiber rovings and the matrix resins (see the chapter on aircraft applications).

Graphite fiber reinforced thermoplastics were used to fabricate exit guide vanes on engines. Type HM-S graphite reinforced polysulfones, polyethersulfones, and polyarylsulfones were evaluated at different temperatures. Bending fatigue tests were an important criterion and the graphite/polyethersulfones showed a gradual shear failure mode as a result of a loss in stiffness and as a result of stress and the number of fatigue cycles. They were reported as superior to graphite/epoxy in this application.[21]

C/G fiber reinforced acrylic polymers have also found use in surgical implants.[22] Reinforced bone cements containing a small amount of randomly dispersed short, high modulus graphite fibers have been developed for use in surgical implants–particularly for total hip prosthesis. Porous surgical polymethyl methacrylate (PMMA) room temperature setting cements and grouting compounds were blended with 1 to 2% by volume of 0.25 inch chopped high modulus graphite fibers (E = 380 - 450 GPa). The reinforced surgical PMMA had a marked increase in tensile strength; a doubling of the elastic modulus; and a 50% increase in fatigue strength.

Fiber reinforced thermoplastic polyester resins are increasing in their industrial applications. A typical product, polybutylene terephthalate (PBT), has been used because of its rapid cycle time in injection molding and its lower cost than some of the more exotic thermoplastics. Studies of the thermal and hydrolytic stability of glass fiber filled PBT has been reported by Kelleher *et al.*[23] They evaluated the effects of molding, regrinding, and reprocessing on the properties of fiber filled flame retardant PBT. This assisted their analysis of coil bobbins, insulators, protector housings, switch components, and connectors. It was shown that processing without thorough drying caused a decrease in tensile and impact properties, as well as an increase in melt flow rate–related to chain scission of the ester bond. In effect, embrittlement by hydrolysis could be expected on these materials on long-time aging under humid conditions at high temperature.

Another current background paper on the processing of highly filled thermoplastics reports on the flow rates and processing of mineral filled thermoplastics being prepared in a kneader-extruder apparatus.[24] Relationships were established between viscosity and shear velocity of highly filled thermoplastics, including glass fibers at different temperatures. As the total picture begins to develop, it is apparent that the matrices available for the utilization of C/G fibers are numerous. Through examination of their strengths and weaknesses, processing characteristics, and salient cost factors, it becomes possible to make good engineering decisions.

Milewski presents a detailed study of short fiber reinforcements for injection molding, extrusion, and transfer molding compounds.[25] Micro-fibers, though still experimental, offer promise in formulating molding compounds and comparisons are drawn between various mineral fibers in the 1 to 3 micron diameter size, and various chopped and milled glass and carbon fibers. The microfibers do not possess the purity and crystalline perfection of a true sub-micron whisker. Packing, wetting, and distribution problems are examined under high volume loadings.

REFERENCES

1. Husman, G.E., and Hartness, J.T., "Polyphenyl Sulfone Matrix Composites," *SAMPE Twenty-fourth National Symposium,* Vol. 24, page 21, San Francisco, California, May 1979.
2. Platzer, N., "Performance Characteristics of Thermoplastics," *Thirty-ninth International Congress of Industrial Chemistry,* Bucharest, Rumania, September 1970.
3. Arkles, B., Theberge, J., Cloud, P., and Gerakaris, S., "Elevated Temperature Performance of Thermoplastic Composites," *Thirtieth Annual Technical Conference,* SPI, RP/C Institute, Section 17E, Washington, D.C., February 1975.
4. Arkles, B., Theberge, J., and Robinson, R., "Carbon Fibers for Discontinuous Reinforcement," *ACS Division of Organic Coatings and Plastics Chemistry,* Vol. 35, No. 2, August 25–29, 1975.
5. Theberge, J., Robinson, R., Badner, J., and Arkles, B., "Carbon Fiber Reinforced Thermoplastics," *SAMPE National Symposium,* Vol. 20, page 72, April 1975; and *Industrial Engineering Chemical Product Research and Development,* Vol. 15, No. 2, pages 100–114, 1976.
6. Theberge, J., "Glass and Carbon Fiber Thermoplastic Composites," *Advanced Composites Technology Conference,* by Technology Conference Associates, El Segundo, California, page 52, March 14–16, 1978.

6a. Valentine, J., "Graphite Fiber Reinforcement Molding Compounds," *SPE, Pactec V,* Los Angeles, California, February 1980.

7. Anon., *Plastics Engineering,* page 48, September 1978.

8. Cloud, P. and Wolverton, M.A., "Predict Shrinkage and Warpage of Reinforced and Filled Thermoplastics," *Plastics Technology,* pages 107–113, November 1978.
9. Hill, H.W., Jr., *ACS Polymer Reprint,* Vol. 20, No. 1, pages 173–176, April 1979; and Campbell, R.W., U.S. Patent 3, 919, 177, to Phillips Petroleum Chem., November 11, 1975.
10. Smarook, W., "Polysulfone Reinforced Foam Molding," *Thirty-second Annual Technical Conference,* SPI, RP/C Institute, Section 11G, Washington, D.C., February 1977.
11. O'Connor, J. and Lewis, A.F., "Enhancement of Carbon Fiber/Nylon Matrix Adhesion Via In-situ Polymerized Coatings," *ACS Division of Organic Coatings and Plastics Chemistry,* Vol. 37, page 751, New Orleans, Louisiana, March 20–25, 1977.
12. Fox, D.W., "Chemistry and Technology of Engineering Plastics," Chapter 30, page 430, Craver and Tess (Editors), *Applied Polymer Science,* Published by *ACS Division of Organic Coatings and Plastics Chemistry,* 1975.
13. Kardos, J.L., Cheng, F.S., and Tolbert, T.L., "Tailoring the Interface in Graphite Reinforced Polycarbonates," *Polymer Engineering and Science,* Vol. 13, No. 6, pages 455–461, November 1973.
14. Laasko, J.H. and Hoggatt, N.T., "Development of a Low Cost Graphite Reinforced Composite," Boeing, D180–18236–5, 164 pages, December 1976.
15. Hoggatt, J.T. and Laasko, J.H., "Fabrication of Low Cost Primary Aircraft Structure with Thermoplastic Composites," *Eighth SAMPE National Symposium,* Vol. 8, pages 459–478, Seattle, Washington, October 1976.
16. May, L.C. and Goad, R.C., "Graphite Reinforced Thermoplastics," *Thirty-first Annual Technical Conference,* SPI, RP/C Institute, Section 17C, Washington, D.C., February 1976.
17. Hoggatt, J., "Aerospace Application of Graphite Reinforced Thermoplastic Composites," *Advanced Composites Technology Conference,* by Technology Conference Associates, El Segundo, California, page 217, March 1978.
18. Anon., *Modern Plastics Encyclopedia,* Vol. 55, page 49, 1978–1979.
19. *Chemical Abstracts,* 88-23904d, Vol. 88 (1978); and CA–87–40411N, Vol. 77, 1977.
20. Lubin, G. and Poveromo, L.M., "Reinforced Thermoplastic Composites for Space Beam Fabrication," *Thirty-fourth Annual Conference,* SPI, RP/C Institute, Section 15E, New Orleans, Louisiana, February 1979.
21. Navak, R. (United Technology Research Center), CA 87–185654d, NASA–CR–135196, 1977.
22. Pilliar, R.M., Bratina, W.J., and Blackwell, R.A., "Mechanical Properties of Carbon Fiber Reinforced Polymethyl Methacrylate for Surgical Implants," *Fatigue of Filamentary Composite Materials,* ASTM STP 636, pages 206–227, 1977.
23. Kelleher, P.G., Bebbington, G.H., Falcone, P.R., Ryan, J.T., and Wentz, R.P. (Bell Telephone Labs), "Thermal and Hydrolytic Stability of Poly(butyleneterephthalate)," *Thirty-seventh SPE ANTEC,* page 527, New Orleans, Louisiana, May 1979.
24. Geisbusch, P. and Menges, G., "Flow Properties of Highly Filled Thermoplastics," *Thirty-sixth SPE ANTEC,* page 113, Washington, D.C., April 1978.
25. Milewski, J.V., "Short Fiber Reinforcements," *Plastics Compounding.* page 17, November/December 1979.

6
Surface Treatments of Carbon/Graphite Fibers and Their Effect on Composites

Several factors influence the choice and application of surface treatments to C/G fibers. One concern is to enhance further processing and handling, by the use of sizing agents on fiber surfaces to reduce the tendency for the fibers to break. Because most carbon fibers now in use are derived from synthetic fiber precursors, the handling of the precursor fibers (such as polyacrylonitrile (PAN) or rayon, requires sizing agents applied at the time of, or shortly after, their issuing and/or handling, during oxidation and carbonization treatments. Subsequent heat treatments will decompose the organic sizing agents, and the problems of resizing must be addressed, with efforts towards facilitating the handling of stiff C/G fibers, whether as monofilaments or as woven goods, as well as with the additional concern for enhancement of the bond between the surface of the carbonaceous fiber and the resin matrix to be applied. It is accepted that good bonds between fiber and resin are essential to achieve good performance of the composites under adverse environmental conditions. With newer carbon fibers derived directly from pitch, problems related to surface treatments are different, in that at the time of dry spinning, one is not dealing with synthetic resin precursors, and the selection of sizing agents or treatments is concerned with further handling of brittle fibers and avoidance of fiber to fiber sticking. Further heat treatment through graphitization removes organic material, requiring surface treatments to enhance subsequent handling and resin adhesion qualities, with fibers derived from synthetic fiber precursors as well as pitch fibers. Some carbon fiber manufacturers leave open the option of surface treatment or no treatment.

C/G fibers have had unique requirements for surface treatment, and while excellent information is available on the silane coupling agents for glass fibers, these data are not necessarily applicable to the C/G fibers. Progress has been made in recent years, judging from the improvement in properties, such as shear strength of the C/G composites, where there have been concerted efforts to enhance the matrix to fiber bond. Another mitigating factor which has an influence upon composite performance is the nature of the chemicals which remain behind on the surface of the fiber after treatment. As discussed earlier, in examining the performance of composites after sustained time at high temperatures, the presence of certain metallic ions, such as sodium, on C/G fibers (for example) has been shown as contributory to more rapid loss in physical properties of the composite at high temperatures. In the following chapters, an attempt will be made to examine those salient ideas, which suggest acceptable methods for the treatment of C/G fibers before they are combined with the resin matrix.

Excellent fundamental studies on the adhesion of synthetic resins to the surfaces of strong fibers have been carried out at the Naval Research Laboratory and Naval Ordnance Laboratory in Washington, D.C. and Maryland. As Zisman[1] points out, it is always desired to optimize the energy of adhesion at the resin/fiber interface. In this way, fibers are made to distribute and carry as much of the stress load as possible. The presence of the fibers can also interfere with crack propagation through the surrounding resin matrix. Although this reference places emphasis upon glass reinforcing fibers, much of the presentation is applicable to other fibers. To assure intimate contact of liquid resin and the entire surface of the solid, it has been felt that the surface tension of the resin should be as high as possible, consistent with maintaining the contact angle between the fiber surface and the resin matrix at or close to 0°. Too often, the resin matrix does not fill completely all the crevices and pores on the surface, including the interfibrillar crevices in fiber-reinforced plastics. The presence of only partially filled surface pores and crevices contributes a significant loss of strength and moisture resistance in the composite material. One objective of surface treatments is to offset or compensate for these situations, either through chemical or mechanical means.

A further difficulty resulting from the presence of voids on fiber surfaces or at the interface with the resin matrix is that the diffusion of water through the resin/fiber composite is facilitated. Good adhesion between the resin and the fiber surfaces reduces this tendency. As will be examined in later chapters, cyclic stressing of advanced composites while submerged in water greatly accelerates their rate of failure. Measures can be taken to improve performance (for example, through good adhesion between carbon fibers and resin matrix).

It should be noted that the possibilities of chemical bonding to carbon surface, particularly if activated by some form of heat treatment, offer more prospects for good adhesion than with glass fibers, particularly if the oxidation treatment has developed functional bonding sites for the synthetic resin matrices used with the carbon fibers.

When one refers to the literature on the role of silane coupling agents for enhancing the bond of resin matrices to glass fibers, it is impressive to observe the considerable research performed in this area. Silane coupling agents, in particular, have been tailored to accommodate a variety of resin systems to glass fibers. Although the same coupling agents are not necessarily applicable to the C/G fibers, the methods of test and the evaluation on the permanence of composites prepared form a sound referral base. Plueddemann's[2] work has been thorough and afford good background material. As he has commented, an organic polymer based on stable covalent bonds, and a mineral based on ionic bonds (glass), are such dissimilar materials that it is difficult to achieve a good water resistant bond between the two. Hybrid organic-inorganic compounds, such as organosilanes and methacrylate-chrome complexes, are used as coupling agents at the polymer/mineral interface. Improved adhesion across the interface is accompanied by improved mechanical properties and water resistance in the composites. It may be necessary to attain a compromise in interface properties of adhesion, because fracture toughness in glass composites may be decreased by improved adhesion; hence, the coupling agent may require chemical modification to accommodate the toughness (which requires a weaker adhesion at the interface) with the good adhesion required for maximum strength.

Silane treated glass fibers have been observed at high magnification in electron micrographs, according to Erikson and Plueddemann.[2A] Optimum performance in composites employed 0.1 to 0.25% silane coupling agent on the glass fibers, though a monolayer would require only 0.02 to 0.04%. Electron micrographs showed a heavy concentration of the silane in the interfilament channels of the glass fibers. In the development of coupling agents for glass, the chemical bonding theory is advanced most frequently, in which chemical functional groups on the coupling agents could theoretically react with the silanol groups on the glass by covalent bonds. In addition, the coupling agent could contain a different chemical functional group which would co-react with the resin matrix. Typical of the silane coupling agents used with glass fiber composites are phenolic resin/glass laminates (gamma-aminopropyltriethoxysilane) and epoxy resin/glass laminates (glycidoxypropyltrimethoxysilane).

Among other theories that are more applicable to the C/G composites are those which suggest the importance of a deformable interface layer that will accommodate the low thermal expansion C/G fibers in the relatively high expansion resin matrix. There are thermal stresses which arise from the wide difference of thermal coefficients of expansion of the materials (resin matrix and fiber) comprising the composite, as well as the fact that the thermal conductivity of the C/G fibers is greater than for the resin matrix (hence, they will more rapidly reach the temperature of the environment).

Scola[3] examines high modulus fibers and the fiber/resin interface in resin composites. The most common surface elements on graphite, other than carbon, are hydrogen and oxygen. Surface functional groups reported include carboxyl (-CO_2H); aromatic and aliphatic alcohols; carbonyl (=C=O); and lactones. In the case of graphite fibers, the surface contains appreciable quantities of oxygen and traces of hydrogen, due in part to the nature of the precursor fiber (rayon or PAN), as well as due to the reactive nature of carbon. Scola reported on the work of Tuinstra and Koenig,[4] which shows that the size (diameter) of the crystallite on the graphite surface is related to the degree of interaction at the fiber/resin interface. Using Raman spectroscopy, they showed that the fiber whose surface structure consisted of crystallites of small diameters in the *a* plane gave composites of the highest shear strength. Data from this paper are shown in Fig. 6-1. Several graphite fibers are included in the development of the data, and while the relationship is established between the apparent diameter of the graphite crystallite and the composite shear strength, surface treatments on the fibers appeared to be random in their influence.

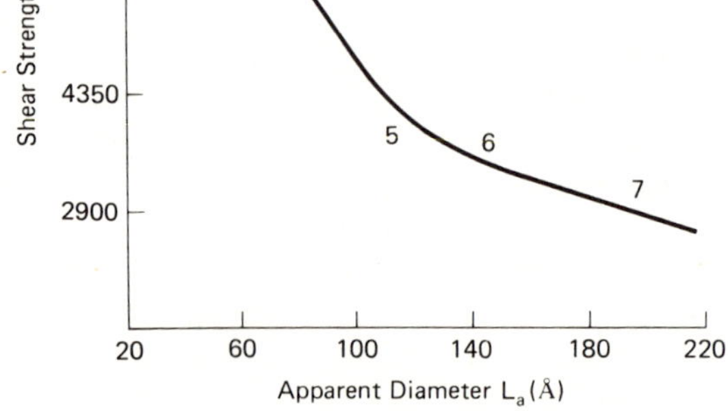

Fig. 6-1. Relationship between composite shear strength and crystallite size from Raman studies. (1) Morganite II (u), (2) Thornel 25 (HNO_3) (3) HMG-50 (u), (4) Thornel 25 (air oxidized), (5) Thornel 25 (u), (6) Thornel 40 (u), (7) Morganite I (u) u = Untreated.[4]

Scola[3] showed further that the fiber surface area (as measured by low temperature absorption of krypton) did not necessarily have a pronounced influence upon the composite short beam shear strength. He uses a series of treated and untreated fibers from Hitco, Courtaulds, Morganite, and Union Carbide in establishing this relationship (see Table 6-1). The composites were prepared from unidirectional graphite fiber: Union Carbide's cycloaliphatic 2256 epoxy and eutectic aromatic amine cure agent based on meta-phenylene diamine. The fiber volume was 55%, and the cure cycle was 2 hours at 80°C and 200 psi, followed by 2 hours at 150°C.

It is interesting to observe that some of the better results in Table 6-1 were attained by manufacturers' proprietary surface treatments. Not all carbon fiber cross-sections are circular, and some of them, from rayon, as well as from PAN, are bilobal.

Because glass fibers are more moisture sensitive than graphite fibers, it would appear that interfacial degradation in glass/resin composites is obtained more easily than in graphite/resin composites. Not to be overlooked, of course, is the chemical resistance of the resin matrices themselves, as those polymers which are known to be very highly resistant to the absorption of water (certain halogenated epoxies, for example) impart this good quality to the retention of bond strength of most fibers used for reinforcement.

The surfaces of C/G fibers affect adhesion of the matrix binder. This is apparent when there are changes in the surface morphology and changes in the chemical functionality. Didchenko[5] discloses valuable information on the pore size distribution of some of the early C/G fibers (for example, "Thornel" 40,

6-1. Comparison of Fiber Surface Area and Composite Shear Strength.[3]

GRAPHITE FIBER	TREATMENT	FIBER DIAMETER (microns)	FIBER SPECIFIC SURFACE AREA (m^2/g)	COMPOSITE SHORT BEAM SHEAR STRENGTH (psi)
U.C. "Thornel" 50	PVA size	6.6	0.59	4000
U.C. "Thornel" 50	Oxidation	6.6	2.5	7000
Hitco HMG-50	None	6.8	0.87	4800
Hitco HMG-50	By manufacturer	6.8	0.66	6500
Hitco HMG-50	Oxidation	6.8	7.3	8500
Morganite I	None	7.5	0.11	4000
Morganite I	By manufacturer	7.5	0.13	9000
Courtaulds B	None	7.5	0.31	4000
Courtaulds B	By manufacturer	7.5	0.39	9000

from a rayon precursor), which is shown in Fig. 6-2. Didchenko reproduced electron micrographs of surface replicas of carbon fibers in the magnification range of 6000 to 20,000. In general, the carbon fibers were unremarkable by the absence of surface detail, whereas the graphite fibers showed well developed surface roughness. Typical porosity data for "Thornel" 40 are reported by Didchenko (data developed from a nitrogen absorption test):

Total pore volume (cc/g, 0–600 Å)	0.0030 ± 0.0002
Surface area (m^2/g)	1.3 ± 0.2
Mercury density test	1.59 ± 0.02
Upper quartile of pore diameter (25% of total pore volume is contained in pores larger than values shown)	473 ± 75 Å
Median quartile of pore diameter (50% of total pore volume is contained in pores larger than values shown)	166 ± 59 Å
Lower quartile of pore diameter (75% of total pore volume is contained in pores larger than values shown)	67 ± 18 Å

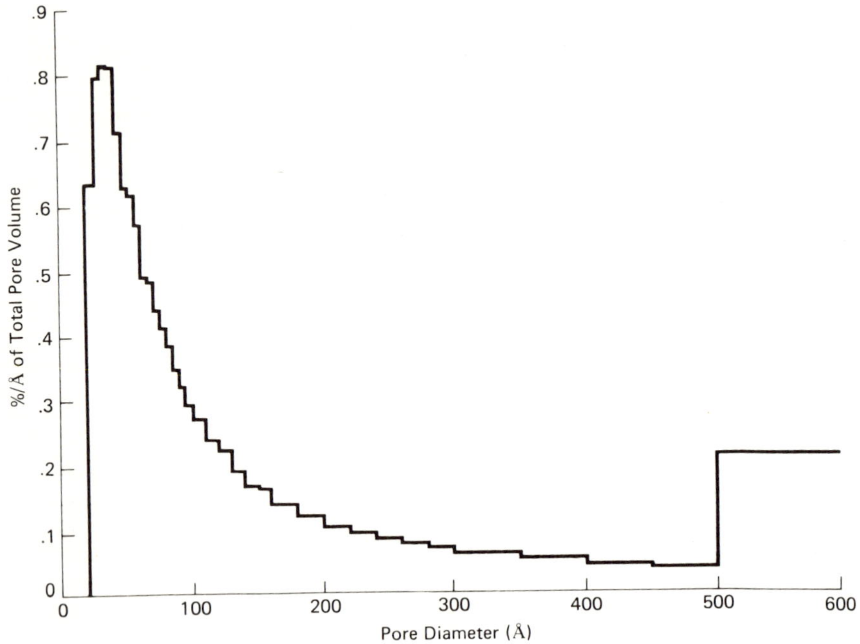

Fig. 6-2. Typical pore distribution for "Thornel" 40.[5]

While the above data are not necessarily generic to graphite fibers, and not necessarily applicable to other C/G fibers, they do suggest the character of pore distribution of fibers prepared from textile precursors (rayon or PAN). In all likelihood, the appearance of extrudable fibers of C/G from pitch is notably different. Their surface roughness will reflect oxidation treatments during thermal processing of the fibers.

Larsen[6] comments that the greatest single problem associated with C/G fibers has been the interface area and the development of strong bonds between the fiber and the resin matrix. Scanning electron microscopic photographs substantiate the existence of crystalline and amorphous regions within the fiber, with the crystalline regions forming continuous or semi-continuous fibrils axially along the fiber. Chemically, the fiber surface contains several oxygen containing functional groups. The surface available for resin bonding consists generally of low energy basal planes, interspersed with edge planes and amorphous regions. Larsen establishes a significantly high proportion of edge displacements, vacancies, and crystal edges on the high strength, intermediate modulus graphite fibers, as compared with the higher moduli graphite fibers. Higher moduli graphite fiber composites have exhibited lower composite shear strength—attributed to lower concentration of reactive sites where resin bonding can occur. The mechanism by which composite shear strength is improved, according to Larsen and others, appears to be related to both the surface area increase and the increase in chemical functionality groups on the fiber surface.

Although it may be considered far-fetched to draw comparisons between the reinforcement of elastomers by finely divided carbon black, with the behavior of C/G fibers in advanced composites for structural and molding purposes, there are elements of similarity which may be examined. In particular, polymer/carbon black surface interactions have been reported voluminously, and some of the conclusions arrived at from these studies will aid developments of structural composites. Kraus[7] poses the question for carbon black filled rubbers: "Is it necessary for polymer molecules to become grafted to the carbon black surface by primary valence bonds to produce reinforcement [in rubber products]?" He presents evidence that the question can be answered in the negative.

It is pointed out that carbon black surfaces contain functional groups capable of reacting with polymer molecules during processing. Among the possible mechanisms by which grafts may be formed on the surfaces of finely divided carbon black are the following:

1. Chemisorption of simple olefins—enhanced in the presence of sulfur.
2. Shear generated polymer free radicals, formed during intensive mixing, which have been shown to graft to carbon black.

3. Hydrogen exchange between rubber and bound hydrogen located at the edges of large aromatic ring molecules making up the layer planes of carbon black (according to Papirer [8]).

Kraus[7] points out further that if the carbon black is heated to 2700°C or higher in an inert atmosphere, as in graphitization, all surface functional groups and hydrogen are removed as highly energetic sites for physical absorption. The graphitization process does lead to formation of a more regular, inert graphite crystal structure, but the gross morphology, surface area, and structure are altered only moderately. It has been demonstrated that inactivation by graphitization of the carbon black has only a modest effect on the properties of rubber.[9]

Table 6-2 lends support to the concept that chemical bonding at the rubber/carbon interface is not a necessary condition for reinforcement. Although the stress for 300% elongation is markedly lower, the ultimate strengths are quite comparable for the materials reported. It is concluded that the highly complex morphology of the reinforcement of rubber by carbon black can be approximately represented by the specific surface area accessible to the rubber molecules and by a measure of the volume pervaded by the primary aggregates ("structure").

The considerable study given to the interaction of polymers to carbon black fillers provides data of fundamental importance to C/G fiber composites. In these circumstances, we are not dealing with finely divided particles and intensive co-mingling, except in the preparation of molding compounds and non-aligned fibers. The specific techniques of C/G fiber surface preparation can be examined in the light of earlier information gleaned from studies of carbon black.

The adhesion of polymers, such as polyesters, epoxies, and polyimides, to the surfaces of high modulus C/G fibers is generally weak, and the surface treatments which have been developed have concentrated upon the enhancement of the bond. In general, an improvement in adhesion to the surface of the carbon fiber is accompanied by an improvement in the interlaminar shear strength of

Table 6-2. Effects of Graphitization of Carbon Black on Rubber Reinforcement.[7]

	ORIGINAL	GRAPHITIZED
Specific surface area (m^2/g) (by N_2 absorption)	116	86
Structure by vehicle demand, linseed oil (cm^3/g)	1.72	1.78
Stress at 300% elongation (MPa)	14.4	3.5
Tensile strength (MPa)	26.2	23.4

the C/G fiber composite. For this reason, data on the interlaminar shear strength of subsequent composites have been presented.

As one examines the treatments discussed below, side effects which may prove harmful to the environmental stability of the composite must be identified. For example, traces of sodium salts left on the surface of the fibers, which may have originated from the original wet spinning process, washing procedures, or chemical treatments, will adversely affect the stability of the C/G fibers at high temperatures of operation, with a rapid loss of physical properties.

CHEMICAL COUPLING AGENTS

The success of using silane coupling agents with glass fibers in preparing reinforced plastics has prompted the application of silanes to C/G fibers.[10,11] In general, the results with silane coupling agents have not proven successful with the C/G fibers. Somewhat better performance was reported for these fibers when using titanate coupling agents.[12] Some advantages of post-baking of the carbon fibers after treating with silane coatings have been reported. For example, poly(dichlorodimethylsilane) applied to a carbon fiber tow from a toluene solution, and subsequently baked in an inert atmosphere at 1000°C, for six hours, offered moderate gains in interlaminar shear strength, when fabricated into composites with epoxy resin matrix.[13] Attention is also called to glow discharge polymerization of thin polyorganosilicon films, particularly hexamethylcyclotrisilizane, for thermally resistant (up to 500°C) films, as a procedure for coating fibers.[14]

A direct correlation between the amount of carboxylic groups on the fiber surface and the efficiency of reinforcement has been shown for carbon fiber reinforced phenolics by Fitzer *et al.*[15] It is also recognized that good adhesion between fiber and matrix before carbonization is disadvantageous from the viewpoint of fracture behavior, resulting in lower strength. This has led to thermoplastic pitch matrices for the carbon fiber/carbon matrix, high temperature composites. Studies were conducted upon "Sigrafil" (West Germany) high strength and high modulus carbon fibers; which were nitric acid etched for this study.

Coatings on carbon fibers for improved adhesions have generally been in the range of 1 to 2% by weight of the fibers (6 to 10 microns in diameter), though when applied to tows, or bundles of fibers, additional polymer film content may be embedded in voids or spaces between individual fibers. When present in this manner, the effectiveness of the coupling agent may be lost. In fact, to be effective, uniformly thin coatings on the fibers are preferred, and attention must be given to the molecular weight and spreading characteristics of the polymer coating materials. The surface coating agent may be best modified to improve its quality as a coating on a surface such as carbon fiber–such modification by the inclusion in its formulation of about 0.5% spreading agent of a solution of a cellulose ester or proprietary polymer such as "Modaflow."

The literature is not very definitive on the subject of polymer coating agents for the treatment of carbon fibers for enhanced adhesion of resin matrices. Phenolic resin coatings, furfuryl alcohol resin coatings, epoxy resin coatings, and polyvinyl alcohol coatings are mentioned most frequently for the C/G fiber epoxy composites, while thin coatings of polyimide resins are cited for the polyimide resin matrices. As to what is "thin" is still open to question. When one examines primers on non-porous adherent bulk surfaces to maximize adhesive bonding, highest shear strengths in lap-shear tests are obtained when such primer coatings are very thin.

Coatings of furan resins and phenolic resins were described by Watts[16] in his patents on carbon fibers. Lubowitz[17] discussed solutions of low molecular weight polyimide prepolymers to treat carbon fibers. Maleic anhydride/styrene copolymers were cited in other efforts to improve shear strength and water resistance,[18] functioning as chemical coupling agents on carbon fibers. The grafting of methyl, ethyl, and n-butyl acrylate polymers on "Toray" T-300 carbon fibers was studied in considerable detail.[19] Using a benzoyl peroxide catalyst for polymerization, the rate of grafting was reported as increasing with an increasing length of ester group in the monomer. Isocyanate prepolymers are also promising for surface treatments, not only because they offer chemical functionality, but also because of their ability to scavenge surface moisture by their NCO groups.

Marks *et al.*[20] investigated several organic coatings on graphite fiber surfaces to enhance interfacial adhesion. "Thornel" 300 graphite yarn was treated with four types of organic resin coatings, with the following results.

SURFACE COATING	TREATMENT	RESULT (SCANNING ELECTRON MICROSCOPY)
SC-1008 Phenolic	Pyrolyzed at 1000°C	Smooth carbonaceous surface
SC-1008 Phenolic + ferrocene (1:1)	Pyrolyzed at 1000°C	Rough, pock-marked surface
"Hercules" H-polyaromatic resin	Pyrolyzed at 1000°C	Rough, pock-marked surface
"Kerimid" 601 polyimide	Manufacturer's cure	Heavy, net-like structure

Unidirectional composites prepared with the above fibers and an epoxy matrix did not indicate much differences in tests at 350°F, when compared with untreated controlled T-300 fiber. All composites showed a 50 to 60% fall in flexural strength when tested at 350° from a wet condition as compared to tests from a dry condition. The "Hercules" H-polyaromatic resin appeared to have a slight edge over the other treatments.

Further studies of the surfaces of graphite fibers have been conducted with the aid of scanning electron microscopy (SEM) and electron spectroscopy for chemical analysis (ESCA).[21] Surface treatments of C/G fibers were evaluated for polyimide, nadic anhydride-epoxy, and styrene/maleic anhydride coatings. Data reported the nitrogen and oxygen enrichment on C/G surfaces as a consequence of these surface treatments. Epoxy resin sizes have been used as coupling agents, as well as bisphenol polyester sizes for improved adhesion to C/G fibers.[21A]

Thermoplastic resin matrices for C/G fibers have specific selections for chemical coupling agents to enhance the bond, though the problems are not as acute for molding compounds with non-oriented short segments of C/G fibers as they are for structural laminated composites with continuous monofilaments. As observed in the chapter on thermoplastic resin matrices for advanced C/G fiber composites, there are many candiate materials which have been used as matrices for C/G fibers. Recommendations for chemical coupling agents to carbon fibers should be obtained from the suppliers of the thermoplastic polymers. One effective study has been published by Kardos[22] on graphite reinforced polycarbonates. He reports that engineering thermoplastics such as polycarbonates, polyethylene terephthalate, and polysulfones exist as small ball-like structures of about 75Å in size when in an amorphous state. When these systems are held near T_g for long periods, morphological changes begin to occur which involve gradual alignment to form lamellae. Preliminary work showed that graphite fiber could function as a nucleating site for polycarbonate resin, and that the properties of the resulting composites were significantly improved. (See the chapter on thermoplastics resin matrices for C/G fiber composites.)

LIQUID OXIDIZING AGENTS

Processes for treating carbon fibers have included concentrated nitric acid, acidic potassium permanganate, sodium hypochlorite, and hypochlorous acid.[23] Boiling nitric acid achieves significant surface activation, though long treatment times are necessary. As indicated earlier, recent studies question the use of sodium salts in surface treatments of carbon fibers. The liquid oxidation treatments generally etch the fiber surface and yield substantial gains in interlaminar shear strength of the C/G composites, with relatively little decrease in the tensile strength of the fibers. An improvement in the adhesion of polymers to carbon fibers has been reported by prior immersion of fibers in a solution of chlorine in water (2g/1).[24] In the patent review in Chapter 1, bromine treatment of carbon fibers was also referenced. Earlier references on Fitzer *et al.* discussed nitric acid treatments of carbon fibers to oxidize the surfaces and enhance adhesion. Pretreatment of pitch fibers in chlorine solution before oxidation greatly reduces the oxidation time.[25]

Goan's[26] studies on graphite fiber oxidation provided guidelines for further research. Although this work was performed in 1969 upon C/G fibers that were then available, including the rayon precursors of Union Carbide, some of the conclusions are applicable to the newer improved fibers of recent years. The program was organized to determine the effect of a given treatment process on the graphite fiber by incorporating the treated fibers into a resin composite and measuring the properties of the composite. The epoxy resin system ERL 2256/ZZL 0820 (see Chapter 3 on thermosetting resin matrices) was employed in preparing the composite. The principal wet oxidizing agent used was 60% aqueous nitric acid. Carbon and graphite yarns from various sources were treated, and thoroughly washed to remove any traces of acid. Table 6-3 gives data from Goan's report.

The shear strengths of the composites listed in Table 6-3 are substantially lower than the values obtained from advanced composites with boron fibers and glass composites. It was also noted that carbon fibers from PAN precursors attained shear strengths over 9000 psi. Herrick[11] has reported on the improvement of interlaminar shear strength of composites through oxidizing treatment of the carbon fiber surfaces. NOL test rings were observed to yield higher shear strengths than those obtained by the short beam interlaminar shear tests, using a 5 to 1 span to depth ratio. More recent work on shear strengths of C/G polyester composites indicated substantial improvements in shear strength, when oxidation was carried to a 0.6 to 0.8% weight loss, which produced pits on the fiber surfaces.[27]

Judd[28] has reported gains in interlaminar shear strength through surface treatment of carbon fibers with a hypochlorite solution followed by soaking in

Table 6-3. Effect of Fiber Oxidation on "Thornel" 50 Composite Properties (Epoxy Resin System ERL 2256/ZZL0820 (27phr).[26]

	COMPOSITE PROPERTIES				STRAND TENSILE STRENGTH	
FIBER TREATMENT	DENSITY (g/cc)	FIBER VOLUME (%)	SHEAR (psi)	STRENGTH (MN/m^2)	(ksi)	(MN/m^2)
None (PVA finish)	1.43	58	3660	25.2	288	1990
Water wash	1.45	61	3820	26.3	220	1520
Nitric acid (60%)	1.46	57	5850	40.3	220	1520
Sodium hypochlorite (5.5%)	1.41	–	4330	29.9	123	850
Chromic acid	1.59	59	3510	24.2	190	1310
Air (400°C)	1.44	57	3790	26.1	178	1230

an ether extract of ferric stearate. Cured with an epoxy resin system, there was a reduction in water absorption of the composite and an increase of interlaminar shear strength to 51 MN/m^2.

Additional work on liquid oxidizing agents for treating C/G surfaces was reported in 1972.[29] In trying to circumvent the substantial loss in tensile strength of fibers during high temperature oxidative exposure, aqueous sodium iodate ($NaIO_4$) treatment was shown to impart carbonyl functionality to high modulus graphite fibers, and hence improve possibilities of improved adhesion. Aqueous OsO_4-$NaIO_4$ dioxane solutions achieved the same results at room temperature.

A new technique for evaluating the oxidizability of carbon fiber surfaces was reported by Kalnin.[30] In this method, the characterization of both the surface area and the polarity is measured by the differential capacitance assumed by a conductive fiber bundle anode in an electrolytic solution. The differential capacitance of carbon fibers increases with progressive anodic oxidation in a manner which permits a quantitative assessment of the susceptibility of the carbon fiber surface to oxidative surface treatments. A C/G fiber bundle is anodically oxidized by transferring to a potassium hydroxide solution, charging a capacitor (250 to 500 microfarads) from a dc source (40 volts) and discharging through carbon fiber/platinum electrodes. Relative differential fiber capacitances were measured after discharges. High modulus (GY-70 Celanese fibers 400 to 500 GPa Young's modulus) were treated in this manner and conditions observed for higher composite short beam shear strength.

GASEOUS OXIDATIVE TREATMENT

Gaseous oxidative treatments produce more severe degradation of fiber tensile strength than do the liquid treatments, without substantial improvement in composite shear strength. Nitric acid treatment on the previously described "Thornel" fibers produced a much greater increase in fiber surface area, while gaseous oxidations achieved minor increases. The composite shear strengths appear improved through an increase in surface area as well as functionality on the surface of the carbon fibers.

Oxygen plasma etching of C/G fibers prepared from PAN precursors has been studied with the aid of scanning electron photomicrographs.[31] In these experiments, more severe etching of fiber surfaces was observed for the intermediate modulus graphite fibers than for the high modulus fibers. In earlier work, ozone treating of graphite fibers was practiced for improved adhesion of polyimides. The process was carried out at temperatures above 600°F.[32] Ozone oxidation was also accomplished by heating carbon fibers in an ozone enriched environment for several hours at 100°C.

PAN-based carbon fibers with various heat treatment temperatures (1100° to 2500°C) were oxidized in carbon dioxide or oxygen at temperatures affording similar reaction rates, before use in advanced composites. The oxidation treatment used affected the fiber surface layers to a depth not exceeding several hundred Angstroms. With iron impurities, fiber smoothing occurred with the carbon dioxide and increased striations with the oxygen oxidation. In prepared graphite fiber/epoxy composites, best shear strengths were obtained with a 5% wear due to oxidation with carbon dioxide and on fibers that had been previously heat treated at high temperatures. Electron micrographs indicated an irregular shear fracture for unoxidized fibers which slipped in the composite, but a clean break of fiber and resin for the oxidized fibers.[33]

Carbon fibers prepared from PAN precursors were graphitized at 2100°C and then further subjected to oxidative treatments to improve adhesion. The matrix used in preparing the composite was an epoxy resin and phenolic resin curing agent. The specific surface area of the fibers was determined by B.F.T. using argon gas. Data were presented as follows.[34] Modulus of fibers was reported as 40 to 42 msi.

OXIDATIVE TREATMENT OF GRAPHITE FIBERS	PERCENT LOSS IN FIBER WEIGHT	SPECIFIC SURFACE (m^2/g)	TENSILE STRENGTH (UNIDIRECTIONAL)
None	0	0.45	270 ksi
20–40 minutes at 550°C	0.8	0.51	280 ksi
15–25 minutes at 600°C	1.3	1.17	280 ksi
10–20 minutes at 650°C	2.5	2.03	320 ksi

When oxidation processes are developed for carbon fibers, industrial manufacturers concentrate upon continuous processes. Exposing the carbon fibers for even short intervals to high temperatures in air, for example, may attain increases in composite shear strength but at a substantial sacrifice in fiber tensile properties. Metallic ions are known to have a strong influence on degradation rates. Hence, quality control of the carbon fibers requires the utmost care during processing on a continuous basis. It should also be noted that "heat cleaning" of graphite yarns, say at around 1000°C in an inert nitrogen or argon atmosphere, is a desirable step before coating with a polymer surface film to improve adhesion. McKague[35] discusses techniques of purification of C/G fibers to remove oxygen, water, etc. under reduced pressure just prior to application of epoxy resin matrices.

GAS PLASMA TECHNIQUES

Recent work suggests that gas plasma techniques for treating organic surfaces, including fibers and films, may be adapted to enhancement of bonds to C/G fiber surfaces. DeLollis[36] has reported on the effects of argon, oxygen, air, and nitrogen gas plasmas on silicone and epoxy resin surfaces. Improvements on bondability have been significant. An approach to a continuous coating of a substrate pulled through a plasma is described by Yasuda.[37] He employed a variety of frequencies up to 13.56 MHz. In the apparatus shown, an R.F. glow discharge was maintained by electrons oscillating between the electrodes and achieving ionizing energies by means of numerous collisions on the substrate. The mounting of permanent magnets behind the electrodes increased the efficiency of plasma polymerization. This particular work was directed towards the polymerization of tetrafluoroethylene.

Goan[26] also described radio frequency plasma oxidation of graphite fibers in oxygen at reduced pressure in the presence of R.F. discharge to activate the fiber surface. The process was not intended for continuous processing and conditions for a typical oxidation treatment were as follows:

R.F. frequency	100 MHz
Discharge potential	330 volts
Discharge current	170 milliamperes
Reaction time	3 hours
Oxygen pressure	0.15 torr

When one examines the effect of bombardment of surfaces with high energy charged particles, the recent work of Professor Tombrello at Cal Tech in which he describes a phenomenon called "enhanced desorption" is encountered. The impact of particles on a surface may release many more atoms than would be expected. Though this observation has been discovered on metal surfaces, it is not unreasonable to expect similar processes on graphite fibers—with an ensuing enhancement and roughness which would aid adhesion of polymers. Scientists are trying to determine how such tiny causes could trigger such large effects.

The irradiation of carbon fibers with fast neutrons ($\sim 7 \times 10^{17}/cm^2$ to $3 \times 10^{20}/cm^2$) contributed to an improvement in the mechanical properties of C/G/epoxy composites.[38] 10 to 20% increase in flexural and shear strength properties was reported.

TREATMENT WITH CHEMICAL SOLUTIONS

Larsen[39] describes a number of solution reduction treatments for C/G fibers for the enhancement of shear strength of the composites. After application of the following solutions and drying, heat treatments were attained at high temperatures (400° to 1600°C) by resistance heating. The data below are from Larsen's paper (Hitco's HMG-50 fibers were treated).

SOLUTION REDUCTION	EFFECT ON COMPOSITE PROPERTIES	
TREATMENT	TENSILE STRENGTH	SHEAR STRENGTH
$FeCl_3$	- 7%	+ 100%
$Fe(C_5H_5)_2$ (ferrocene)	+28%	+ 80%
Phenylated polyquinoxaline (PPQ, 0.1% solution)	+ 0%	+ 170%

The improvements in shear are dramatic, and for the ferrocene it is noteworthy that there was no loss in fiber strength. However, the surface contamination of iron compounds on the carbon fibers would contribute to instability at high operating temperatures as composites. The polyphenylquinoxaline resin was chosen for its high carbon content and low weight loss on thermal decomposition (see the chapter on high temperature resin matrices).

Cleaning carbon fibers in chemical solutions under more moderate conditions than above have been reported.[40] Carbon fibers from PAN precursors were impregnated with 1% sulfuric acid and heated for 3 minutes at 750°C for cleaning purposes. These fibers, prepared into composites with an epoxy/anhydride resin system, showed improvement in tensile and shear strength over composites prepared from similar carbon fibers with no cleaning preparation. Others have recommended oxygen aerated solutions of 5% sulfuric acid to treat the surface of carbon fibers to promote adhesion.[41] The roughening of carbon fiber surfaces by 0.1% copper acetate solution, drying, and then heating in air at 200° to 500°C, is reported as promoting improved adhesion.[42] Note, however, that traces of copper acetate accelerate decomposition of carbon fibers at higher temperatures.

VAPOR PHASE DEPOSITIONS AND TREATMENTS

By developing other crystalline structures on the surface of carbon fibers, the mechanical adhesion of the resin matrix may be improved. This process is sometimes referred to as "whiskerizing," as typified by the deposition of silicon carbide fibers on the carbon fiber surface.[43] Bond strengths superior to nitric acid or oxygen treatment of carbon fibers were reported.[44] Composites with shear strengths well in excess of 12,000 psi were obtained. By exposing the fibers to sources of silicon and carbon at 1100° to 1650°C, the whiskers are formed. The crystalline growths must be controlled in size and volume, to avoid reducing the high concentration of C/G fibers required for advanced structures. Another approach to vapor phase deposition was the application of titanium boride and carbides to surfaces of graphite fiber to assist retention of metal coatings such as aluminum or copper.[45] (See the chapter on matrices for over 300°C service for further details on chemical vapor deposition.)

The vapor phase decomposition of methane on hot, electrically heated carbon fibers has been known for some time. There has been a pronounced improvement in shear strength of composites with the deposition of carbon on the surface of graphite fibers, with only minor loss of tensile strength of fibers. The methane is decomposed from a nitrogen mixture at a temperature of 1200° to form a carbon coating.[46] Iron carbonyl, $Fe(CO)_5$, has also been decomposed on the surfaces of hot C/G fibers in a similar manner to leave deposits of pure iron to serve for mechanical bonds to resin matrices. Molten aluminum has also been flame sprayed on carbon fibers to effect metallic coatings on the fiber surface, the oxidation of which would aid adhesion of resin matrices.

Strong acidic chemicals, such as Lewis acids, BCl_3 and BF_3, have been applied in vapor phase to carbon fibers with the objective of improving the shear strength of the resultant composites. Ferric chloride, decomposed at high temperatures, and tin tetrachloride have been reported with good results, though not without loss of fiber tensile strength.

ELECTROPOLYMERIZATION ON GRAPHITE FIBERS

A novel approach to the preparation of C/G fibers for inclusion into composites was through interphase modification of the fibers by electrochemical polymerization, by Subramanian and Jukubowski.[47] "Thornel," "Fortafil," and Hercules PAN-based carbon fibers were used as electrodes. Monomers examined included several types of acrylic, styrene, vinyl acetate, and acrylonitrile monomers. In several instances, solvent-electrolyte systems were used with solvents such as dimethyl sulfoxide and dimethyl formamide, which were compatible with dilute aqueous salt solutions. In some cases, visual deposits of polymer coatings could be observed within a few seconds. There were moderate improvements in the interlaminar shear strength and the impact strenght of composites prepared from the carbon fibers onto which the polymers had been grafted by electroplating. Most interesting results were observed on the aziridinyl monomer, with its large improvement in shear strength of composites with high fiber loading.

Typical results indicating shear strength versus volume fiber loading with with graphite fibers and different polymer graftings (Fig. 6-3) are reproduced from Fig. 3 of the subject reference of Subramanian and Jukubowski.[47] Contacts were established to the graphite fiber tows through copper discs, and a

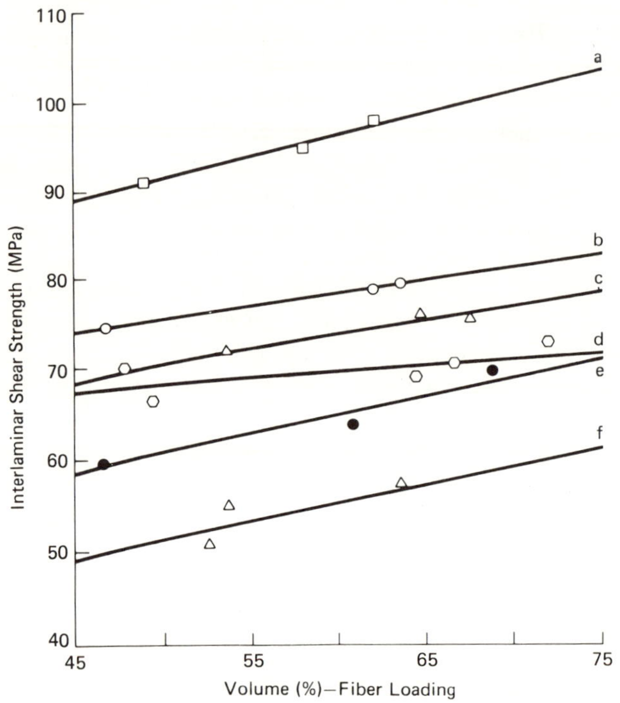

Fig. 6-3. Shear strength of composites prepared from graphite fibers: (a) Hercules AS and Hercules AU coated by electropolymerization of various monomers, (b) acrylic acid, (c) DAA, (d) untreated Hercules AU, (e) MMA, and (f) styrene-acrylonitrile (2.5 s polymerization time).[47]

typical electropolymerization voltage was 10 volts—though plating variables and their effect on properties of ultimate composites were not reported. Electropolymerization through C≡C and C≡N bonds are reported to be significant to high temperature resistant polyimide and polyquinoxaline composites, by improving shear strength and toughness. Electropolymerization conditions were not optimized with respect to composite properties, though the potential of interphase modification of the graphite fiber and resin matrix was pointed out.[48]

Intermediate modulus C-fibers ("Fortafil" CG-3-30 $\times$ 10^6 psi) free of surface treatment, were run continuously through an electrolyte cell and coated with a variety of co-polymers carrying carboxyl functional groups. Best results were reported with copolymers of vinyl methyl ether and maleic anhydride. Deposits were made for solutions of 2.5-10% solids at 10-50 volts, for 1 to 10 minutes—followed by washing. Improvements of interlaminar shear strength were reported (Ref. 49).

REFERENCES

1. Zisman, W.A., "Surface Chemistry of Plastics Reinforced by Strong Fibers," *I.E.C. Product Research and Development,* Vol. 8, No. 2, pages 98–111, June 1969.
2. Plueddemann, E.P., *Interface in Polymer Matrix Composites,* Academic Press, New York, 1974.
2A. Erikson, F., "Historical Background of the Interface", Chapter 1 in Plueddemann, E.P. (Editor), *Interface in Polymeric Matrix Composites.* Academic Press, New York, 1974.
3. Scola, D.A., "High Modulus Fibers and the Fiber-resin Interface in Resin Composites," Chapter VII, page 217, in Plueddemann, E.P. (Editor), *Interface in Polymer Matrix Composites.* Academic Press, New York, 1974.
4. Tuinstra, F. and Koenig, J.L., *Journal of Composite Materials,* Vol. 4, page 492, October 1970.
5. Didchenko, R., "Carbon and Graphite Surface Properties Relevant to Fiber Reinforced Composites," AFML–TR–68–45, page 16, February 1968.
6. Larsen, J.V., "Surface Treatments and Their Effects on Carbon Fibers," U.S. N.O.L., Silver Spring, Maryland, 1971.
7. Kraus, G., "Reinforcement of Elastomers by Carbon Black," *Rubber Chemistry and Technology,* Vol. 51, No. 2, page 297, May/June, 1978 (reprinted from *Die angewandte Makromolekulare Chemie 60/61,* pages 215–248, 1977.
8. Papirer, E., Vort, A., and Given, P.H., *Rubber Chemistry Technology,* Vol. 42, No. 1, page 1200, 1969.
9. Brennan, J.J., Jermyn, T.E., and Boonstra, B., *Journal of Applied Polymer Science,* Vol. 8, page 2687, 1964.
10. Prosen, S.P., Duffy, J.V., Erickson, P.W., and Kinna, M.A., *Twenty-first Technical Conference,* SPI, RP/C Institute, Section D, Washington, D.C., February 1966.
11. Herrick, J.W., "Resin Fiber Interactions in Graphite Fiber Reinforced Composites," SPI, RP/C Institute, *Twenty-third Annual Technical Conference,* Section 16A, pages 1–8, Washington, D.C. February 1968.
12. Monte, S.J., Sugarman, G., and Seeman, D.J., "Titanate Coupling Agents," *Thirty-second Annual Technical Conference,* SPI, RP/C Institute, Washington, D.C., February, 1977.
13. CA87–202637v (Nippon Carbon Co., Ltd.), 1977.
14. Kryszewski, M., Wrobel, A.M., and Tyczkowski, J., "Plasma Polymerized Thin Polyorganosilicon Films," *ACS Division of Polymer Chemistry, Polymer Reprints,* Vol. 19, page 457, September 1978.
15. Fitzer, E., Geigl, K., and Huttner, W., "Influence of C-fiber Surface Treatment on Mechanical Properties of C/G Composites," Fourteenth Biennial Conference on Carbon, American Carbon Committee, page 236, Penn State University, June 1979; and *Proceedings of the Fifth Converence on Carbon and Graphite,* Vol. 1, page 405, London, 1978.
16. Watts, J.C., U.S. Patent 3,367,812, 1968.
17. Lubowitz, H., French Patent 1,580,456, to TRW (September 5, 1969).
18. Riess, G., Bourdeaux, M., Brie, M., and Jouguet, G. (France), CA-84–45084F (1974).
19. Fujisawa, T., Tanaka, A., Hashimoto, K., and Kobayashi, M., "Studies of Graft Polymerization on Carbon Fibers," *A.C.S. Polymer Preprints,* Vol. 20, No. 1, page 853, April 1979.

20. Marks, B.S., Mauri, R.E., and Bradshaw, W.G., "New Graphite Surface Treatments," *Twelfth Biennial Conference on Carbon,* American Carbon Committee, page 337, Pittsburgh, Pennsylvania, July 1975.
21. Chen, W. and Wightman, J.P., "The ESCA Analysis of Graphite Fibers," *Fourteenth Biennial Conference on Carbon,* American Carbon Committee, page 107, Penn State University, June 1979.
21A. Toray Industries, British Patent 1,543,099, March 28, 1979
22. Kardos, J.L., Cheng, F.S., and Tolbert, T.L., "Tailoring the Interface in Graphite-reinforced Polycarbonate," *Polymer Engineering and Science,* Vol. 13, No. 6, page 455, November 1973.
23. Barr, J., "Treatment of C-fibers to Improve Shear Strength," U.S. Patent 3,791,840, February 12, 1974.
24. CA-76-60438C, 1972; and German Patent 2,112,455, 1972.
25. Union Carbide Corp., British Patent 1,538,042, January 10, 1979.
26. Goan, J.C., "Graphite Fiber Oxidation," Naval Ordnance Laboratory TR-153, November 28, 1969.
27. Clark, D. and Watt, W., CA-84-45083E, 1976.
28. Judd, J., British Patent 1,479,445, July 13, 1977 (CA-88-23853-1978).
29. Dauksys, R.J., "Coupling of Epoxy Polymers to Graphite Fibers," A.F.M.L. 72-23, May 1972.
30. Kalnin, I.L., "Oxidizability of Carbon Fiber Surfaces," *Fourteenth Biennial Conference on Carbon,* American Carbon Committee, page 218, Penn State University, June 1979.
31. Nyo, H., Heckler, D.L., and Hoernschemeyer, P.L., "Characterizing the Structures of PAN Based Carbon Fibers," *SAMPE Twenty-fourth National Symposium,* Vol. 24, page 51, May 1979.
32. CA-73-15644u, 1970.
33. Molleyeyre, F. and Bastick, M., CA-87-24026K (1977) "Processing of Carbon Fibers by a Gas Phase Oxidation," *Carbon '76, International Carbon Conference,* 2nd Preprint, 1976, pages 500-503 (held at Bad Honnef, Germany).
34. Kobets *et al.,* "Effect of Surface Treatment on Properties of Carbon Fibers" (translated from Russian), *Polymer Mechanics,* Vol. 4, No. 4, January 1979.
35. McKauge, E.L., "Graphite Filament Epoxy Matrix Interface Study," *Sampe Journal,* Vol. 5, page 49, April/May, 1969.
36. DeLollis, N.J., "Activated Gas Reactions with Silicone and Epoxy Resins," *SAMPE Journal,* Vol. 15, No. 3, page 10, May/June 1979.
37. Yasuda, H. and Morosoff, N., "Tandem Plasma-polymerization Apparatus for Continuous Coating of Fibers and Films," *ACS Polymer Preprints,* Vol. 19, page 488, September 1978.
38. McKague, E.L., Bullock, R.E., and Head, J.W., "Improved Mechanical Properties of composites Reinforced with Neutron Irradiated Carbon Fibers," Chapter 2 in Hilado (Editor), *Carbon Epoxy Systems,* Technomic, Westport, Connecticut, 1974.
39. Larsen, J.V., "Surface Treatments and Their Effects on Carbon Fibers," Paper and Bibliography from Naval Ordnance Laboratory, Maryland, 1971.
40. CA-81-153748X (1974), Japan Kokai "Cleaning Carbon Fibers," Vol. 74, No. 50, page 296, May 15, 1974, to Mitsubishi Rayon Co., Ltd.
41. CA-87-15299Q, 1977.
42. CA-76-86634P, 1972 (Ger Off, 2,119,805, November 11, 1971).

43. Milewski, J.V. and Anderle, J.V., "Interstitial Growth of Silicon Carbide Whiskers in Carbon and Graphite Fibers," *GTC Technical Report 127.7-1,* September 1967 (N.O.L. Contract N60921-67-C-0293).
44. CA71-125352, December 1969.
45. Lachman, W.L., Penty, R., and Jahn, A., U.S. Patent 3,894,863, July 15, 1975.
46. Byrne, J. and Ledbetter, H.D. (Dow Chemical), "Composite Materials Comprising Epoxy Resin Matrix and Carbon Fibers," U.S. Patent 3,853,610, December 10, 1974.
47. Subramanian, R.V. and Jukubowski, J.J., "Electropolymerization on Graphite Fibers," *Polymer Engineering and Science,* Vol. 18, No. 7, page 590, May 1978.
48. Subramanian, R.V. and Jukubowski, J.J., "Interphase Resin Modification with Graphite Composites," *ACS Organic Coatings and Plastics Chemistry Preprints,* Vol. 40, page 688, April 1979.
49. Subramanian, R.V., Sundaram, V., and Patel, A., *Thirty-third Annual Technical Conference,* SPI, RP/C Institute, Section 20F, Washington, D.C., February 1978.

7 Mechanical and Physical Properties of Carbon/Graphite Composites

The mechanical and physical properties of C/G fiber composites have been examined for some years by many qualified scientists and engineers. It is a complex field, the analysis of which is directly applicable to advanced composites of carbon and graphite fibers and matrices of organic resins, as well as metal matrices and carbonaceous binders. Before these high performance composites can be widely accepted in long-term, critical high stress applications and under adverse environmental extremes, reliable means must be established to characterize fundamental properties and behaviors. Many of the properties characterization methods for composites have been extrapolated from the metal industry. In consequence, methods of analysis have been developed which reflect the homogeneity of the metals. These methods cannot anticipate the numerous variables in the composite constituents or in the structure of the composite—variables which have significant influence on the behavior of the composite structure. This chapter will attempt to direct attention to some of the programs and activities in this country and abroad which have addressed themselves to the fundamental mechanical and physical properties of advanced composites.

Current research efforts on advanced composites may be broadly classified into: 1) mechanical responses, such as deformation processes without any damage considerations, and 2) detailed studies of various failure modes that occur in the composite during application of stress. It should also be remembered that stresses which arise may not only be applied externally (mechanically or even electrically, as noted in the chapter on electrical properties), but also may be applied from internal considerations due to differences in thermal properties of the composite components. More knowledge of material damage at microscopic level is needed. This will assist understanding of the various modes and of the way in which physical damage (as distinct from intrinsic material behavior), such as micro-cracking, affects the overall behavior of the composite. Sih[1] has proposed continuing symposia on subjects, such as micro- and macrofracture, statistical analyses, nondestructive testing methods, prediction of fractures, limitations of conventional linear fraction mechanics, environmental influences on composites, and design and technical considerations.

Experience of the metal industry in developing fracture testing, particularly in cooperation with the American Society of Testing Materials (ASTM) and Committee E-24, affords a background which will assist similar efforts for the composite materials. In 1957 and 1958 repeated failures through brittle fracture of high strength metal castings of the Polaris missiles led to government sponsorship of fracture testing and evaluation through ASTM. An impressive list of standards has appeared, starting in 1968 for the first fracture mechanics-related method for measuring toughness. Today we recognize that the onset of fracture instability is the result of interaction of the crack driving force, which is due to the structural configuration and stress field, and the crack resistance of the material, which is the inherent material toughness property. When the crack driving force equals or exceeds the crack resistance, instability ensues.[2] Other recent fracture mechanics-related methods include characterizing the life of the material from a subcritical crack growth through the fracture event. Examinations of recent programs on composite materials indicate similar concern, as will be brought out in further references. Advanced composites, more than metals, lend themselves to design modifications which, in ameliorating stresses and crack growth, will have less tendency to catastrophic failure and will demonstrate improved fatigue strength.

POLYMER FRACTURE

Before examining the considerable efforts which have accrued in studies of fracture phenomena in advanced composites, with emphasis upon C/G fiber reinforcements, attention should be given to the mechanics of polymer fracture. An enlightening analysis has been conducted by Knauss.[3] The importance of incorporating rate-dependent material properties into the analysis of crack propagation in polymers is emphasized. Atomistic and molecular concepts are used in the analysis of fracture by Knauss, rather than phenomenological failure investigations. As a consequence of the high thermal mobility of the polymer chain segments above the glass transition temperature, orientation tends to precede bond rupture. Radical chain ends formed during rupture above the glass transition temperature (T_g) have a high chance of recombination with other radical chain ends during the time-dependent flow process. Relaxation and creep processes occur at an increased rate the higher the temperature is raised, and close to, but above T_g, the rate of relaxation may change by a factor of 10 per degree centigrade. Straining of the material near a crack tip generates heat, and is dependent on the rate of straining. The low thermal conductivity of polymers will not permit a rapid dissipation of heat generated. Under cyclic loading the heat build-up in the crack tip region may be quite significant. Also covered in this review are effects of environment on crack growth in polymers. For example, chemically active gases or liquids may cause highly strained materials at the tip of a crack to degrade more rapidly than the material in an unstrained condition.

The complexity of stress distribution in polymers arising from processing methods introduces other factors which influence the fracture strength of the polymers, which are functioning as the matrices in advanced composites. Because the forming and curing of the matrix base takes place above the T_g, followed by cooling down to remove parts from a mold or fixture, there are numerous instances where stresses and strains have been retained by the part or component below T_g, which influence subsequent resistance to fracture. While the presence of reinforcing C/G fibers will do much to alleviate this condition, studies of the polymer "per se" will assist in determining the more dangerous stress concentrations, particularly where a complex geometry is involved in the final part.

Transparent polymer models with good stress optical sensitivity would permit examination through polarized light and determination of sites for potential fracture. The empiricism of this method is not as rigorous as a mathematical analysis, but for complex configurations, designers gain an advantage. This technique has been used successfully on industrial models and products. When the critical areas subject to fracture have been "pinpointed," more scientific studies may be focused on those regions. The composite, when fabricated, will have had the benefit of an optical preview. Benefits will be apparent, even though there will not be an exact relation between the physical properties of the composite structure and the transparent medium.

Developments in two dimensional photoelasticity and related optional methods for studying crack tip stress and deformation fields have been examined.[4] The studies embraced two dimensional problems involving dynamic photoelasticity and plastic zone development for both small and large scale plastic yielding. A method described for estimating stress intensity on three dimensional problems, was reported by using stress-freezing photoelasticity. In photoelastic studies of polysulfone and polycarbonate models, it was felt that birefringence was strongly thickness related, and studies were reviewed on crack tip stress fields.

Also more recent techniques permit microtoming of thin sections (typically 0.25 mm thick), which if removed from a fracture sensitive area of the composite will provide means of ascertaining stress concentrations, and material distribution (fiber and resin) through optical analyses. Microtoming has merit for "in-situ" tests of materials and should enjoy increasing utilization. Further, and the comment will be repeated again, the final proof is the in-service application of the composite structure under the temperature and actual stress conditions in service. Only then will a complete confidence level be established in the application of C/G composite structure.

The degradation of polymers, which leads to a loss of desired properties, occurs when a polymer is affected by mechanical action, heat, radiation, or the chemical action of such agents as oxygen, water, chemicals, or atmospheric pollutants, either singly or in combination. Polymers are usually shaped by flow resulting from the application of some degree of shearing, generally in the presence of heat. It has been recognized that shearing breaks polymer chains to yield radicals, at the same time reducing the molecular weight of polymers. Zhurkov[5, 6] studied the production of radicals in polymers as a result of bond rupture by infra-red and by electron paramagnetic resonance. Dunn[7] detailed the influence of metal catalysts on oxidative degradation of polymers used in the manufacture of synthetic rubbers, listing the order of activity of metal ions in contributing to degradation. For synthetic rubbers, the following ions were listed in descending order of activity: Co^{2+}, Fe^{3+}, Cu^{2+}, Na^{+}, Mn^{2+}, etc.

Of particular importance is the role of the matrix and the interface in isolating individual fiber fractures by inhibiting the formation of a single crack, which will contribute to composite failure. Mullin[8] studied the influence of fiber property variations and drew the following conclusions.

1. Reinforcing C/G fibers are the primary load carriers and their strain energy density and volume fraction are critical parameters in the energy per unit volume of composite to initiate cracks.
2. Identified are two sources of energy absorption at the fiber fracture site: Mode I (cleavage normal to the fiber) and Mode II (interfacial failure). The first is less effective than the interfacial failure in absorbing energy.
3. Mode I energy absorption is inversely related to the volume fraction of fibers and Mode II energy absorption is independent of volume fraction of fibers. Both modes are inversely related to the fiber diameter.
4. There are distinct changes in local fracture modes when surface treatments are applied to the fibers. (See the chapter on fiber surface preparatory measures.)

FRACTURE MECHANICS OF COMPOSITE MATERIALS

Materials used in advanced composites today are predominantly thermosetting resins and high temperature resistant thermoplastics with graphite, glass, boron, or aramid filaments at nominal fiber volumes of 50 to 60 percent. Their high modulus, high strength, and low density offer advantages to the designer concerned with lightweight high performance structures. Furthermore, there is a versatility available in design and fabrication, unavailable in conventional metals. Composite properties reflect the type of fiber or fibers used, the orientation of the fibers, and the type and proportion of resin matrices which are used. This flexibility makes available a wide latitude of design expediencies to achieve maximum structural efficiency. This wide latitude of material selection compounds the problems associated with fracture mechanics, because one is not necessarily dealing with a homogeneous entity, though in some instances this assumption is made to facilitate the mathematical analysis and model study. As pointed out already, the metal industry in recent years has been engaged in active study of fracture testing and evaluation, and many of these procedures are applicable to the fracture mechanics of composite materials.

The subject materials covered under the aegis of fracture mechanics is extensive and while there is an obvious interdependency of the various disciplines, because the main focal point is the successful utilization of advanced composites, emphasis of the reports may differ. The subject matter lends itself to categorization as follows.

1. Fracture mechanics of anisotropic materials, with attention given to the distribution of reinforcing fibers; locations of the initial failure mode; influence of size and location of voids, holes, and cracks; and factors concerned with the growth rate of such defects.
2. Toughness of advanced composites is important, wherein the work or energy per unit volume to cause fracture is examined. It is here that reports begin to place emphasis upon remedial actions such as modification of the properties of the resin matrix or the developments of hybrid constructions among the fibers to improve impact strength.
3. Fatigue and creep phenomena are fundamentally time dependent and must be understood to predict long time service performance of the advanced composites. The environmental exposures play an important role in affecting fatigue behavior—for example, the derogation of composites under cyclic fatigue stresses during high humidity exposure is of serious concern. Concomitantly, the influence on defect propagation is examined under fatigue conditions.

There are influences which arise from manufacturing methods and procedures, as well as from aging of raw materials and of finished components. Where appropriate these factors are brought out in the discussions, though the designer relies heavily on the experiences of the manufacturing personnel to utilize proven, acceptable techniques to produce satisfactory products. In these instances strong reliance is placed upon the data offered by materials suppliers, who have examined and tested extensively the physical properties and thermal properties of their materials. As fracture mechanics brings out, the physical properties alone are not adequate to develop full confidence levels in the acceptance of composites, hence in this chapter the failure modes and performance of engineered structures are studied. Much of the work on fracture mechanics of composites has been distributed among the glass fiber, boron fiber, aramid fiber, and C/G fiber utilization, and the reports strictly on carbon and graphite fibers alone are limited. However, the procedures and analyses which will be examined are applicable to most types of reinforcements.

Zeben[9] emphasizes the importance of understanding fracture behavior, because of the statistical variability in material strength and observation of fracture phenomena that are significantly different from those of metals. His paper discusses three main approaches: 1) classical fracture mechanics (CFM) on a macroscopic level, treating composites as anisotropic materials; 2) treatment of composites as heterogeneous materials and applying CFM to the problems of crack propagation in the matrix, fibers, and interfaces separately; and 3) the use of approximate models to represent the major effects of heterogeneity and to simplify the analysis. Zeben felt that available experimental data for unidirectional and multidirectional laminates are not sufficient to permit definitive decisions regarding the applicability of analytical approaches. The available data for the fracture of laminates having sharp center notches indicate that the behavior of these materials is more complex than observed in metals. In view of the heterogeneity of advanced composites and the multiplicity of damage modes that can occur at a micromechanical level, designers for the time being must rely on experimental data for reduction in strength caused by notches and holes. Macroscopic classical fracture mechanics may apply in a limited sense to unidirectional materials having a notch parallel to the fibers, it was concluded.

Freeman and Kuebler[10] presented a detailed review of mechanical and physical properties of advanced composites which are reproduced in Table 7-1. They summarize information on production composite materials parallel (0°) and perpendicular (90°) to fibers available in 1973. There have been further developments in recent years which have yielded higher moduli C/G fibers, improved heat resistant matrices, improved fiber reliability, availability of matrices with lower moduli and greater strain capability, and hybridization among the reinforcing fibers. In the meantime, there are reference points in the table for comparison purposes. Data are also included on thermal properties of advanced composites, summarized in Table 7-2. The low coefficients of expansion of the graphite fibers (A, HM, VHM–distributed by Hercules and manufactured by Courtaulds in 1973) and their good thermal conductivity relative to other fibers are quite noteworthy and of advantage to aerospace and industrial applications.

Freeman and Kuebler point out that the maximum strain criteria is generally used with advanced composites for two important reasons:

1. Elastic properties of oriented laminates can be predicted from measured lamina elastic constants.
2. Failure of any ply in a laminate (commonly referred to as "first ply failure" or the appearance of crazing in the matrix) is analagous to yielding in metals. First ply failure generally results from low transverse elongation in a highly orthotropic lamina or large Poisson's ratio mismatch between the lamina (Poisson's ratio for most unidirectional composites lies between 0.21 and 0.28). While higher elongation is generally associated with lower heat resistant resins, this performance must be weighed against having the first ply failure strains above the static design limit load.

Sharp, crack-like flaws in fiber composite structures may be present for many reasons, ranging from material irregularities, improper fabrication, and adverse environmental exposures to in-service damage of the structure. One of the purposes of fracture mechanics is to study such flaws and to determine how to design and apply advanced composites so that the appearance of flaws is minimal, and if they occur, for whatever reason, to suggest measures to contain the flaws and avoid rapid and catastrophic failures. For example, in August 1979 there was held in England, under the auspices of the Royal Ocean Racing Club in Plymouth, the annual yacht race to Ireland and back. As reported by a national weekly magazine published in the United States, an inordinately high number of lightweight carbon-fiber rudders snapped during the severe storm which accompanied the racers.[11] As admitted by one of the leading yacht designers, R. Holland, "We have not had enough experience to really know how to use (carbon fiber) properly yet." While boat designers do not have the background of the aerospace industry, the previous terse statement expresses the need for detailed study of fracture mechanics of fiber composites. The other side of the story would be a reference to another national weekly publication, which directed attention to the fact that extensive use of C/G fiber composites were being planned for new generations of aircraft.[12] These designs have had the benefit of many thousands of hours in-service application on existing commercial aircraft, with notable success.

Ply stacking or the alignment of unidirectional fibers in a prescribed direction appear frequently in the analysis of fracture mechanics on the subject of fiber

Table 7-1. Mechanical Properties of Composites (60% Fiber Volume).[10]

REINFORCEMENT	ULTIMATE TENSILE STRENGTH (ksi)		ULTIMATE COMPRESSIVE STRENGTH (ksi)		INITIAL TENSILE MODULUS (msi)		ULTIMATE TENSILE STRAIN (%)		COMPOSITE DENSITY (lb/in.^{3})
	0°	90°	0°	90°	0°	90°	0°	90°	
A–graphite fiber	220	7–12	220	40	18	1.5	1.2	0.5–0.9	0.056
HT–graphite fiber	220	6–12	200	35	22	1.3	0.8	0.5–0.9	0.057
HM–graphite fiber	130	5–10	130	30	27	1.3	0.5	0.3–0.8	0.059
VHM–graphite fiber	200	2–3	97	27	45	0.9	0.5	0.2–0.3	0.058
Aramid 49	220	3–4	45	15	13	0.8	1.5	0.2–0.3	0.050
S-glass	260	5–9	100	23	6–8	2.0	3.0	0.3–0.4	0.072
Boron fiber	230	9–13	400	30	30	3.0	0.7	0.3–0.4	0.075

Table 7–2. Thermal Properties[10]

MATERIAL	COEFFICIENTS OF THERMAL EXPANSION (10^{-6}/in./in./°F)			SPECIFIC HEAT	THERMAL CONDUCTIVITY (BTU-ft/hr/ft^2/°F)		
	UNIDIRECTIONAL	TRANSVERSE	ISOTROPIC		UNIDIRECTIONAL	TRANSVERSE	ISOTROPIC
Type A Graphite fiber	-0.1	14	0.5–1.5	0.17	6–10	0.4	1–3
HM Graphite fiber	-0.5	15	0–0.5	0.17	28–35	0.5	6–12
VHM Graphite fiber	-0.8	17	-0.5–0.5	0.17	70–75	0.6	14–18
Aramid (Type 49)	-2.0	30	-0.5–0.5	0.33	1	0.1	0.1
S-glass	3.5	11	6	0.18	2	0.2	0.2
Boron	2.5	8	2–3	0.20	1	0.6	0.8
Aluminum	–	–	12–14	0.23	–	–	–
Epoxy-resin	–	–	30–50	0.45	–	–	0.2

composites. Bending properties and equations are derived from homogeneous, isotropic beams, and hence the flexural strength of fiber composites is not always easy to interpret. Whitney *et al.*[13] pointed out that the heterogeneous nature of laminating materials requires that bending stiffness, which is a function of ply stacking sequence, must be utilized in determining flexural properties. Tensile modulus of symmetrical laminates, on the other hand, is independent of stacking sequence. Chamis[14] notes that the measured values of flexural strength tend to be greater than longitudinal tensile or compression strength of unidirectional composites, suggesting that one of the highly stressed outer plies fails first.

McGarry[15] observed the influence of stacking sequence on longitudinal splitting in specimens having sharp edge notches. The splitting was most severe when the 0° layers (0° fibers parallel to main axis of test specimen) were stacked in the center of the laminate, with 90° layers on the outside. The situation was improved through separation of the 0° layers by intercalation with the 90° layers.

Halpin[16] emphasizes the design philosophy for high-modulus, fiber composite technology. He points out that, because current research efforts indicate that stress levels required for a safe-life fatigue, for restricted environmental degradation on the composite filaments, and for controlled stiffness properties, are in the vicinity of the stress level predicted for first ply failure, design should anticipate the first occurrence of mechanical damage. As will be pointed out in test methods, acoustic emission evaluations, correlated with mechanical tests, are becoming more valuable in such studies. Halpin states: "Design ultimate loads shall result in a stress that does not exceed the design ultimate stress for the laminate used, where design ultimate stress is the maximum laminate stress attainable without the rupture of any lamina." Using glass fiber reinforcement, he illustrates (see Fig. 7-1) the influence of direction of fiber reinforcement upon the shape of stress-strain curve (0°; 0°/90°; 0°/±45°/90°; and 90° lay-ups are shown). Boron and graphite fibers are relatively fatigue insensitive filaments, and no fatigue damage is realized below the first ply failure. Lamination techniques make it possible to meet conflicting stress-strain demands at different points in the structure and are utilized in design of aircraft and automotive components (see the chapter on applications).

The extension of linear elastic fracture mechanics (LEFM) to the study of crack growth in more general composites has been pursued by various investigators. Konish and Cruse[17] suggest that the procedures of LEFM, suitably modified for anisotrophy, may be applied to the study of crack extension in a graphite/epoxy laminate. They treated the fracture strength of a given composite laminate as a function of the fracture properties of an individual ply and the relative orientation of those plies. Crack growth parallel to the material fibers

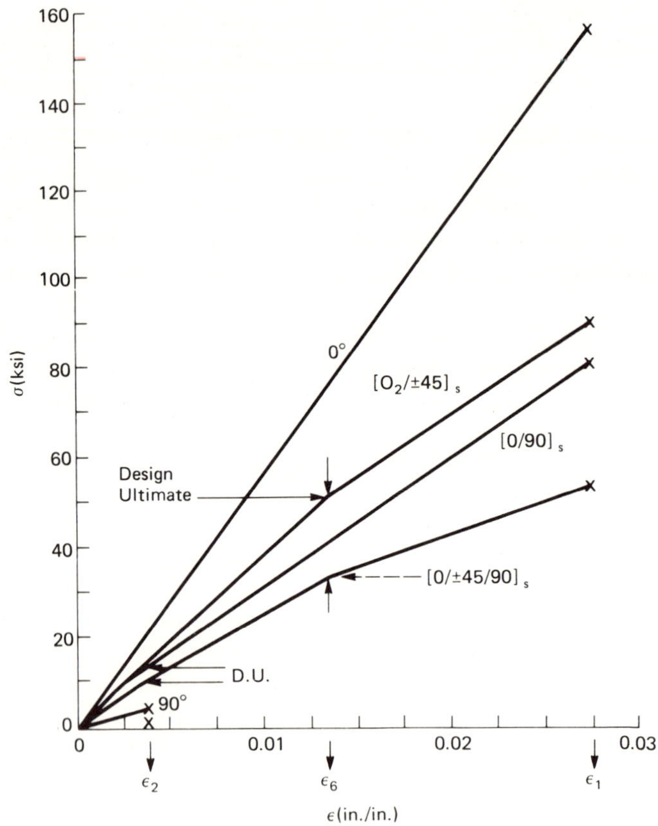

Fig. 7-1. Influence of lamination ply lay-up on stress-strain curves of glass/epoxy.[16] *(Courtesy of Technomic Publishing Company, Conn.)*

was treated as matrix fracture and crack growth across fibers was treated as fiber fracture. Predicted fracture strengths of angle-ply graphite/epoxy laminates, shown in Fig. 7-2 as presented by Konish and Cruse,[17] had experimental verification at several points (0°, 45°, and 90°). It should be noted, however, that modes of crack growth in C/G composites show a dependence upon stacking sequence.

Static and fatigue behavior of angle-ply laminates are shown for several advanced fiber composites, including hybrids.[18] The graphite fibers used in this study ("Thornel" T-300) were laid up in conjunction with epoxy and polyimide resins. Other fibers included were glass, boron, and polyaramid fibers. Most filamentary composites are supplied as prepregs and used in the form of angle-ply laminates such as the quasi-isotropic (0°/±45°/90°) lay-up, where the state

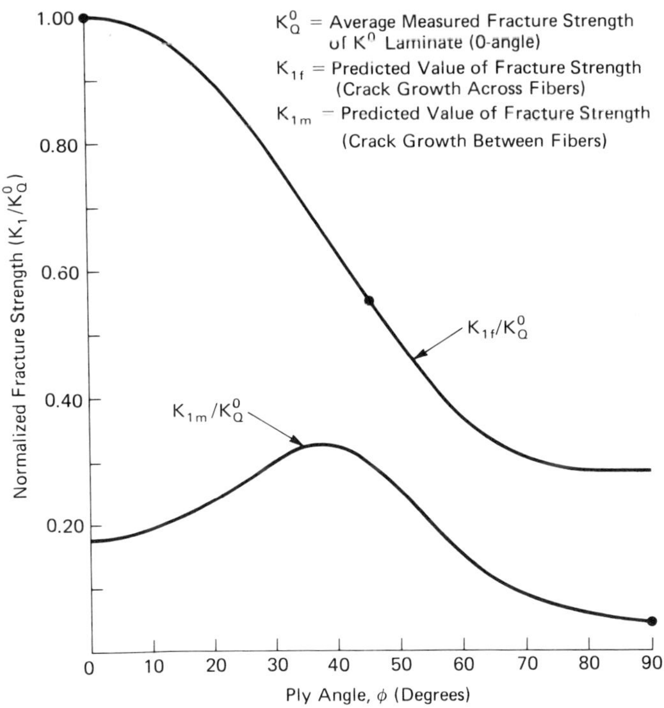

Fig. 7–2. Predicted and experimental fracture strengths of single angle-ply graphite/epoxy laminates.[17]

of the stress can be varied depending upon the preferential orientation for the principal stress. The more anisotropic lay-up ($0^\circ_2/\pm45^\circ$) would be used in situation requiring high tensile strength in one direction and shear resistance in the transverse direction, such as in jet engine blades.

Daniel[18] also reviews the behavior of laminates with cutouts. Geometric discontinuities such as notches and cutouts make stress analysis of composite structures more complex and failure modes more difficult to predict. As seen in the previous reference, modified LEFM procedures, developed for isotropic materials, permit analysis of holes or cracks in graphite/epoxy composites. Stress concentration and strength reduction were studied for boron/epoxy specimens with holes. The influence of laminate lay-up is indicated whereby the strength is decreased with higher values of the stress intensity factor.

With a preponderance of 0° fibers parallel to the main axis, higher strengths are obtained, though enough 45° plies will reduce the stress intensity factor. Daniel offers a strength reduction ratio (the ratio of the strength of a laminate

Table 7-3. Influence of Laminate Construction and Stacking Sequence on Tensile Behavior of Boron Fiber/Epoxy Laminates and Graphite/Epoxy Laminates With Holes.[18]

LAY-UP	MODULUS OF ELASTICITY GPa	(10^6 psi)	STRAIN CONCENTRATION FACTOR	TENSILE STRENGTH MPa	(ksi)	MAXIMUM STRAIN INS/IN.	RATIO OF STRENGTH OF LAMINATE WITH HOLE TO LAMINATE WITHOUT HOLE*
(0/±45)0/90)s	115	(16.7)	3.37	291	(42.2)	0.0080	0.435
(±45/0_2/0)s	127	(18.4)	4.02	426	(61.7)	0.0076	0.529
(0_2/±45/0)s	134	(19.4)	3.50	498	(72.2)	0.0087	0.617
(0/±45/90)s	79	(11.5)	3.08	183	(26.5)	0.0090	0.394
(±45/±45)s	19.9	(2.88)	2.06	121	(17.5)	0.027	0.909
(±45/0/±45)s	59	(8.6)	2.46	206	(29.9)	0.009	0.546
(0/±45/0/90)s**	70	(10.2)	3.58	165	(23.9)	0.0095	0.389

*Specimens were 10 inches wide, 26 inches long, and had 1 inch diameter holes.
**Graphite/epoxy composite.

with a hole to that of a similar laminate without a hole). In general, he observed that the higher the strain concentration factor, the lower the strength reduction ratio. Stacking sequences were also observed to affect failure modes significantly in the hole vicinity.

Graphite/epoxy plates with holes ranging in diameter between 0.008 and 1.0 inches showed a response similar to the quasi-isotropic boron/epoxy specimens discussed above. Maximum measured strains at the hole boundary reached values of up to twice the ultimate strain of the unnotched laminate. For some of the specimens studied ($0_2 \pm 45$), there was a critical hole diameter (1.6 mm) below which the C/G laminates were notch-insensitive.

Daniel[19] has reported on the deformation and failure of graphite/epoxy laminates with crack lengths of 0.25 to 1 inch. Failure at the tip of the crack takes the form of a damage zone consisting of ply subcracking along the fiber direction, fiber breakage, and local delaminations. Comparisons of results for cracks and holes showed that, in these specific tests, strength reduction was nearly independent of the notch geometry.

A program was initiated by Walter and June[20] to develop engineering data necessary to design advanced composite structures to satisfy stringent damage tolerances. The T-300 graphite fiber and tetrafunctional epoxy resin system (tetraglycidyl ether of methylene dianiline with DADS as a curing agent was selected for study). S-glass prepregs were used for some hybrid adaptations, which demonstrate less brittle failure modes than the all-graphite laminates. The slower, more controlled failure mode by the glass hybrid laminates provided higher damage tolerance. The relationships between static strength and fracture strength for fiber controlled laminates were examined for several quasi-isotropic laminates. Care was taken to fabricate the panels from the same batches of material, using the same processing specifications. Test specimens 3 inches wide and 16 inches long were fabricated with no damage and with damage sizes of 0.10 to 1.0 inch width. Results are shown in Fig. 7-3 for a balanced laminate consisting of 24 plies with the following angle-ply sequence $(0°/\pm45°/90°)_{3S}$. Results shown were also corrected with a small increment added to the damage size. When corrected in this manner, fracture strength followed the direction of LEFM.

Fracture strength was also influenced by the stacking sequence. Stacking sequences with 90° and 0° plies separated by 45° plies had highest fracture strength. There was approximately 30% difference in strength between the strongest and weakest laminate stacking sequence. It has been pointed out that the relationship between fracture strength and the static strength could be defined as a function of the laminate critical strain. This was further studied by substituting glass plies (equivalent strength, 1/3 modulus, and greater weight) for the primary load carrying graphite plies. Substantial improvements were obtained

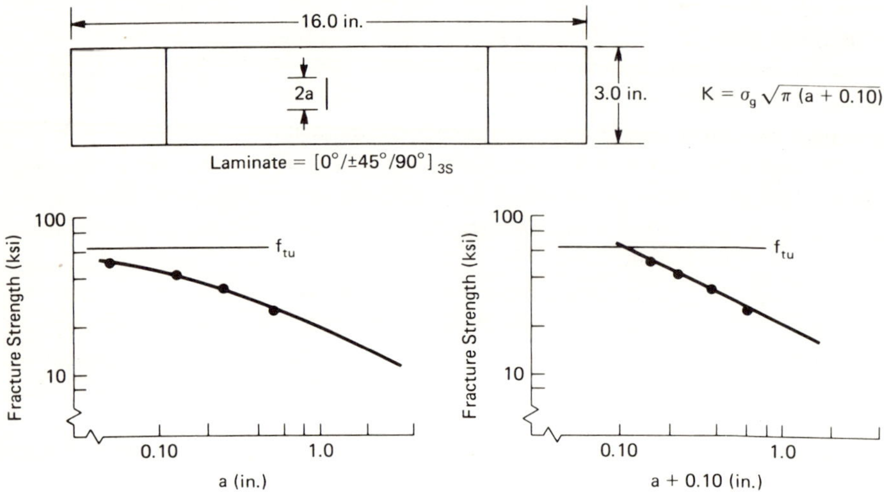

Fig. 7-3. Fracture strength performance.[20]

in the hybrid laminates with regard to fracture energy over the all-graphite laminates attributable to the large strain capability of the glass, enabling the unidirectional plies to break away from graphite in a high stress region and behave as a ply of unnotched material. These data demonstrate improvements in fracture strengths, though for the maximum stiffness criteria other factors enter the picture. The general survey of results of elastic stress analyses of cracks and the stress-intensity-factor of representing results has been presented by Paris and Sih.[21] For in-depth studies of fracture toughness, this publication is recommended for composites.

Laminates of composites can be tailored to meet specific requirements. Stacking the plies symmetrically and maintaining overall balance is necessary to avoid interactions which produce dimensional instability, such as bending and twisting. Symmetry requires a mirror image of ply stacking about the midplane. Laminates with alternate layers of 0°, 90°, +45° and −45° plies are illustrated in Fig. 7-4. They are referred to as quasi-isotropic (i.e., the properties are equal in all directions in the plane of the laminate). Properties may be varied depending on the distribution of C/G fibers.

The variations in tensile strength and modulus of elasticity for angle-ply lay-up of typical *high strength* carbon fiber/epoxy advanced composites, are shown in Figs. 7-5 and 7-6 as a function of the percentage composition of 0°, ±45°, and 90° plies. The charts are based on a carbon fiber composite with an allowable unidirectional composite tensile strength of 160 ksi, and tensile modulus of 20×10^6 psi.[22]

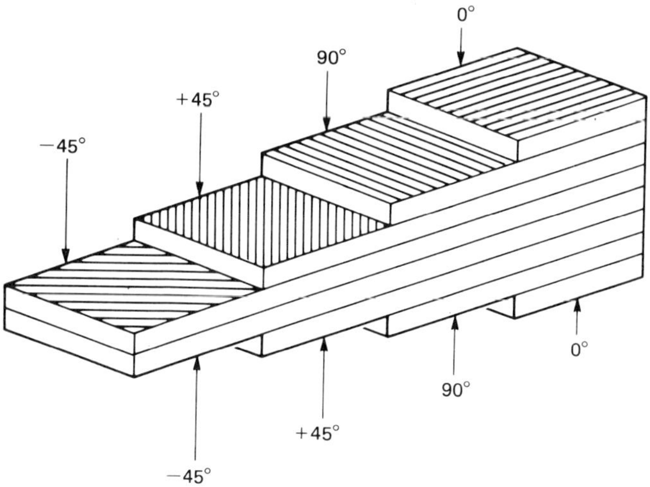

Fig. 7-4. Quasi-isotropic lay-up $(0/90°/\pm 45°)_S$.

Similar charts for a family of 0°, ±45°, 90° plies are available in *Advanced Composites Design Guide*[22] for intermediate strength C/G fiber/epoxy composites and high modulus C/G fiber/epoxy composites for both room temperature and 350°F. Referring to the Figs. 7-5 and 7-6, the high value of strength and modulus in each series lies at 100% 0° plies and no ±45° plies. For the three classes of carbon-fiber composites reported, the following data will permit comparisons on the charts for tensile properties and compressive properties.

	LONGITUDINAL TENSILE STRENGTH ksi		LONGITUDINAL TENSILE MODULUS msi		COMPRESSIVE STRENGTH ksi	
	At R.T.	At 350°F	At R.T.	At 350°F	At R.T.	At 350°F
High strength carbon fibers	180	180	21	19	180	70
High modulus carbon fibers	110	95	25	25	100	60
Intermediate strength carbon fibers	160	65	17	16	160	65

An additional chart is reproduced in Fig. 7-7, showing the coefficients of

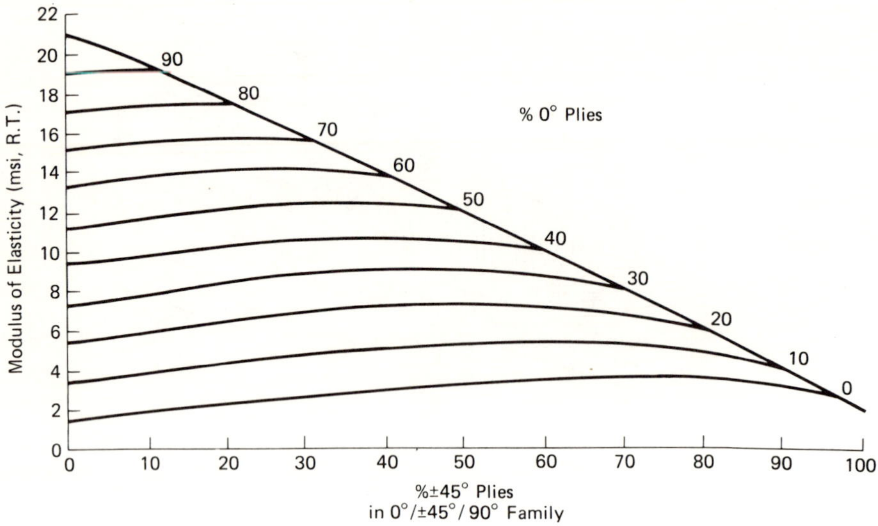

Fig. 7-5. High strength graphite/epoxy.[22]

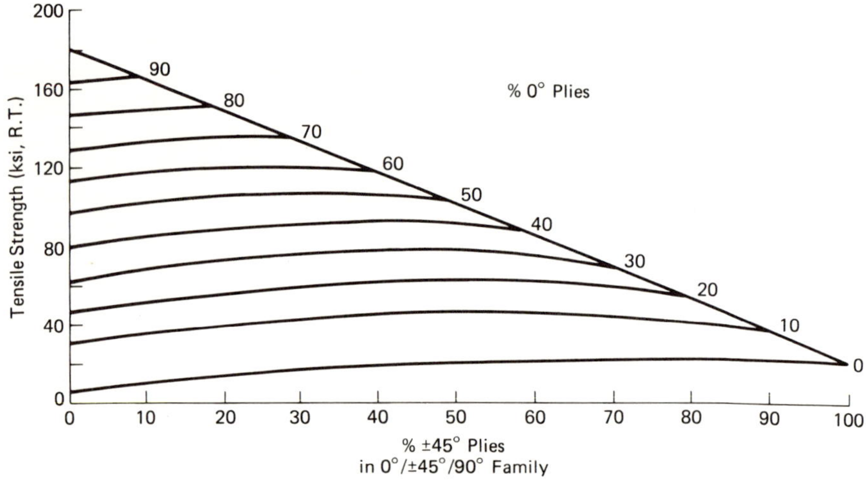

Fig. 7-6. High strength graphite/epoxy.[22]

thermal expansion for a series of laminates with different distributions of 0°, ±45°, and 90° plies. The value of the coefficient of thermal expansion (C.T.E.) for the composite containing no ±45° and no 0° plies is basically the tranverse C.T.E. perpendicular to the fiber direction. The chart shown is for the high strength carbon-fiber composite with its highest C.T.E. at 16 micro-inches/inch/°F,

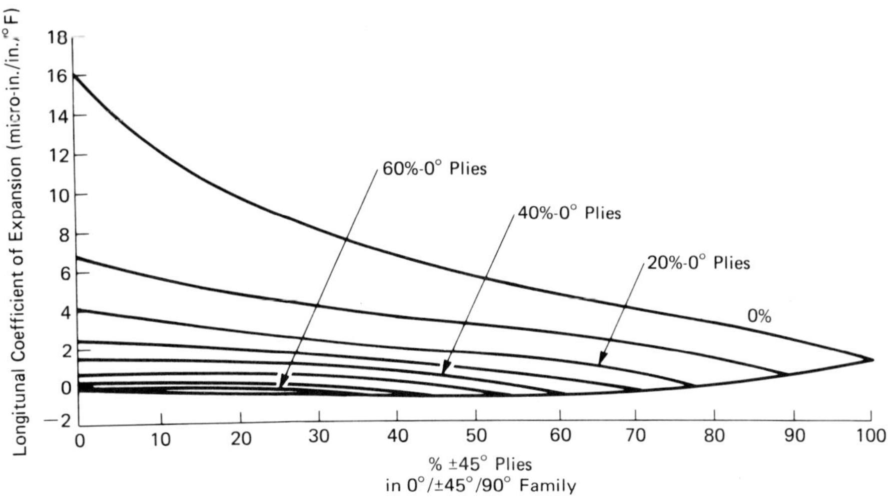

Fig. 7-7. High strength graphite/epoxies.[22]

while for the high modulus fiber composite this was 20 micro-inches/inch/°F, and for the intermediate strength carbon fibers the value was 15 micro-inches/inch/°F. It should be noted how rapidly the addition of 0° plies reduces the coefficient of thermal expansion.

Further tables from this advanced composite design manual are given as Tables 7-4 and 7-5, illustrating the design strengths and elastic properties of (0°, ±60°) composites and unidirectional composites prepared from high strength, high modulus, and intermediate strength C/G (60% volume of fiber). While the commercial sources were not identified, the epoxy resin matrix appeared in prepregs that were procured, suitable for low pressure (100 psi) and high temperature (350°F) cure.

Some excellent design guides were offered by Kliger[23] in the use of C/G fiber composites. While many of these guidelines have already been noted, key design features are noted below from his paper.

1. Use of balanced and symmetric lay-ups.
2. Reinforcement of attachment areas.
3. Use of hybrids for cost effectiveness and property modification. Carbon fibers are brittle, strain sensitive materials, with tensile strains close to 1.0%. For improved impact, glass and aramid fibers are recommended.
4. Use the 0°/±45°/90° design curves which are available for different percentages of plies.

5. Overdesign and do not overreact to micro-cracking and acoustic sounds—which are equated to yielding in metallic structures.

In a paper by Beaumont and Phillips[24] we are reminded of the influence of surface treatments on carbon fibers on tensile strength of notched composites. The untreated carbon fiber specimens show long fiber pull-outs at the time of tensile failure (matrix-epoxy/MNA/BDMA system). Very much shorter fiber pull-outs were observed for surface treated carbon fibers. Notch insensitivity was associated with long fiber pull-out lengths or strips of composite at the tips of the cracks. A considerable difference was demonstrated in the shear strength of composites, with surface treatments offering a marked improvement in shear over the untreated fibers. Composites prepared from the latter fibers would be expected to shear at crack tip before crack propagation occurs.

Opto-acoustic techniques were developed by Wolff *et al.*[25] to study the effects of matrix micro-cracking on the dimensional stability of angle-ply graphite fiber reinforced tubular laminates. As they point out, micro-cracking in composites may occur in advanced composites by reasons of:

1. Breaks and splits in the fiber bundles;
2. Cracks in the resin matrix, occurring between fibers;
3. Cracking between plies, or delamination;
4. Debonding or fiber pull-outs (as when shear stress exceeds the interfacial shear strength); and
5. Growth of pre-existing flaws or microvoids.

The effects of micro-cracking are numerous, including reduced strength and stiffness factors in the composite, increased permeability of gases and moisture, and decreased dimensional stability, as observed by changes in thermal expansion coefficients and net dimensions. Laser interferometry was employed to measure dimensional changes and acoustic emission was combined with optical signals to record the audible cracking that occurred during test (see the chapter on test methods). From the analytical techniques employed, precise lamina properties of the GY70/934 materials were developed. Data included the following for the GY70/934 lamina:

Modulus of elasticity	44.0×10^6 psi
Transverse modulus of elasticity	$.97 \times 10^6$ psi
Coefficient of thermal expansion	$-.57 \times 10^{-6}$/in./in./°F

Recent papers by Sinclair and Chamis[26, 27] have made useful contributions to fracture studies of fiber composites. They have presented a mechanics theory for predicting the physical, thermal, hygrometric, and mechanical properties of unidirectional intraply composites, which is also applicable to the flexural stiffness of intraply hybrid angle-plied laminates. With the assumptions that the

Table 7-4. Basic Properties of C/G Fiber Composites (0 ± 60°).[22]

				INTERMEDIATE STRENGTH GRAPHITE/EPOXY		HIGH STRENGTH GRAPHITE/EPOXY		HIGH MODULUS GRAPHITE/EPOXY	
				R.T.	350°F	R.T.	350°F	R.T.	350°F
Design strengths (typical)	Longitudinal tensile, ultimate	(ksi)	F_x^{tu}	60.0	45.0	65.0	62.0	42.0	35.0
	Transverse tensile, ultimate	(ksi)	F_y^{tu}	48.0	36.0	52.0	43.0	36.0	24.0
	Longitudinal compression, ultimate	(ksi)	F_x^{cu}	64.0	24.0	69.0	26.0	38.0	22.0
	Transverse compression, ultimate	(ksi)	F_y^{cu}	55.0	21.0	59.0	22.0	33.0	19.0
	In-plane shear, ultimate	(ksi)	F_{xy}^{su}	31.0	23.0	40.0	33.0	26.0	24.0
	Interlaminar shear, ultimate	(ksi)	F^{isu}	–	–	–	–	–	–
	Ultimate longitudinal strain	(μ in./in.)	ϵ_x^{tu}	8,840.0	7,800.0	7,900.0	9,150.0	4,410.0	3,980.0
	Ultimate transverse strain	(μ in./in.)	ϵ_y^{tu}	7,060.0	6,360.0	6,430.0	6,590.0	3,820.0	2,750.0
Elastic properties (typical)	Longitudinal tension modulus	(msi)	E_x^t	6.75	5.95	8.09	6.78	9.42	8.73
	Transverse tension modulus	(msi)	E_y^t	6.75	5.95	8.09	6.78	9.42	8.73
	Longitudinal compression modulus	(msi)	E_x^c	6.75	5.95	8.09	6.78	9.42	8.73

Table 7-4. Continued

				INTERMEDIATE STRENGTH GRAPHITE/EPOXY		HIGH STRENGTH GRAPHITE/EPOXY		HIGH MODULUS GRAPHITE/EPOXY	
				R.T.	350°F	R.T.	350°F	R.T.	350°F
	Transverse compression modulus	(msi)	E_y^c	6.75	5.95	8.09	6.78	9.42	8.73
	In-plane shear modulus	(msi)	G_{xy}	2.58	2.26	3.03	2.57	3.56	3.31
	Longitudinal Poisson's ratio		ν_{xy}	0.307	0.318	0.312	0.322	0.324	0.317
	Transverse Poisson's ratio		ν_{yx}	0.307	0.318	0.312	0.322	0.324	0.317
Physical constants (typical)	Density	($lb/in.^3$)	ρ	0.055	0.055	0.056	0.056	0.058	0.058
	Longitudinal coefficient of thermal expansion	(μ in./in./°F)	α_x	1.79	1.90	1.21	1.10	1.27	0.984
	Transverse coefficient of thermal expansion	(μ in./in./°F)	α_y	1.79	1.90	1.21	1.10	1.27	0.984

Table 7–5. Unidirectional C/G Fiber Composites.[22]

				INTERMEDIATE STRENGTH GRAPHITE/EPOXY		HIGH STRENGTH GRAPHITE/EPOXY		HIGH MODULUS GRAPHITE/EPOXY	
				R.T.	350°F	R.T.	350°F	R.T.	350°F
Design strengths (typical)	Longitudinal tensile, ultimate	(ksi)	F_L^{tu}	160.0	144.0	180.0	180.0	110.0	96.0
	Transverse tensile, ultimate	(ksi)	F_T^{tu}	7.0	4.0	8.0	4.0	4.0	3.1
	Longitudinal compression, ultimate	(ksi)	F_L^{cu}	160.0	65.0	180.0	70.0	100.0	60.0
	Transverse compression, ultimate	(ksi)	F_T^{cu}	25.0	15.0	30.0	12.0	20.0	8.0
	In-plane shear, ultimate	(ksi)	F_{LT}^{su}	10.0	4.0	12.0	6.8	9.0	3.6
	Interlaminar shear, ultimate	(ksi)	F^{isu}	13.0	8.0	13.0	8.0	10.0	6.5
	Ultimate longitudinal strain	(μ in./in.)	ϵ_L^{tu}	9,500.0	9,100.0	8,700.0	9,650.0	4,600.0	4,000.0
	Ultimate transverse strain	(μ in./in.)	ϵ_T^{tu}	4,500.0	4,300.0	4,750.0	4,100.0	2,500.0	3,700.0
Elastic properties (typical)	Longitudinal tension modulus	(msi)	E_L^t	17.0	16.0	21.0	18.7	25.0	24.1
	Transverse tension modulus	(msi)	E_T^t	1.7	1.0	1.7	0.87	1.7	0.85
	Longitudinal compression modulus	(msi)	E_L^c	17.0	16.0	21.0	18.7	25.0	24.1

Table 7-5. Continued

				INTERMEDIATE STRENGTH GRAPHITE/EPOXY		HIGH STRENGTH GRAPHITE/EPOXY		HIGH MODULUS GRAPHITE/EPOXY	
				R.T.	350°F	R.T.	350°F.	R.T.	350°F
	Transverse compression modulus	(msi)	E_T^c	1.7	1.0	1.7	0.87	1.7	0.85
	In-plane shear modulus	(msi)	G_{LT}	0.65	0.36	0.65	0.32	0.65	0.50
	Longitudinal Poisson's ratio		ν_{LT}	0.21	0.21	0.21	0.21	0.30	0.30
	Transverse Poisson's ratio		ν_{TL}	0.021	0.013	0.017	0.010	0.020	0.011
Physical constants (typical)	Density	(lb/in.^{3})	ρ	0.055	0.055	0.056	0.056	0.056	0.056
	Longitudinal coefficient of thermal expansion	(μ in./in./°F)	a_L	0.3	0.5	-0.21	-0.005	-0.3	-0.01
	Transverse coefficient of thermal expansion	(μ in./in./°F)	a_T	14.4	20.7	16.0	21.8	19.5	23.0

model has uniform parallel-piped cells with the properties of the primary fiber composite, and that the cell bodies are contiguous and completely bonded, and further that the uniform average stress on the orthogonal planes produce an average uniform strain, and that there is uniform average temperature, moisture, and heat conductivity, they have reduced the properties of unidirectional composites to the expression:

$$P_{HC} = P_{PC} + V_{SC}\,(P_{SC} - P_{PC})$$

where P denotes the property of interest, V denotes composite volume ratio, subscript HC denotes hybrid composite property, subscript PC denotes primary composite property and subscript SC denotes secondary property, and they have developed a group of equations which are used to compute hybrid properties. The basic data on unidirectional composites which are used in their calculations are contained in Table 7-6. The principal epoxy resin used was a tetrafunctional glycidyl ether as a matrix. As discussed earlier in this chapter, there are other influences, such as stacking sequences, discontinuities, and heterogeneities, which affect results, but therein was an attempt to relate a large amount of data on different fibers and offer a starting position for analysis.

Fracture modes in angle-ply composites have been studied at NASA and Lewis Research Center where the mechanical behavior of high modulus graphite fiber/epoxy matrix composites (Modmor I-GR/ERLA-4617) were subjected to off-axis tensile loading.[27] The major results of the investigation by Chamis and Sinclair were summarized in their conclusions, as follows.

1. Predominant fracture modes:
   a. Longitudinal tensile (fiber breaks) near 0° load angle
   b. Intralaminar shear (matrix shear fracture) in the 5° to 20° angle range
   c. Transverse tensile (matrix tensile fracture) in the 45° to 90° load angle
   d. Mixed mode (intralaminar shear and transverse tensile) in the 20° to 45° load angle range.
2. Stress-strain curves to fracture are linear (see Fig. 7-8).
3. Results by linear composite mechanics were in good agreement with measured data for predicting mechanical response to off-axis tensile loads of high modulus fiber/epoxy matrix composites which exhibit linear stress-strain to fracture.
4. Plotting procedures were developed to identify which stress dominates off-center tensile fracture.

Chamis and Sinclair also used NASA's strain analysis computer program (NASTRAN) to examine out-of-plane twisting and bending eccentricities and

Table 7-6. Unidirectional Composite Properties: Typical Room Temperature Values (60% Volume Fiber).[27]

				LONGITUDINAL (0°) PROPERTIES				TRANSVERSE (90°)	
				TENSION		COMPRESION		TENSION	
MATERIAL	EPOXY MATRIX RESIN	FORM	DENSITY g/cm^3	STRENGTH (ksi)	MODULUS (msi)	STRENGTH (ksi)	MODULUS (msi)	STRENGTH (ksi)	MODULUS (msi)
Graphite									
A-S	3501	Tape	1.54	220	20	170	16	9.0	1.3
HM-S	934	Tape	1.63	175	32	90	25	12.5	2.0
HT-S	5208	Tape	1.55	215	25	155	24	13	1.3
T–300	5208	Tape	–	210	20	210	20	6.5	1.5
GY–70	934	Tape	1.69	85	40	75	38	6.0	1.2
Aramid 49	–	Tape	1.38	220	12	40	11	4.1	0.8
S-2 class	1002S	Tape	1.82	180	6.3	110	6.0	–	–

Table 7-6. Continued

FLEXURAL PROPERTIES		INTERLAMINAR (SHORT BEAM) SHEAR	IN-PLANE (INTRALAMINAR) SHEAR PROPERTIES			PLY	PREP REG
STRENGTH (ksi)	MODULUS (msi)	STRENGTH (ksi)	STRENGTH (ksi)	MODULUS• (msi)	POISSON'S RATIO	THICKNESS (mils)	COST ($/LB)
225	17	14	8.7	0.83	0.25	5–8	45
150	28	10.5	10.4	0.85	0.20	7–8	75
245	23	16.5	10.5	0.85	0.21	7–8	65
260	20	14	9.0	0.95	0.21	5–7	45
135	38	7.5	14.0	0.60	0.25	6–8	60
90	11	7.0	8.7	0.3	0.34	5–7	45
170	6.0	10.5	12.0	–	0.3	10	6.2

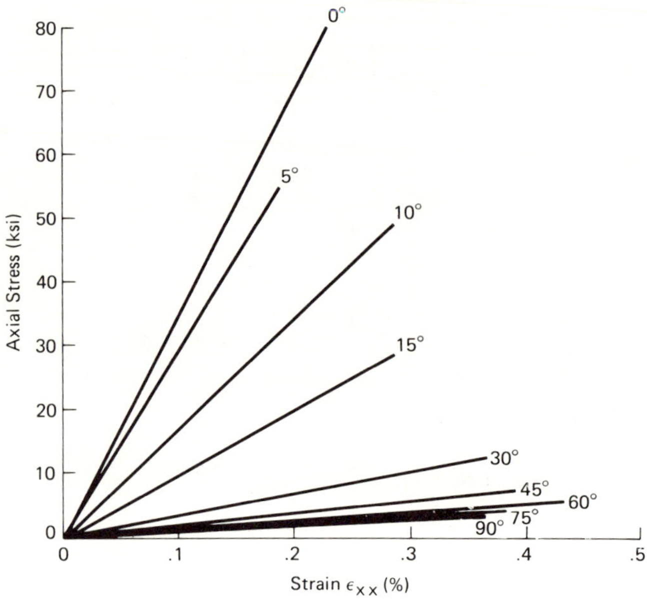

Fig. 7-8. Stress-strain curves for specimens subjected to tensile loading at various angles.[27]

their effects on axial strain variations and fracture stress of off-axis fiber composites.[28]

Also representative of the considerable work on hybrid laminates was a study of the effects of thermal cycling on composites prepared with graphite/epoxy hybridized with lower cost, high strength, lower stiffness aramid fibers ("Kevlar" 49)/epoxy or S-glass/epoxy.[29] Specimens were subjected to a tensile pre-load between 50 and 70% of the static strength of uncycled specimens. One hundred thermal cycles were applied to specimens between room temperature and -100°F and between room temperature and 280°F. Surviving specimens were tested statically to failure. These properties compared with static properties obtained by testing similar specimens without cycling. However, the ultimate strains of the hybrids are significantly lower than for their single material equivalents, suggesting that strengths may be influenced by the lower strain capability of the plus or minus 45° graphite/epoxy plies. The epoxy resin matrix used in all plies was the cycloaliphatic ERLA-4617.

Henriksen and Thornton[30] examined the material damage of graphite/epoxies under tensile loading. The graphite/epoxy systemT300/5208 epoxy was fabricated into tensile test specimens with and without circular flaws (see Table 7-7).

Table 7-7. Tensile Properties of Composite Specimens.[30]

COMPOSITE SYSTEM	WITHOUT HOLE	WITH HOLE*
(0)	227,000 psi	169,000 psi
(0/±15/0)	115,480	104,650
(0/±30/0)	139,910	89,300
(0/±45/0)	116,790	64,500
(0/±60/0)	100,360	76,080

*Area based on specimen without hole.

Specimens contained 68.3 fiber volume, 30.4 resin volume, and 1.3% void. Tensile test specimens were 8 plies in thickness, 1 inch wide, and with a 3/16 inch diameter circular hole at the center.

TOUGHNESS OF ADVANCED COMPOSITES

Toughness studies on advanced composites depart significantly from the standard methods, such as Izod and Charpy tests (ASTM D256-72), performed upon isotropic materials. With the highly diverse materials which characterize composite structures, toughness considerations are more complex by reason of the many variables which arise from the choice and application of materials, as well as the manufacturing, processing, and environmental conditions that occur. Among homogeneous plastic materials, toughness is identified with higher elongation materials, and large areas under tensile stress-strain curves. Such considerations must be tempered by design factors which are frequently directed for advanced composites of C/G towards optimum stiffness and exacting dimensional tolerances. The heterogeneous character of advanced composites requires examination of impact strength or fracture toughness to assess material toughness. Instrumented Charpy (or Izod) impact testing has been improved, yielding interesting data on fiber-reinforced composites.

Chang[31] points out the continued usefulness of Charpy and Izod impact strength in characterizing high impact materials. The toughness of a solid material is commonly defined as the work or energy per unit volume required to cause fracture. LEFM describes the release of elastic strain energy as

$$G = 0.5\, P^2\, \frac{\partial(1/k)}{\partial a}$$

where P = the applied load,
k = the ratio between P and elongation, and
a = one-half of the crack size normal to direction of applied load

Since fracture toughness is the work rate with respect to crack length to extend a crack, fracture toughness represents a critical value, such as G, upon fracture. The changed length divided by applied load is called the "compliance" of the specimen. Chang describes three modes: (a) direct opening mode, (b) edge sliding mode, and (c) tearing mode. Fracture toughness can be given in terms of G_{ic} (inch-lb/inches^2). Alternately, fracture toughness of a material can be given by the stress intensity factor K_{ic} in units of ksi/inches^2, whereby in a plain stress condition the following holds:

$$EG_{ic} = K_{ic}^2$$

where E is the modulus of elasticity.

Chang also discusses hybrids and the introduction of S-2 glass fibers and polyaramid fibers in place of some of the "brittle" graphite fibers in advanced composites.

Sih *et al.*[32] performed tests upon unidirectional directional fibers reinforced composites with the crack running parallel to the fibers (see Fig. 7-9). The current theory of fracture mechanics is shown to apply with good accuracy for glass fiber composites, though with modification for graphite fiber reinforcement. For specimens with various crack lengths, compliance calibration curves were obtained for various fiber-volume concentrations of glass and graphite

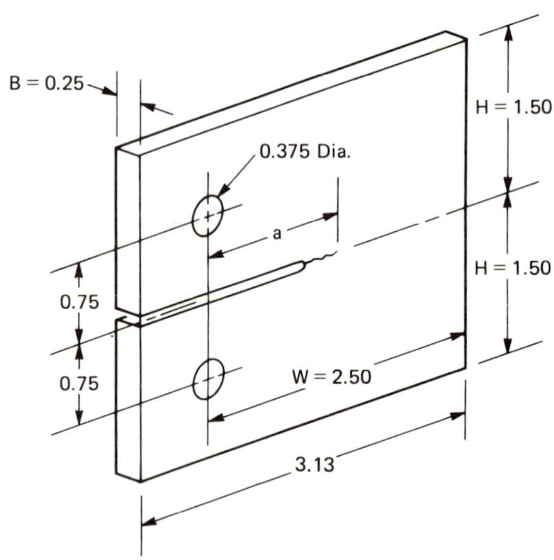

Fig. 7-9. Edge-notched tension specimen.[32]

composites. Specimens were cut and prepared, as in the previous curve, with a jeweler's saw. Compliance is plotted against the ratio of crack length and width of specimens (see Fig. 7-10). It is significant that Sih *et al.* indicate that there exists an optimum filler volume fraction for which the composite achieves maximum fracture toughness.

Data were presented to show significant improvement in the impact resistance of composite materials through the use of hybrid construction (i.e., graphite with aramid or with glass fibers). Beaumont *et al.*[33] used instrumented Charpy tests to study a number of materials. They stressed the importance of separating the initiation energy from the propagation energy. Some of the hybrid constructions were more effective in resisting impact damage than thicker all-epoxy composites.

There are other ways of classifying the processes which affect toughness of composites. Cooper and Piggott[34] offer the following:

1. Among the processes which lead to energy dissipation during crack propagation are those which arise from fiber movements relative to matrix. It is

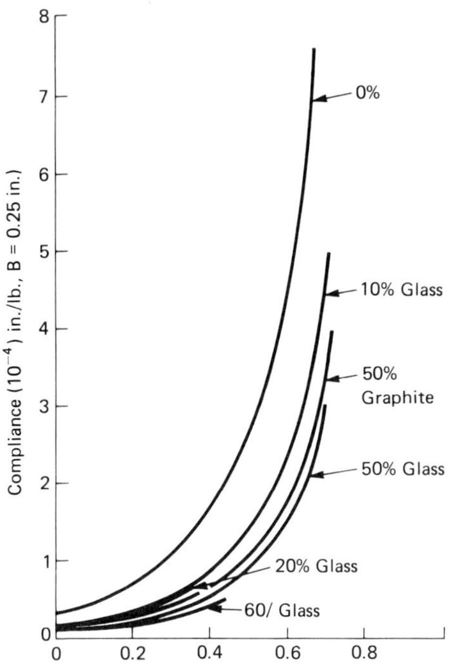

Fig. 7-10. Compliance calibration curves for various fiber-volume concentrations of glass and graphite composites.[32]

shown that stress redistribution makes the greatest contribution to toughness when the crack opening has to be minimized.

2. There are indirect processes which enhance the toughness of materials, such as interactions of brittle fibers with relatively ductile matrices which result in non-linear stress-strain relationships. There are synergistic benefits which may arise from such interactions.
3. Stress concentrations at fiber ends and fiber breaks can contribute to early failures in the matrix.

The importance of distinguishing between interface controlled failure and matrix controlled failure by microscopic examination of fracture surfaces was emphasized by Hahn.[35] Furthermore his analysis predicts that the composite-to-matrix strength ratio in transverse tension and shear can be greater than unity and can increase with constituent modulus ratio and the fiber content. The use of strong and ductile matrices may improve fiber efficiency.

The resistance of graphite/epoxy sandwich panels to low energy, foreign object impact damage was studied on 104 rectangular sandwich panels[36]. Face sheets of graphite or S-glass/epoxy were fabricated over a 1 inch thick "Nomex" core. A 2 inch diameter steel ball was dropped at heights ranging from 1 to 7 feet for the graphite panels and up to 14 feet for S-glass panels. Graphite faced sandwich panels exhibit marked susceptibility to impact, attributed to the low strain to fracture of graphite composites. S-glass sandwich panels sustained significantly higher impact energy levels than the graphite composite faced specimens.

Impact tests upon composites have "real-life" significance when the velocity of impact, as well as the stress upon the composite at the time of impact, are examined. Sayers[37] presents the results of such an examination upon carbon fiber reinforced composites. Using "Hyfil" carbon fibers (28×10^6 psi modulus) and an epoxy resin, he fabricated test specimens for interlaminar shear strength (ILS), which in turn were tested under impact conditions. The interlaminar shear fixture maintained a spacing of 1.50 inches between 0.25 inch rollers, and the span to thickness ratio was 5.0. Because interlaminar shear cracks have occurred on C/G fiber composite blades of a turbofan engine in a fraction of a millisecond, the approximate time to maximum load was recorded in milliseconds for the impact tests. Higher values for ILS were reported for the unidirectional composites, as compared to the $0°$–$90°$ cross-plied. Weight of falling platform was 4.3 pounds and striker radius 0.1 inch (see Table 7-8).

More recent tests of impact on composites were performed by Rhodes *et al.*,[38] conducting tests at higher impact velocity than shown in Table 7-8. Tests were conducted upon several laminates of C/G fiber/epoxy system with different ply orientation and thicknesses from 0.56 to 0.75 cm. Panels were damaged by

Table 7-8. Variation of ILS with Time to Failure.[37]

UNIDIRECTIONAL COMPOSITES			0°–90° CROSSPLIED		
DROP HEIGHT (in.)	TIME TO MAX. LOAD-MICROSEC.	ILS RANGE lb/in.^{2}2	DROP HEIGHT (in.)	TIME TO MAX. LOAD-MICROSEC.	ILS RANGE lb/in.^{2}
0 (Static)	55×10^3	6400–7500	0 (Static)	40×10^3	3950–5000
3–6	1.4–0.6	5500–6400	2–4	0.9–0.6	3550–4300
6–27	0.6–0.4	4530–6100	4–18	0.6–0.3	3070–4500
27–39	0.4–0.3	4650–5650	18–39	0.3–0.2	2400–3700

(Reprinted Courtesy of Technomic Publishing Co. Conn.)

aluminum spheres 1.27 cm diameter which were propelled normal to specimen surfaces at velocities of 30 m/second to 140 m/second. The first visible evidence of impact damage was a small crack on the side opposite the impact location. Ultrasonic inspection showed internal damage due to low velocity impact with no surface damage. Further inspection of specimens showed these internal damages were principally intraply cracking and delamination. There was no appreciable difference in damage of laminates with woven fabric reinforcements versus unidirectional tape lay-ups (see Fig. 7-11 relating damage area in cm^2 to projectile velocity in m/second).

A symposium on foreign object impact damage to composites was held by the ASTM in Philadelphia in 1973 by the committee D-30 on high modulus fibers and their components. Husman *et al.* examined the damage inflicted by spherical steel projectiles measuring 0.77 inch and 0.250 inch diameter at velocities in the range of 200-500 feet per second.[39] Results suggest that residual strength can be predicted as a function of kinetic energy of impact by conducting two experiments: static tension test on an unflawed specimen and static test upon a coupon previously subjected to a single-point impact. Specimens were 12 ply thick with a thickness of 0.1 to 0.144 inch. All of the epoxy laminates displayed local crushing, cracking, and delamination. When a thermoplastic matrix was used, there was a denting mode and crushing, but no delamination.

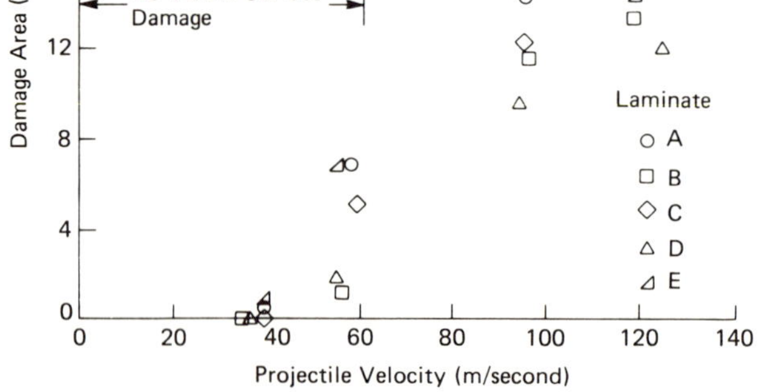

Fig. 7-11. Damage area in five test laminates as determined by C-scan inspection for different impact velocities.[38]

The role played by the matrix in controlling the impact properties of advanced composites has been recognized in several papers suggesting the use of toughened matrices. The attainment of higher impact epoxies, for example, has been reported in considerable detail through appropriate chemical modifications. Epoxy matrices of higher elongation, accompanied by an increase in toughness, are attained by:

1. Selection of curing agents, which in forming part of the polymer backbone will impart increased flexibility to the epoxy resin. For example, diamine terminated long alkyl chains; polyamide curing agents; or dianhydrides with long chain pendant groups (i.e., dodecenyl succinic anhydrdide).
2. Special epoxy resins synthesized with higher elongation in mind—for example, epoxy-urethane copolymers or special chain extended cycloaliphatics, such as Union Carbide Corp.'s 4289.
3. Plasticizers such as polysulfides or dibutyl phthalate which are formulated into the composition. These do not have the permanence of flexibility of the internally modified epoxy resins.
4. Additions of small amounts of a rubbery phase such as amine-terminated (ATBN) or carboxyl-terminated butadiene/acrylonitrile (CTBN) synthetic rubbers to the epoxy resin.

In effect, the above information will remind the reader that there are chemical methods for altering the resin matrix to attain greater elongation and tougher resins. What is to be identified, however, is the correct amount of alteration to achieve the increased strain required by the structure. An increase in elongation, usually accompanied by some loss in tensile strength, will yield a matrix with lower heat resistance and, in some instances, poorer environmental resistance. The cooperative efforts of the several disciplines will yield the tougher system for the advanced fiber composites.

There are several literature references which have identified the need for tougher resin matrices for advanced composites. In sporting goods applications of C/G fiber composites, Segal and Boyle[40] described the utilization of epoxy resins of varying moduli to improve impact strength C/G fiber composites. Interlaminar fracture energy was significantly increased by the addition of elastomeric toughening agents to the epoxy resin matrix.[41] Improvements were obtained by formulating the epoxy matrix with 8% CTBN 1300 $\times$ 13 ("Hycar") and 5.4 percent "Hycar" 1472. It should also be noted that the addition of such synthetic rubber additives to epoxy adhesives has greatly improved the peel strength and toughness of adhesive products. It is also recognized that the viscosity of the epoxy resin is considerably increased by such additions which will have a bearing on the wetting out and adhesion to the C/G fibers.

ATBN synthetic rubber derivatives were dispersed in epoxy resin matrices by Allred *et al.*[42] to toughen the matrix. Stress-strain curves of standard epoxy resin systems with different percentages of ATBN concentration are reproduced in Fig. 7-12. Results suggest that the toughened matrix suppresses the growth of flaws initiated at the interface, thereby effectively increasing the critical flaw size and hence the transverse strength of the unidirectional composite.

Hunston[43] reviewed the fracture toughness of composites and showed a preference for discrete elastomeric particles in the resin matrix to enhance toughness. He employed a width tapered double cantilever beam specimen to check the propagation of cracks between the layers, demonstrating a significant improvement in fracture energy of the composite with an increase in fracture energy of the resin matrix. He also presented some data on the high fracture energy of experimental phthalocyanine resins with high values of T_g (375°C).

Improvement in performance of high modulus graphite/epoxy tubes, 4 inches in diameter, was reported when graphite fiber with a 34×10^6 psi modulus was substituted for an 80×10^6 psi modulus graphite fiber.[44] The tube fiber volume was decreased 2%, with an increase in resin weight to 42%, resulting in a more stress resistant material and a better quality end product. While in this example the resin matrix remained the same, it should be noted that fiber modification as well as matrix modification offer alternatives to improved design and fracture toughness. All graphite/epoxy torque tubes were observed in some studies to delaminate severely under low velocity, low energy impact tests.[45] There is an increasing interest in impact tests of finished components of advanced fiber composites which are under stress, and aircraft manufacturers engaged in producing such components are applying service stresses to the materials at the time

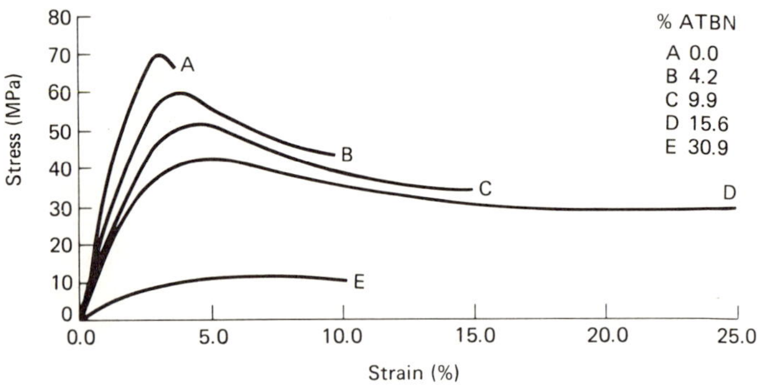

Fig. 7-12. Stress strain of epoxy (modified with ATBN rubber) matrices.[42]

of impact study. This study has contributed to the sum of the knowledge to improve design and material selection.

Cracking within the composite structure may also arise as a consequence of thermal cycling. The large difference in thermal expansion between the graphite fiber and its near zero C.T.E. parallel to fiber direction, and the resin matrix C.T.E. (about 30×10^{-6}) in./in./°F), contribute to matrix cracking. As shown in the photomicrograph (Fig. 7-13), after 25 cycles from -320°F to 212°F exposure to a graphite fiber/phenolic composite section, the cracks are quite visible. On the other hand, with higher elongation thermoplastic matrices, such as polyphenylsulfones, such cracks are much less prevalent. Because of the

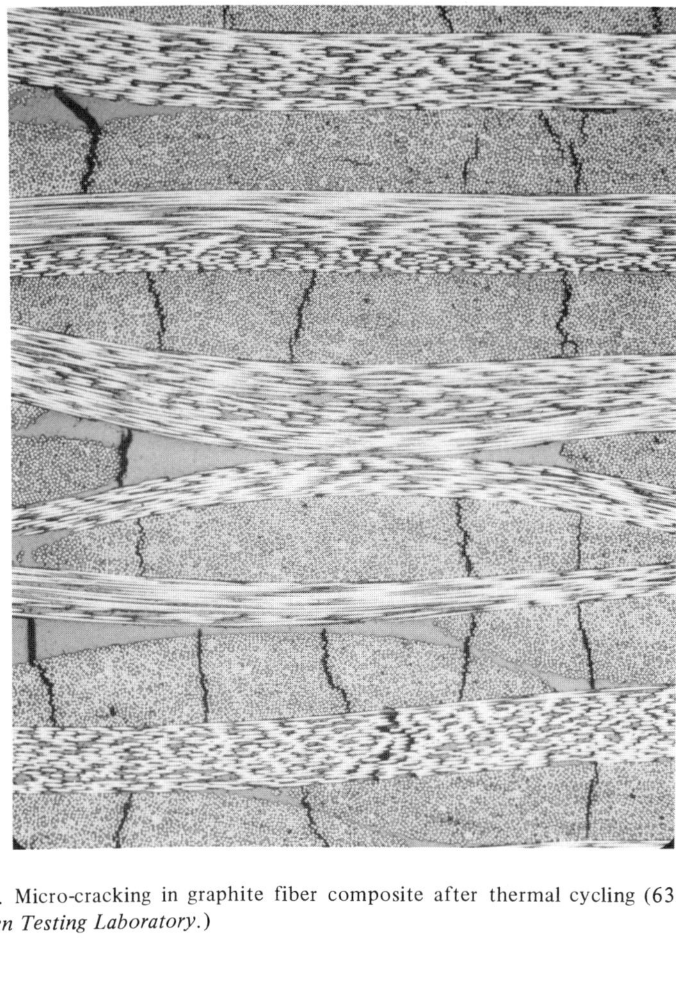

Fig. 7-13. Micro-cracking in graphite fiber composite after thermal cycling (63X). (*Courtesy Delsen Testing Laboratory.*)

audible sounds of cracks during cool down from cure temperatures during fabrication, it is quite likely that some advanced composites develop matrix cracking internally, at this time.

The effects of thermal cycling on graphite-polyimides have been reported by Hyer *et al.*[46] The development of resins which retain their strength at elevated temperatures has been accomplished in the higher heat resistant non-metallics, such as the polyimides. As a structure is repeatedly heated and cooled, the question arises as to whether or not the changes in material properties are reflected in changes in the load deformation behavior from one cycle to the next. We have already seen the effect of "thermal" spikes on changes in moisture absorption (see the chapter on environmental influences). Unsymmetric laminates were examined and changes in strain as a function of temperature was used as a measure of thermal deformation. It was recommended that optical gages be used at higher temperatures because of questions on strain gage attachments. Specimens were fabricated from HTS graphite fiber/PMR-15 polyimide resin. In the 180°C temperature cycling there was no evidence that thermal cycling affected the temperature-thermal deformation from one cycle to another.

FATIGUE AND CREEP PHENOMENA

Fatigue and creep phenomena are time dependent and have a vital role in determining the performance of advanced fiber composites. The repetitive character of fatigue stressing, whether by tensile, compressive, or flexural loading, is of fundamental interest to designers, particularly on advanced composites where the diverse nature of materials produce varying results. It may be reasoned that in-service testing offers "real-life" load conditions which yield, through failure analysis, data which will dictate the improvements which can be made in composites. In most cases, however, the time required to generate meaningful results may run into months or years, and hence laboratory accelerated fatigue testing is used to attain early data. Creep phenomena characteristic of the behavior of viscoelastic materials under stress is usually performed upon the resin matrices of fiber filled, isotropic components where the amount of resin present is the major component. Of interest to manufacturers of advanced composites are those resin systems which exhibit low creep at elevated temperature. In fact, standard ASTM test methods of heat distortion usually exhibit creep deformation processes under static loads and under an increasing temperature, with an arbitrary deformation limit defining acceptable performance. The interest in the fatigue resistance of C/G composites stems from data which indicate the superiority of these systems to other materials of construction (see Fig. 7-14).[47]

There are two ways of looking at creep phenomena: creep as a stress relieving process and hence capable of forestalling an early failure (more possible in an

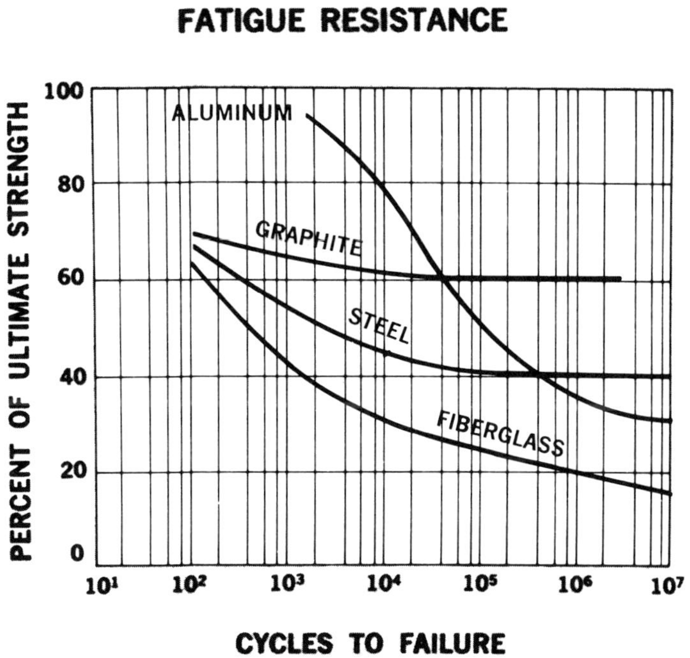

Fig. 7-14. Comparison of fatigue strengths of graphite, steel, fiberglass, and aluminum. Note the greater strength of graphite after 10,000,000 cycles.[47]

isotropic material), or creep under stress with its attendant dimensional instability and adverse effect upon performance. In advanced fiber composites, the high stiffness of C/G fibers, particularly, provide the over-riding consideration and the transient nature of creep in the resin matrix is usually small. However, the fact that stress concentrations may produce creep in the resin will contribute to the reduced volume of resin, as, for example, at fiber crossovers. Factors such as these contribute to the heterogeneity of composite make-up. However, as observed in earlier references to fracture mechanics, those assumptions made in LEFM, in treating some laminates as homogeneous bodies, have not been too much in error.

It appears proper to introduce here environmental factors and their effect on fatigue of composites, because they may prove severe and lead to catastrophic failures. Water and high humidity affect most applications, and in consequence, a separate chapter has been devoted to this important subject. When the emphasis is on fatigue behavior, pertinent references will be examined in this chapter. There are advantages cited sometimes for composites exposed to water gain, particularly by those expressing acceptance of the plasticizing and stress relieving effect of moisture on the resin matrix in composites. On the other hand, there are those who properly point out the accelerating influence of moisture on fatigue behavior.

Glass laminates continue to provide excellent reference points when examining the technology of carbon and graphite fiber composites. Information in the typical fatigue curves (stress/number of cycles–S/N curves) shown in Fig. 7-15 is from an early paper by Davis *et al.*[48] Of note is that there is a higher fatigue strength of the unidirectional glass laminates over other forms of construction and that laminates flexurally loaded in a modified Sonntag machine have 10% high fatigue strength than in axial fatigue with alternate tension and compression loading. Since this data were presented the C/G composites have entered the picture and demonstrate improved fatigue strength in advanced composites.

Santelli and Simon[49] tested graphite-epoxy composites for their static shear and flexural fatigue properties. Specimens included loaded and unloaded shear and flexural specimens in both wet and dry environments. Significant deteriora-

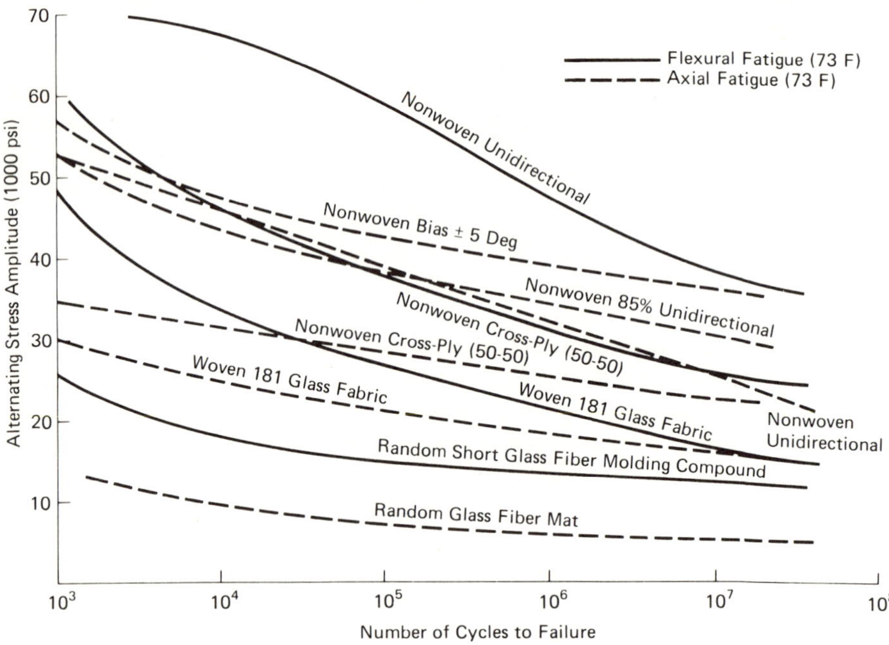

Fig. 7-15. Non-woven, unidirectional glass fiber reinforced plastics provide much better fatigue strength than any other form of glass fiber reinforced plastics.[48]

tion took place on those specimens flexurally loaded in water. At the time of these tests (1970) "Modmor II" graphite fibers were selected in preparing test samples with several epoxy resin binders. The cycloaliphatic ERLA-4617 withstood stresses under water exposure better than other commercial epoxy resins.

Di Benedetto[50, 51] examined the fatigue of graphite filled Nylon 66, showing that the distribution of fatigue lines as a function of stress is related to both the flaw growth rate and flaw distribution. It was established that progressive fatigue damage, slowly building up to some critical value, is the principal mechanism of failure in fiber reinforced plastics. Composite samples contained 10% by volume of 6 micron Hercules AS graphite. Fatigue crack growth in Nylon 66 is a function of stress intensity factor and water content. In the case of Nylon, up to 3% water content has water tightly bound chemically (1 water molecule per 2 amide groups), and fatigue crack rates were 2.5 times lower than for dry specimens. However, for higher amounts of water, up to 8%, in which water is loosely bound, the polymer is weakened and fatigue crack propagation rates significantly higher than those for the dry nylon specimens.

Owen[52] made a detailed examination of the fatigue strength of high modulus and high tensile strength C/G composites. Except at low tensile stresses, axial fatigue strengths are governed by static compressive and tensile strengths, and at ambient temperatures appear to be relatively insensitive to attack by moisture or oil. He employed the C/G fibers of Courtauld's ("Grafil") in twist free tows containing 5,000-10,000 individual filaments of 8 to 9 microns mean diameter. Typical physical properties of the commercial fibers were as shown below.

	DENSITY (g/cm^2)	ULTIMATE TENSILE STRENGTH–(GN/m^2)	YOUNG'S MODULUS (GN/m^2)
"Grafil" A or A-S	1.75	1.90–2.60	180–220
"Grafil" HT or HT-S	1.80	2.10–2.80	225–280
"Grafil" HM-S	1.95	1.80–2.40	350–410

Data are presented by Owen for both epoxy resin matrices and polyester resin matrices. Fatigue specimens were loaded axially at stress levels within the scatter band of static test results of composites. The tests were run at frequencies of 7000 cycles/minute (without noticeable temperature rise) and at 100 cycles/minute, with little difference in results. The resulting S/N curves were more nearly horizontal than for other common structural materials. Typical results are shown in the accompanying Fig. 7-16, for unidirectional, untreated high-modulus fibers with a standard epoxy/MNA/BDMA resin matrix system. Fiber volume fraction was 60% and test frequency 7000 cycles/minute. Specimens

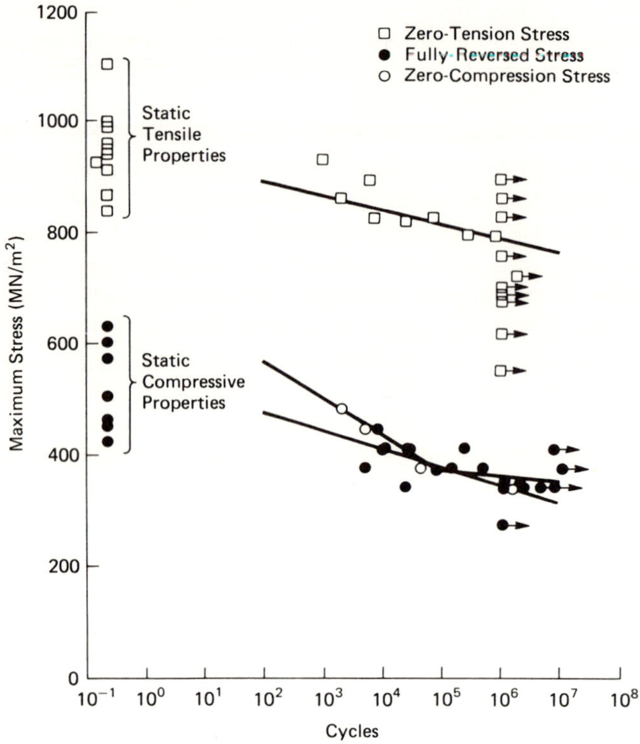

Fig. 7-16. Axial load fatigue results for unidirectional, wet lay-up, surface-treated, high-modulus fiber composites in Shell Epikote 828/MNA/BDMA epoxy resin system; volume fraction 62%; 7000 c/min; compressive stress plotted positive.[52]

were loaded axially. Under these test conditions, specimens which had been immersed for two months in distilled water did not show much difference in fatigue strength. As will be noted further in this chapter, reverse stress loading in flexure shows a much greater influence of water on fatigue strength. This technique leads more rapidly to results in fatigue comparisons and although pure stresses involve shear, compression, and tensile, the flexural fatigue tests are more representative of the engineering applications for advanced composite fiber structures.

Stinchcomb *et al.*[53] examined the effects of frequency on the mechanical response of composite materials. Fatigue damage consists of various combinations of matrix cracking, debonding, delamination, void growth, and fiber breakage. This work reported primarily on the behavior of boron fiber in aluminum and in epoxy matrices. Strain controlled cyclic tests were conducted on specimens in the frequency range of 0.5 to 45 Hz. They observed an unusual non-linear response to the frequency of the load (or strain) oscillation, for which no explanation was offered. They observed a reduction in stiffness early in the fatigue life of the materials under test.

Sumsion and Williams[54] used Hercules prepreg tapes of AS/3501 epoxy resin systems (fiber properties were given as tensile strength 3.09 GN/m^2, a modulus of 222.5 GN/m^2), 1/8 inch specimens were prepared with 0° and plus or minus 45° fiber orientation. Torsional fatigue tests were conducted at 1 Hz under controlled torsional strain and flexural fatigue studies were conducted at 30 Hz in cyclic, fully reversed bending under constant maximum load. Typical results are shown in the Figs. 7-17 and 7-18, where stiffness in torsion changes were pronounced on specimens pre-soaked in 74°C water. Later studies show that flexural (fully reversed cycles) fatigue testing during continuous water immersion lowers strength more rapidly than pre-soaking, and also show the probability of unreliable moisture data in pre-soaked specimens. Some of the interesting observations by Sumsion and Williams are noted in their paper:

1. Both unidirectional (0°) and cross-plied (±45°) graphite epoxies undergo fatigue damage, as evidenced by reduction in torsional stiffness. Cyclic torsional stresses were applied: 5760 psi (40 MN/m^2) for the unidirectional and 11,200 psi (77 MN/m^2) for cross-plied material. See decrease in torsion stiffness as function of applied stress.
2. Limiting stiffness in cross-plied specimens that were torsion tested, corresponded to a limited torsional stress equal to 15% of torsional fatigue

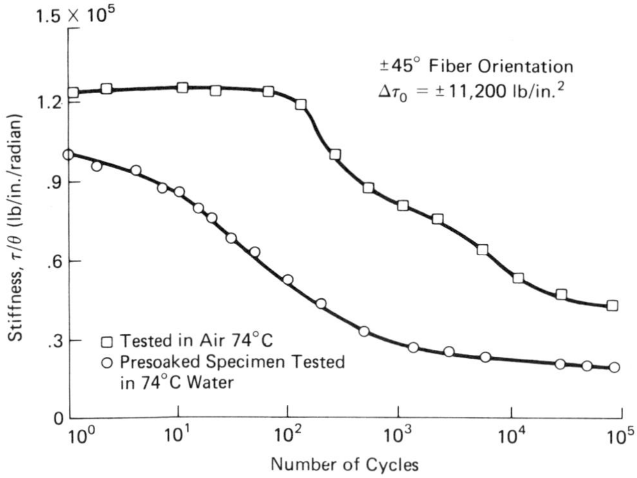

Fig. 7-17. Torsion Fatigue of ±45 deg fiber-oriented graphite-epoxy composite. Decrease in stiffness (τ/θ) with number of cycles (log N) as a function of temperature and water.[54]

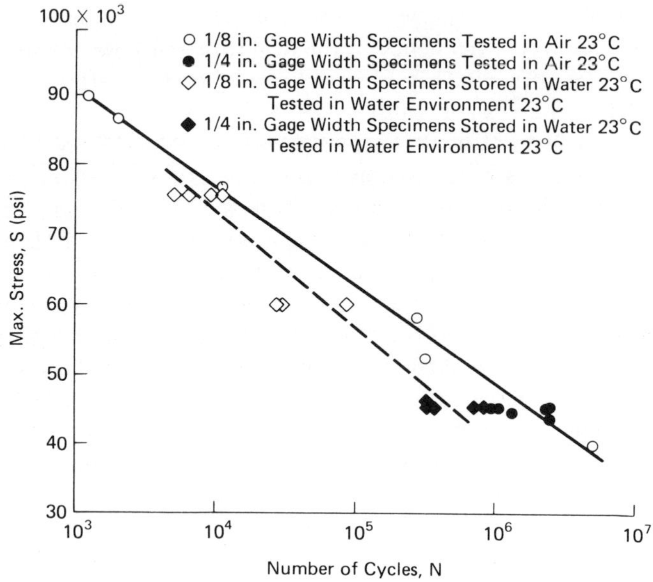

Fig. 7-18. Flexure fatigue tests at ±45 deg fiber-oriented specimens. All bend tests at room temperature.[54]

strength. Reduction in torsional stiffness caused by stress cycling was accompanied by reduction in failure strength in bending.

3. Graphite-epoxy specimens showed significant flexural fatigue damage in both air and water, when subjected to fully reversed plane bending. (See data in chapter on influence of environment on fatigue strength of composites.)

The effect of compressive loading on the fatigue properties of a quasi-isotropic graphite fiber/epoxy composite is discussed by Ryder and Walker of Lockheed.[55] Tests were conducted on unnotched laminates, uniformly prepared from T300/934 (tetrafunctional epoxy) matrix cured at 350°F. Fatigue tests were conducted at 10 Hz under constant amplitude loading. Tension-tension and tension-compression fatigue stress life curves were qualitatively evaluated. Care was taken in preparation of uniform test samples, which were machined and carefully tabbed with grips. Fiber volume range was 60.3-66.4%. Test laminate panels were non-destructively inspected for defects by ultrasonic scan procedures. The chosen stacking sequence was (0/45/90/-45/-45/90/45/0) for the plies. Tension-tension fatigue results and tension-compression fatigue results are shown in the Figs. 7-19 and 7-20.

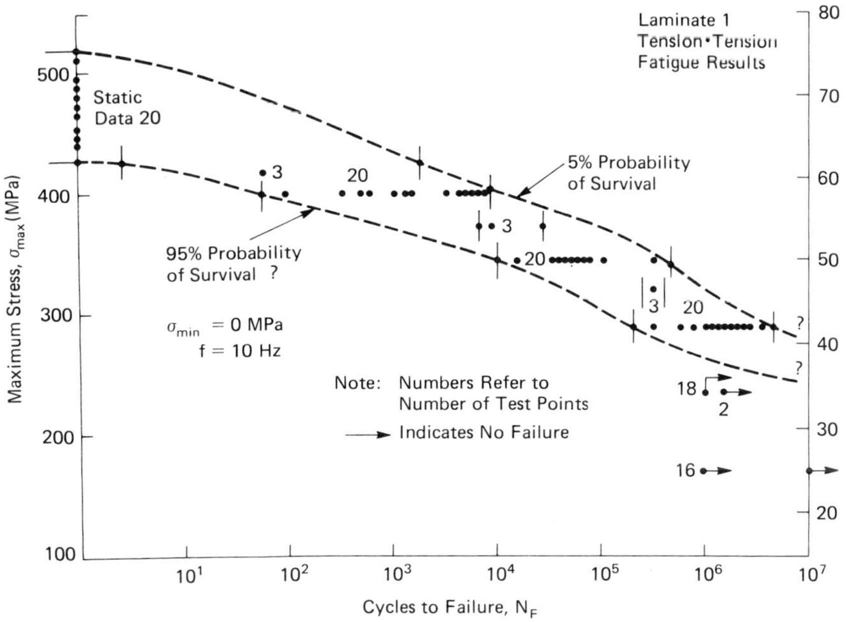

Fig. 7-19. Tension-tension fatigue results at σ_{min} = 0 MPa.[55]

For the tension-tension (T-T) tests failure was defined as the breakage of the coupon; for tension-compression (T-C) tests failure was defined as coupon breakage or inability to sustain compression load due to severe delamination, usually in the two -45° layers. Temperature rises in coupons during fatigue loading were more apparent in the T-C coupons and in the range of 8° to 15°F. Residual strengths were statistically tested in tension and compression on unfailed samples.

Additional data on axial fatigue behavior of both notched and unnotched graphite/epoxy composites has been reported by Ramani and Williams (NASA-Ames).[56] Tests were conducted on AS/3501 graphite/epoxy laminates from which test specimens were laid up on a $(0° \pm 30°)_{3s}$ basis. Apparent fatigue limits, defined as endurance limit for 5×10^6 cycles in tension (T-T) and compression-compression (C-C) cycling, occurred at about 60% of the respective static strengths. Fatigue limit is closer to 70% for $(0/\pm45/0)_{2s}$ laminates. The specimens had a thickness of 2.5 mm and an average fiber volume of 63%. In notched studies, fatigue was investigated in AS/3501 and T300/934 laminates containing sharp notches and circular holes. At stress levels of 50% of the unnotched strength, results demonstrate the insensitivity of these laminates to T-T fatigue. Laminates of $(0/\pm45/90)_{2s}$ were more susceptible to fatigue damage in

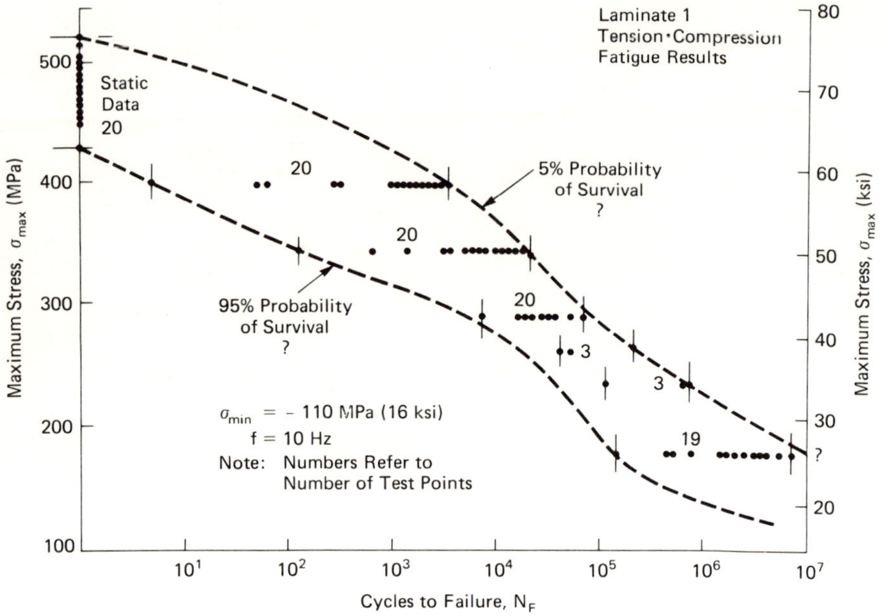

Fig. 7–20. Tension-compression fatigue test results at σ_{min} = –110 MPa (–16.0 ksi).[55]

both the notched and unnotched conditions. Progressive damage accumulation from edge notches into the specimen occurs under T-C fatigue. Resistance to damage was observed by changes in the crack opening displacement. Damage propagation from notches and holes in T-T fatigue occurred exclusively along the vertical stress direction, between axial splits at hole edges.

The Goodman chart (Fig. 7-21) expresses overall results for T-C stresses in the form of a constant life diagram. It will be noted that a maximum in fatigue properties occurs at a positive value of mean stress. The skew symmetry in fatigue arises from the low compressive strength of the unrestrained test specimens. S/N curves for T-T (R = 0.1) and C-C (R = 10) cycling are shown also. Porter[57] also concluded that for a wide range of flaw types in C/G composites there exists a relationship between initial static strength and cycle life, allowing for the use of proof loading techniques as a qualification procedure.

Kunz and Beaumont[58] investigated graphite fiber/epoxy resin composites as notched beams under cyclic compressive loading. Specimen configuration was 5 inches by 0.5 inch by 0.25 inch with a sharpened .05 inch notch. Fatigue testing cycled sinusoidally in zero-compression, constant load control, at 1 to 3 Hz. Axial cracking is the primary fracture mechanism in unidirectional composites, while in cross-ply composites, splitting between transverse plies pro-

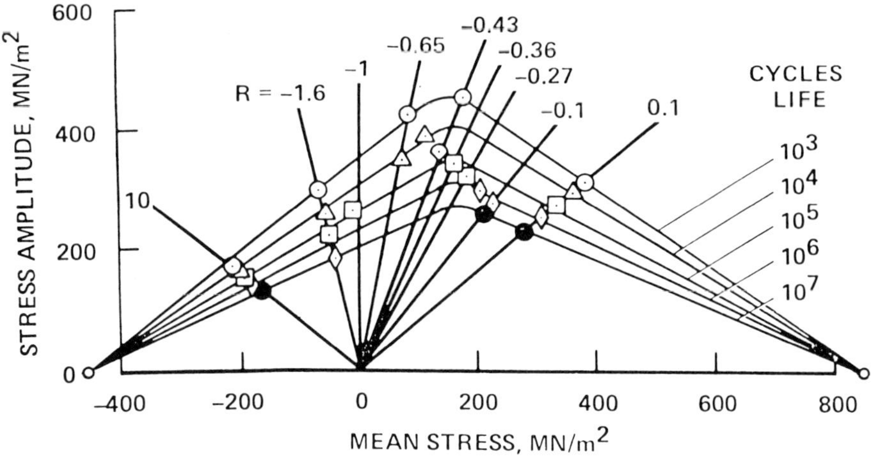

Fig. 7-21. Goodman diagram for $(0°/\pm30°)_{3s}$ graphite /epoxy.[56]

motes crack extension. The authors felt that retardation of micro-fracture processes may be achieved by considering:

1. Composites with high interfacial shear strength are less susceptible to axial cracking along the fiber-matrix interface.
2. High interlaminar shear strength minimizes splitting between and within the transverse plies of a cross-plied composite.
3. There is an optimum bond strength for which axial cracking is minimized and transfibrile crack extension is prevented.

Further data showed that immersion in saline solution greatly increases axial splitting in all composites and enhances transverse ply splitting. This is a significant observation in view of some of the present uses of C/G fiber composites in marine applications. Figure 7-22 reproduced from Kunz and Beaumont's paper depicts some of the failure modes in compression. The possible planes for micro-buckling and shearing displacements are illustrated.

Compressive tests have been made at various temperatures on unidirectional carbon fiber reinforced plastics, with loads parallel to the fiber (longitudinal and compression) and with loads normal to the fibers, while the specimens were constrained.[59] Below 100°C both types of test yielded similar strengths and the failure mode in both was shear through fibers and matrix. Above 100° the failure mode changed to one associated with fiber micro-buckling in the longitudinal compression case. Courtaulds HM-S and HT-S fibers were used with an epoxy matrix system of standard bisphenol A epoxy cured with methyl nadic

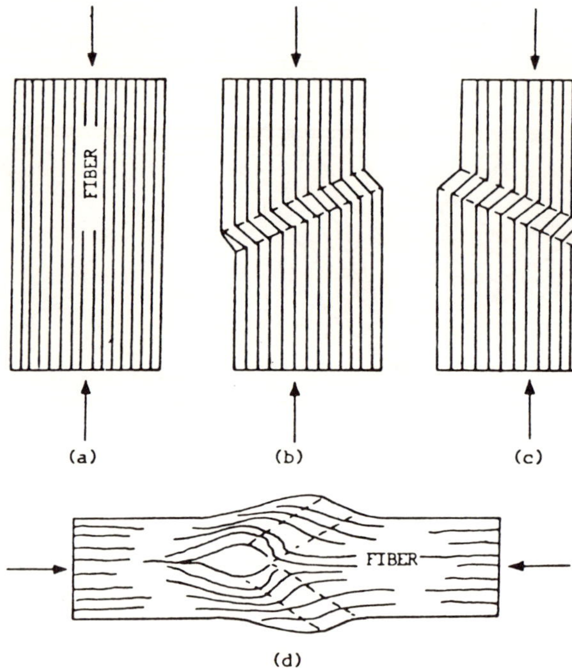

Fig. 7-22. (a) Fiber reinforced material under compression. (b *and* c) Possible planes for microbuckling; shearing displacement is permissible across these planes. (d) Microbuckling may occur on two different planes if axial cracking occurs.[58]

anhydride, and benzyl dimethyl amine (100/85/1 weight ratio). Variation of compressive strength with fiber volume fraction is shown in Fig. 7-23. At a constant fiber volume of 0.62 fraction, temperature tests showed a rapid fall off in compressive strength above 100°C. If the epoxy resin system was not fully post cured, low compressive strength could be expected at this temperature. Ewins *et al.* prepared good micro-photographic studies of the various fractured compressive specimens.

Greszczuk[60] reviewed compressive strength and failure modes of unidirectional carbon fiber composites, in an effort to reconcile compressive strengths lower than theoretical values predicted by the micro-buckling theory. The compressive shear strength of the composites was found to increase with higher values of shear modulus of the resin and with increasing Young's modulus of fibers. Factors contributing to lower compressive strengths were misaligned fibers, unbonded fibers, and round ended fibers or other than flat ended test specimens. A number of the tests were conducted upon extruded 0.078 inch diameter graphite rods inserted in resin matrices of different strengths.

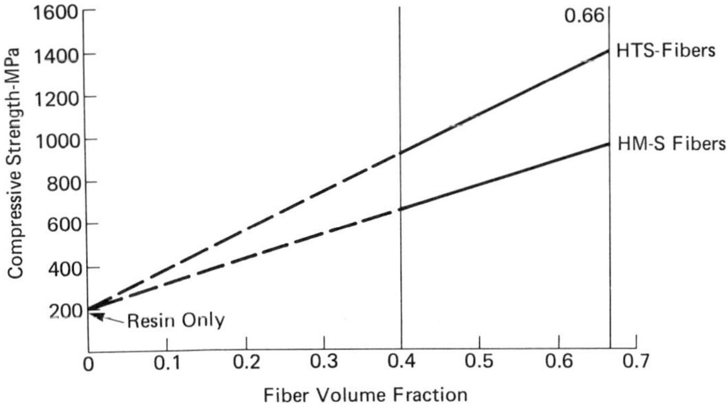

Fig. 7–23. Variation of compressive strength with fiber volume fraction (room temperature).[59] With permission of the A.I.A.A.

MISCELLANEOUS ENVIRONMENTAL FACTORS

Although a relatively limited space has been left in this chapter for environmental factors, the reader is urged to examine the chapter on this subject. Studies have focused upon the effect of water and water vapor upon C/G composites at various temperatures. In time, as applications extend to other fields of interest, such as chemical industries and marine environment, programs will be developed upon the then favored C/G fiber composites and their hybrids to examine performance under these adverse conditions. For the aerospace and automotive areas, moisture is of basic concern, as well as radiation in outer space. In the chapter on electrical applications, the influence of lightning strikes and their effect on advanced composites is considered.

Loughlin[61] reported on the average results at cryogenic temperatures for several advanced fibers, which were tested at 4°K. The matrix resin used was a blend of "Epon" 828 and 1031, cured with methyl nadic anhydride. Unidirectional composites were prepared. Flexural testing was conducted with a 16 to 1 span thickness ratio and interlaminar shear was conducted at a 6 to 1 span to thickness ratio. Data reported were as follows, for properties at 4°K.[61]

FIBER	PERCENT FIBER VOLUME	PERCENT VOIDS	DENSITY (g/cc)	FLEXURAL STRENGTH (ksi)	FLEXURAL MODULUS (msi)	INTER-LAMINAR SHEAR STRENGTH (psi)
S-glass	60.3	1.20	1.98	600	8.1	26,800
Modmor II-Carbon	66.3	2.04	1.60	200	22.5	7,900
Boron	70.5	4.10	2.16	217	31.0	6,500

Studies were also conducted on boron fibers in an epoxy matrix and in aluminum matrix at cryogenic temperature of 76°K.[62] Using degradation of modulus as a criterion, the data on low-cycle fatigue indicate that the cryogenic temperatures have a negligible effect on the tensile or compressive fatigue performance of the boron fiber composites.

During the screening of high-strength graphite/epoxy systems, an unusual type of failure was observed in an epoxy matrix with HT-S and "Modmor II" fibers. Interfiber shear specimens failed in cleavage as an extension of the notch when tested at -320°F. All other graphite/epoxy systems failed in the normal shear failure mode. Examination under high magnification showed a clear separation of resin and fiber for the above specimens, attributed to incompatibility at the interface, and labelled as cryogenic embrittlement of this particular C/G fiber/epoxy composite.[63]

Before advanced composites can be employed to replace aluminum for structural uses in space, they must be exposed on extended flights (as in orbitting vehicles) for long periods.

There is ample evidence that advanced fiber composites possess better fatigue strength and superior vibration dampening qualities to metals. This superior performance has been borne out in service applications of aircraft components. Exposure in outer space is fraught with many conditions which can only be properly evaluated when the composites have experienced the environment. Felbeck[64] at the University of Michigan, for example, is conducting studies on the effect of such exposures on the mechanical properties of graphite/epoxy composites. Spacecraft will have active lifetimes of 2 to 10 years in the space environment. Factors such as strong ultra-violet radiation will have a catalytic effect upon the properties of some resin matrices. The vacuum of outer space may enhance the rate of loss of resin rich outer surfaces on composites, which may be affected by the ionizing radiation from the sun and the charged particle layers surrounding the earth. The inclusion of advanced fiber composite components–both with metal matrices and with resin matrices in the space programs–will enrich our ability to design and utilize these materials.

REFERENCES

1. Sih, G.C., "Summary of USA-USSR Symposium on Fracture of Composite Materials," *Composites Technology Review,* Vol. 1, No. 2, page 15, Winter 1979.
2. Kaufman, J.G., "Committee E-24 on Fracture Testing," *ASTM Standardization News,* Vol. 7, No. 4, page 8, April 1979 (this issue reviews current ASTM activities on fracture mechanics and testing of metals).
3. Knauss, W.G., "The Mechanics of Polymer Fracture," *Applied Mechanics Review,* American Society of Mechanical Engineers, January 1973 (221 references).

4. Smith, C.W., Chapter 1 in *Experimental Techniques in Fracture Mechanics,* Vol. 2, Kobayashi (Editor), Society for Experimental Stress Analysis, 1975.
5. Zhurkov, S.N., Zakrevskii, V.A., and Korsukov, V.E., *Journal of Polymer Science,* Vol. A-2, No. 10, page 1509, 1972.
6. Zhurkov, S.N., "Kinetic Concept of Strength of Solids," *International Journal of Fracture Mechanics,* Vol. 1, page 311, 1965.
7. Dunn, J.R., "Unsolved Problems in Polymer Degradation," *ACS Rubber Chemistry and Technology,* Vol. 51, No. 4, page 686, 1978.
8. Mullin, I.V., "Influence of Fiber Property Variation on Composite Failure Mechanism," *Analysis of Test Methods for High Modulus Fibers and Composites,* ASTM–STP, Vol. 521, pages 349–366, 1973.
9. Zeben, C., "Fracture Mechanics and Composite Materials: A Critical Analysis," *Analysis of Test Methods for High Modulus Fibers and Composites,* ASTM–STP, Vol. 521, pages 65–97, 1973.
10. Freeman, W.T. and Kuebler, G.C. (Hercules, Inc.), "Mechanical and Physical Properties of Advanced Composites," *Composite Materials Testing and Design,* ASTM–STP, Vol. 546, pages 435–456, 1974.
11. *Newsweek,* page 26, August 27, 1979.
12. *Business Week,* page 36D, July 30, 1979.
13. Whitney, J.M., Browning, C.E., and Main, A., "Analysis of Flexural Test for Laminated Composite Materials," *Composite Materials Testing and Design,* ASTM–STP, Vol. 546, pages 30–45, 1974.
14. Chamis, C.C., "Micromechanics Strength Theories," Chapter 3 in *Composite Materials,* Vol. 5, page 141, Broutman, L. (Editor). Academic Press, New York and London, 1974.
15. McGarry, F.J. and Mandell, J.F., *Fracture Toughness of Fiber-reinforced Plastic Laminates,* Chemical Society, Faraday Division, Nottingham, England, September 1972.
16. Halpin, J.C., "Structure–Property Relations and Reliability Concepts," in Hilado, C.J. (Editor), *Carbon Reinforced Epoxy Systems,* page 85. Technomic, Westport, Connecticut, 1974.
17. Konish, H.J. and Cruse, T.A., "A Study of Fracture Phenomena in Fiber Composite Laminates," AFML-TR-73-145, Vol. III, 112 pages, September 1973.
18. Daniel, I.M., "Effects of Material, Geometric, and Loading Parameters on Behavior of Composites," *Thirty-fourth Annual Technical Conference,* SPI, RP/C Institute, Section 20B, February 1979.
19. Daniel, I.M., "Strain and Failure Analysis of Graphite/Epoxy Plates With Cracks," *Experimental Mechanics,* Vol. 18, No. 7, page 246, July 1978.
20. Walter, R.W. and June, R.R., "Designing for Improved Fracture Strength in Advanced Composites." *SAMPE Quarterly,* Vol. 9, No. 3, page 26, April 1978; and *SAMPE Twenty-third National Symposium,* page 877, Anaheim, California, May 1978.
21. Paris, P.C. and Sih, G.C., "Fracture Toughness Testing and Its Applications," *ASTM Symposium,* pages 30–80, STP 381, (sponsored by ASTM and NASA), Chicago, Illinois, June 1964.
22. Lackman, L. *et al., Advanced Composites Design Guide,* Third Edition, Vol. 1, AFML Contract to Rockwell F.33615.71.C.1362, 1973.
23. Kliger, H., "C-fibers–Designer's Tool for High Modulus Parts," *Plastics Design Forum,* May/June 1977.

24. Beaumont, P.W. and Phillips, D.C., "Tensile Strength of Notched Composites," in *Carbon Reinforced Epoxies,* page 5, Hilado, C. (Editor). Technomic, Westport, Connecticut, 1974.
25. Wolff, E.G., Eselun, S.A., and Neubert, H.D., "Microcracking Effects on Dimensional Stability," *SAMPE Twenty-fourth National Symposium,* Vol. 24, page 1299, San Francisco, California, May 1979.
26. Chamis, C.C. and Sinclair, J.H., "Mechanics of Intraply Hybrid Composites–Properties, Analysis, and Design," *Thirty-fourth Annual Technical Conference,* SPI, RP/C Institute, Section 20E, New Orleans, Louisiana, February 1979.
27. Sinclair, J.H. and Chamis, C.C., "Fracture Modes in Off-Axis Fiber Composites," *Thirty-fourth Annual Technical Conference,* SPI, RP/C Institute, Section 22A, New Orleans, Louisiana, February 1979.
28. Chamis, C.C. and Sinclair, J.H., "Effects of Eccentricities on Fracture of Off-axis Composites," *Polymer Engineering Science,* Vol. 19, No. 5, page 337, April 1979.
29. Liber, T.M., Daniel, I.M., and Chamis, C.C., "Static Behavior and the Effects of Thermal Cycling in Hybrid Laminates," *Thirty-second Annual Technical Conference,* SPI, RP/C Institute, Section 15D, February 1977.
30. Henriksen, M. and Thornton, H.R., "Material Damage of Graphite/Epoxy Composites Under Tensile Loading," *SAMPE Quarterly,* Vol. 10, page 15, January 1979.
31. Chang, C., "Toughness of Advanced Composites." *Twelfth Biennial Conference on Carbon,* American Carbon Committee, page 281, Pittsburgh, Pennsylvania, July 1975.
32. Sih, G.C. *et al.,* "Fracture Mechanics for Fibrous Composites," ASTM–STP 521, pages 98–132, 1973.
33. Beaumont, P.W.R., Riewald, P.G., and Zweben, C., "Methods for Improving Impact Resistance of Composite Materials," ASTM–STP, 568, pages 134–158, 1975.
34. Cooper, G.A. and Piggott, M.R., "Cracking and Fracture of Composites," in *Advances in Research on Strength and Fracture Materials,* page 580, Taplin, M.R. (Editor). Pergamon Press, New York, 1977.
35. Hahn, H.T., "Effect of Constituent Properties on Composite Strength," *SAMPE Twenty-fourth National Symposium,* Vol. 24, page 1470, San Francisco, California, May 1979.
36. Oplinger, D.W. and Slepetz, J.M., "Impact Damage Tolerance of Graphite/Epoxy Sandwich Panels," ASTM–STP, Vol. 568, pages 30–48, 1975.
37. Sayers, K.H. and Harris, B., "Interlaminar Shear Strength of a Carbon Fiber Reinforced Composite Material Under Impact Conditions," in *Carbon Reinforced Epoxy Systems,* page 80, Hilado, C.J. (Editor). Technomic, Westport, Connecticut, 1974.
38. Rhodes, M.D., Williams, J.G., and Starnes, J.H., "Low Velocity Impact Damage in Graphite Fiber Reinforced Epoxy Laminates," *Thirty-fourth Annual Technical Conference,* SPI, RP/C Institute, Section 20D, New Orleans, Louisiana, February 1979.
39. Husman, G.E., Whitney, J.M., and Halpin, J.C., "Residual Strength Characterization of Laminated Composites Subjected to Impact Loading," ASTM–STP, Vol. 568, pages 92–113, 1975.
40. Segal, C. and Boyle, G., "Graphite Fiber Composites," *SPE ANTEC,* March 4, 1969 (CA71-4062K-1968).
41. Bascom, W.D. and Bitner, J.L., "Interlaminar Failure of Organic Matrix Composites," *SAMPE Twenty-fourth National Symposium,* Vol. 24, page 1592, San Francisco, California, May 1979.

42. Allred, R.E., Street, H.K., and Martinez, R.J., "Improvement of Transverse Composite Strength," *SAMPE Twenty-fourth National Symposium,* Vol. 24, page 31, San Francisco, California, May 1979.
43. Hunston, D. (Naval Research Laboratory), "Fracture Toughness of Composite Matrix Resins and Structural Adhesives," *SAMPE Workshop,* Palo Alto, California, 1979.
44. Camahort, J.L., Buckley, K.W., Slavin, F.E., and Fahrni, R.E., "High Modulus Graphite/Epoxy Tubes," *SAMPE Twenty-second National Symposium,* No. 9, page 358, 1977, Sunnyvale, California.
45. Figge, I., Henshaw, J., and Roy, P.A., "Low Weight Drive Shafts–Impact Resistance," STP 546, *ASTM on Composite Materials,* page 651, 1974.
46. Hyer, M.W. and Hagaman, J.A., "Effects of Thermal Cycling on Thermal Deformation of Graphite Polyimide," SESA Paper R79-114, San Francisco, California, May 1979.
47. Hercules, Inc., *"Magnamite" C-fiber Bulletin,* 1978.
48. Davis, J.W., McCarthy, J.A., and Schurl, J.N. (3M Company), "Fatigue Resistance of Reinforced Plastics," December 1964.
49. Santelli, M.L. and Simon, R.A., "Fatigue Testing of Graphite Fiber Composites," *Twenty-sixth Annual Technical Conference,* SPI, RP/C Institute, Section 10C, Washington, D.C., 1971.
50. Benedetto, A.T., "Fatigue and Crack Propagation in Graphite Fiber Reinforced Nylon," *Thirty-fourth Annual Technology Conference, SPE,* page 103, April 1976.
51. Benedetto, A.T., "Fatigue Crack Growth in Crystalline Polymers," *Journal of Material Science,* Vol. 14, No. 10, page 2482, October 1979.
52. Owen, M.J., "Fatigue of Carbon Fiber Reinforced Plastics." Chapter 8, in *Fracture and Fatigue,* pages 341–369, Broutman, L.J. (Editor), Academic Press, New York and London, 1974.
53. Stinchomb, W.W., Reifsnider, K.L., Marcus, L.A., and Williams, R.S., "Effects of Frequency on the Mechanical Response of Two Composite Materials to Fatigue Loads," *Fatigue of Composite Materials,* ASTM–STP, 569, pages 115–129, 1975.
54. Sumsion, H.T. and Williams, D.P., "Effects of Environment on the Fatigue of Graphite-epoxy Composites," *Fatigue of Composite Materials,* ASTM–STP, 569, pages 226–247, 1975.
55. Ryder, J.T. and Walker, E.K., "Effect of Compression on Fatigue Properties of a Quasi-isotropic Graphite/Epoxy Composite," *Fatigue of Filamentary Composite Materials,* ASTM–STP 636, pages 3–26, 1977.
56. Ramani, S.V. and Williams, D.P., "Notched and Unnotched Fatigue Behavior of Angle-ply Graphite/Epoxy Composites," NASA-TMX-73, 191, December 1976.
57. Porter, T.R., "Evaluation of Flawed Composite Structure Under Static and Cyclic Loading," *Fatigue of Filamentary Composite Materials,* ASTM–STP, 636, pages 152–170, 1977.
58. Kunz, S.C. and Beaumont, P.W.R., "Microcracking Growth in Graphite Fiber-epoxy Resin Systems During Compression Fatigue," *Fatigue of Filamentary Composite Materials,* ASTM–STP, 569, pages 71–91, 1975.
59. Ewins, P.D. and Ham, A.C., "Nature of Compressive Failure in Unidirectional Carbon Fiber Reinforced Plastics," *AIAA/ASME/SAE–Fifteenth Structures, Structural Dynamics and Materials Conference,* Paper 74-353, Las Vegas, Nevada, April 1974.

60. Greszczuk, L.B., "Compressive Strength and Failure Modes of Unidirectional Composites," ASTM–STP, Vol. 521, pages 192–217, 1973.
61. Loughlin, J.R.M., "Low Temperature Properties of High Performance Composites," *Twentieth Annual Technical Conference,* SPI, RP/C Institute, Section 18D, Washington, D.C., February 1975.
62. Kasen, M.B., Schramm, R.E., and Read, D.T., "Fatigue of Composite at Cryogenic Temperatures," ASTM–STP, 636, pages 141–151, 1977.
63. Hertz, J., Christian, J.C., and Varias, M., "Advanced Composite Applications for Spacecraft and Missles," AFML-TR-71-186, page 282, March 1972.
64. Felbeck, D.H., "U.S./RD" Publication, page 3952, Washington, D.C., April 1978.

8
Electrical Properties and Electrical Applications of Carbon/Graphite Fibers

The electrical conductivity of C/G fibers has been known for a long time, and no better examples can be found than Thomas Edison's early electric light filaments, as described in Chapter 1. The availability of improved grades and consistency in C/G fibers has rekindled interest in electrical applications, though not to the same extent as in structural uses. The subject is complex due to the many-faceted nature of the products that are obtained during processing and heat treatment of carbon fibers. The soft or amorphous carbons, which are capable of graphitization, and the hard or vitreous carbons, which are non-graphitizable, represent two distinct forms with differences in electrical properties. Our interest lies in the fibrous products, available as carbonized materials from organic precursors, which may also be heat treated to the layered hexagonal crystal structure of graphite, with its improved electrical conductivity over amorphous carbon. As *electrical conductors,* C/G fibers have been of interest to many industries, as indicated in the following paragraphs.

The passage of electrical current through C/G fibers has been used as a source of heat.[1, 2] Electrically conductive graphite fibers are made available as yarns, braids, felts, and woven or knit fabrics. Because of their flexibility, as compared to fine electrically resistive metallic conductors, they were designed into heating pads, panels, clothing, etc. It was also convenient to laminate conductive C/G fibrous products into non-metallic tooling, to achieve a heat source by passing electrical current through the C/G fibers within the tool, to assist in the cure of resin impregnated adhesive and molding compounds. These applications were well conceived, though, at their early introduction, they suffered because of the inconsistencies of the early C/G fibrous products; the difficulties of establishing reliable and consistent electrical contact with the fine fibers comprising the filaments; and the denigrating effect on the fibers of high temperature service as a heat source.

Electrical Conductivity

The electrically conductive nature of C/G fiber products has prompted their consideration as electromagnetic shielding from high frequency electrical sources, much in the same manner as copper or aluminum screening for this purpose. Lightweight C/G fibers may be integrated into a non-conductive molding or laminate in an inconspicuous manner to form a composite with potential "built-in" shielding qualities. A requirement in the design of electrical or electronic equipment is to protect the equipment from the effects of interference due to other electronic devices and to prevent the equipment from propagating interference. A measure of the ability of a material to control the passage of radiated electromagnetic energy is its shielding effectiveness. Carri[3] reported on the electrical conductivity and shielding effectiveness of graphite/epoxy laminates, both protected and unprotected with conductive aluminum overcoats. Figure 8-1, reproduced from this report, compares the variation of magnetic shielding effectiveness with frequency for a 2 mil aluminum panel and a 12 and 24 ply of unprotected graphite/epoxy. The panels were fabricated from Hercules AS/3501-5A, graphite/epoxy prepreg. It was noted that there are about three orders of magnitude difference between the epoxy/graphite composites (higher electrical resistivity) and aluminum metal, and hence the C/G fiber panels were not as effective as the aluminum. Sprayed-on aluminum coatings on C/G composites were also examined to improve shielding. Aluminum coated glass fibers are also available from Hexcel Corp. for shielding. The objectives of Carri's tests (Grumman Aerospace) were to examine the vulnerability of aircraft electrical systems to lightning and nuclear electromagnetic pulses. Induced cable currents and voltages from a lightning strike could upset aircraft electrical systems. Based on results of lightning strike model attachment tests, there appears to be no difference between a conventional metal and graphite/epoxy aircraft with respect to lightning attachment points.

Studies of the electrical conductivity of C/G monofilaments have been made to examine the influence of different processing measures and heat treatments upon the fibers. Relationships have been established between the modulus of elasticity of the fiber and the electrical resistivity.[4] McMahon[5] has also reported on a curve (Fig. 8-2) establishing this relationship. At Delsen Testing Laboratory, studies were conducted on the feasibility of using C/G fibers as strain gages, with the objective of introducing them into glass fabric laminates as a means of measuring strain. Electrical resistivity (ohms) can be obtained upon a unit length of tow or yarn. At elongations of 0.66%, the electrical resistance change was

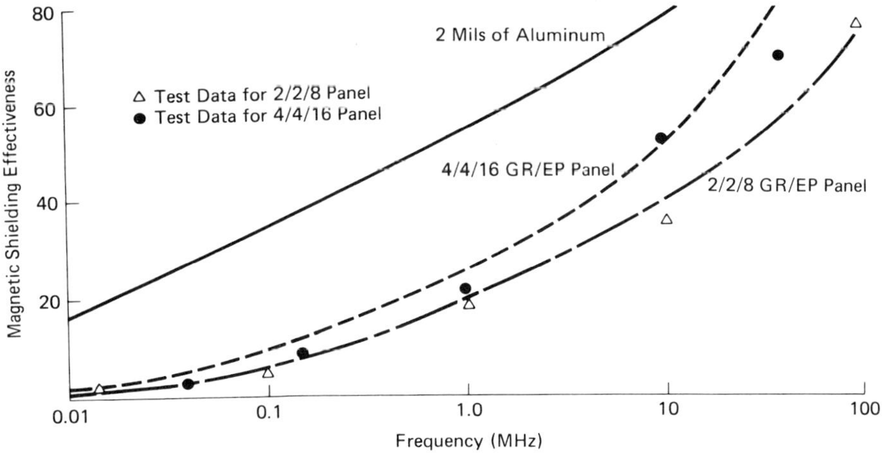

Fig. 8-1. Magnetic shielding effectiveness for 2/2/8 and 4/4/16, unprotected graphite/epoxy panel, and a 2 mil aluminum panel.[3]

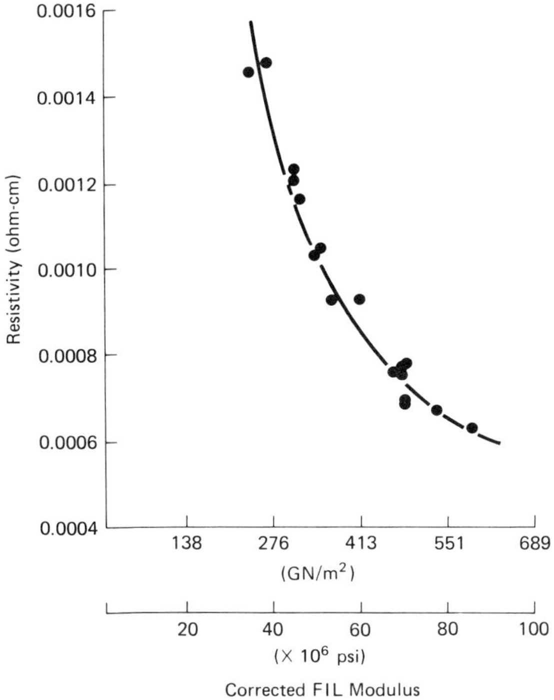

Fig. 8-2. Electrical resistivity as a function of modulus for graphite bundles originating from PAN precursor.[5]

approximately 10%. However, returning to zero elongation showed a permanent electrical resistance increase, reflecting fiber breakage within the tow. Examination of tows from various manufacturers and lot to lot variations pointed up inconsistencies in conductivity, due to a number of processing variables for the fibers.

The subject of the effect of lightning strikes on advanced graphite composites has prompted considerable study by aircraft companies. The data obtained and their relationship to C/G fiber conductivity are handled in a separate section of this chapter.

A novel approach to the preparation of superconductors has been reported by Woollam[6] and by Lin *et al.*[7] Lin *et al.* reacted niobium pentachloride with carbon fiber in the presence of hydrogen and nitrogen at 1873°K(1600°C). A film of niobium carbonitride about 1 micrometer thick was formed on the 7 micrometer carbon filament. The carbon filament gave the strength and support required by the niobium carbonitride. The fibers came in a yarn containing 720 filaments, and the result was a multifilamentary superconductor with superconducting transition temperatures in zero field of about 16°K; critical fields at 4.2 K of 11.0 telsa (110,000 gauss) and critical current densities over 10^6 amp/cm^2. Design problems of the superconductors are aided by the reinforcing presence of very fine filaments, as available from C/G as supportive carriers.

Woollam also reported the high degree of anisotropy in the structure of single crystal graphite is evident in its electrical and thermal properties. For example, the ratio of electrical conductivity parallel to the layers, to the conductivity perpendicular to the layers, is highest at low temperatures, and can be as high as 10^5 at cryogenic temperatures.

APPLICATIONS

There are noteworthy examples of the utilization of carbon and graphite fibers in product and design applications where emphasis is placed upon the electrical properties of the material. In examining some of these applications, the intrinsic qualities of the fibers are utilized just as they are provided by the manufacturers. This is in contrast to structural applications, where research and development have defined more sophisticated requirements.

Tooling

The role of carbon fibers as a heating source in non-metallic tooling has been updated by Bushman.[8] The concept has been used in the past in preparing non-metallic tools for aircraft fabrication, where production quantities are limited and the rapid availability of non-metallic tooling was important. Among the tools which benefit from integral heating elements are bonding fixtures, where components are kept close together under pressure as they are being bonded with a heat-set adhesive; tools for radiating sufficient heat to accelerate the drying and curing of coatings; tools for thermal forming of thermoplastic sheet stock; and tools for the curing of thermosetting matrix prepreg (either of glass or of C/G fiber) into finished components. Bushman presents an up-to-date bibliography on the heating of non-metallic tooling and tooling applications, principally in the aerospace industry. Resistance wires, such as "nichrome," or heating cartridges have been inserted or used in conjunction with metallic tooling and, in a few instances, with plastic tooling. High temperatures on the surface of the metallic wires have decomposed most organic plastics. In this connection, it has been hoped that the use of C/G fibers as conductors of electricity and as sources of heat may dissipate heat more rapidly to their immediate surroundings, particularly if contiguous to glass fibers. Such bonded and laminated structures, with electric conducting and heat dissipating layers of C/G fibers near the operating surfaces, have been the basis of such tooling. Electrical connection to electrically conductive fabrics has not been easy, and the usual measures of mechanical clamping between metal electrodes, flame spraying of conductive metallic electrodes of copper or aluminum onto exposed ends of C/G fibers, or the use of silver filled electrically conductive epoxy adhesives or coatings have been tried with varying degrees of success. Because the electrical resistance has been somewhat variable, such installations have been equipped with a source of variable voltage to ensure a uniform current flow and heat dissipation. Insulative layers of impregnated glass fabrics are important as shields to avoid contact by individuals with exposed C/G fibers conducting electricity. It should also be noted that there are other compelling reasons for use of C/G fibers in non-metallic tooling aside from electrical conductivity. Among these are the low thermal expansion coefficients of C/G fiber laminated structures.

Lightning Conductive Characteristics of C/G Composites

The use of C/G fiber composites in aerospace structures has prompted studies of the effects of lightning strikes and the influence of the electrically conductive properties of carbon and graphite fibers. It has been felt that the optimum lightning protection design for graphite composite aircraft structures is to fully utilize their electrical conductive properties.[9] The relatively low thermal conductivity of the graphite composites (as compared to aluminum) permitted the use of an infra-red scanning process to record lightning current flow patterns. It was observed that fastener hold areas of graphite composites have the highest concentration of lightning current density and are the most vulnerable. The average peak current for unit cross-section area of the graphite fiber composite panels was 64 kiloamps/cm^2. Among the design objectives were considerations of a separate lightning conductive joint to avoid alteration of structural integrity. From the viewpoint of antenna radiation effects, Kung and Amason[9] noted that, at frequencies above 100 MHz, the graphite fiber composite ground plane is similar to an aluminum metal ground plane. In further work at McDonnell Douglas, macroscopic electrical response of aluminum and graphite/epoxy panels subject to a lightning environment, was investigated by Skouby[10] for different potentials of lightning strikes on attachments. Despite property differences between the metal and the graphite/epoxy composites, there were no significant functional differences observed in response to the applied lightning potential.

All present protection systems and conductive materials on aircraft require local repair after major lightning strikes. High current testing has shown that a reasonable cross-sectional area of graphite/epoxy structure can carry lightning current without damage, but the development of electrically conductive joints that will resist damage is desirable. Any arcing or streamering in the vicinity of fuel vent outlets must be avoided.[11]

A graphite/epoxy design of an advanced fiber composite aileron for the L1011 aircraft was evaluated under an idealized lightning waveform for swept stroke tests (see Figs. 8-3 and 8-4) performed upon 12 inch by 15 inch panels. The decision was made to include lightning strike protection on the advanced composite aileron. The composite aileron showed 28% weight savings over the aluminum counterpart.[12]

Lightning, radio, radar, and man-made noise are among the energy sources that can couple into an aircraft. This energy, in turn, can induce currents and voltages potentially harmful to the aircraft's electrical and electronic systems. Fig. 8-5, from Force's paper,[13] demonstrates the shielding effectiveness of graphite/epoxy laminates as a function of frequency and thickness. For comparison, a 0.04 inch aluminum would offer considerably better shielding.

Coupling of Graphite/Epoxy Composites With Metals

Galvanic corrosion rates of selected aerospace alloys coupled to graphite/epoxy composites have been studied in aerated neutral (pH=7) 3.5% NaCl solutions.[14] These data are important to the assemblies of graphite/epoxy composites to aluminum alloys. Typical data appear in Figs. 8-6 and 8-7, which show the galvanic current (I_g) versus time for three different alloys used in aircraft, when they are coupled to graphite/epoxy composites, as per the schematic diagram showing the measurement of galvanic current. The authors examined 23 alloys coupled to the graphite/epoxy composites, and grouped results into three classes.

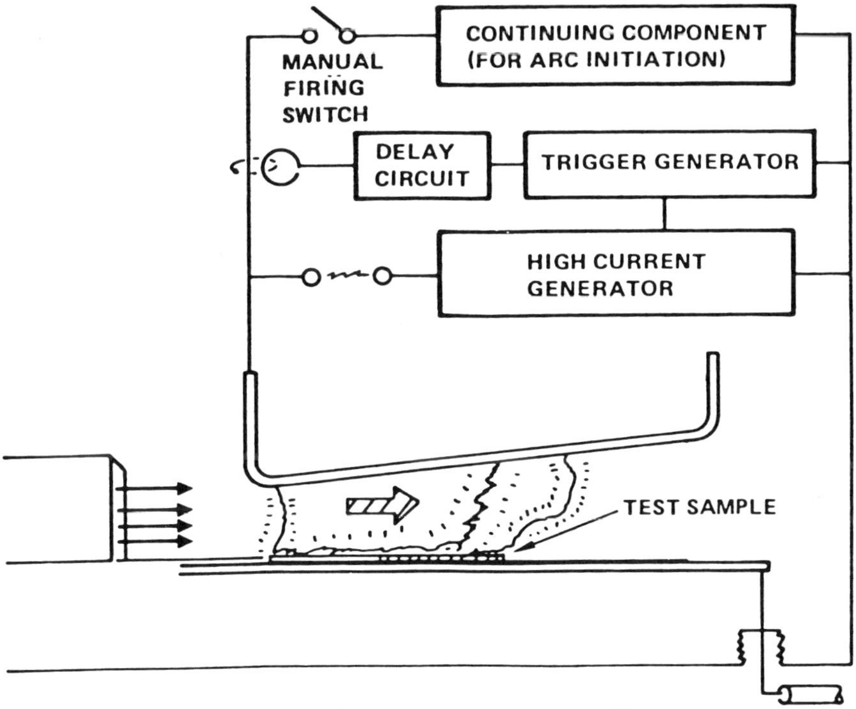

Fig. 8-3. Swept stroke test arrangement.[12]

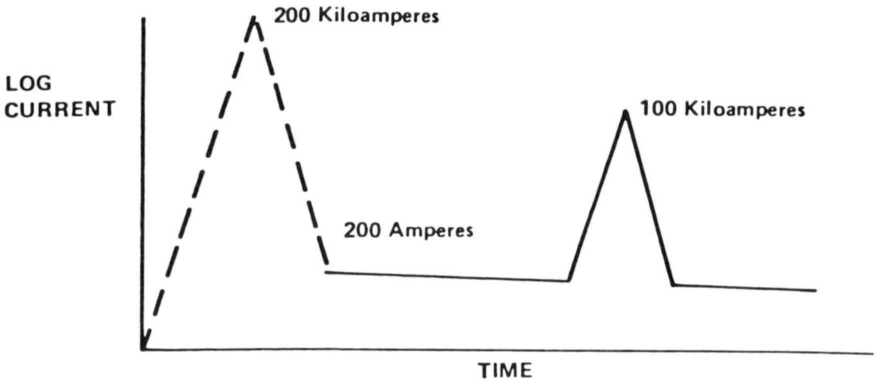

Fig. 8-4. Idealized lightning waveform for swept stroke tests.[12]

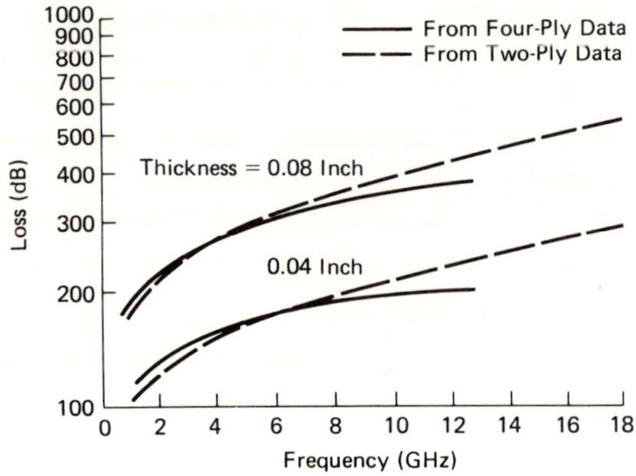

Fig. 8-5. Calculated graphite epoxy plane wave shielding, symmetric cross-ply laminates.[13]

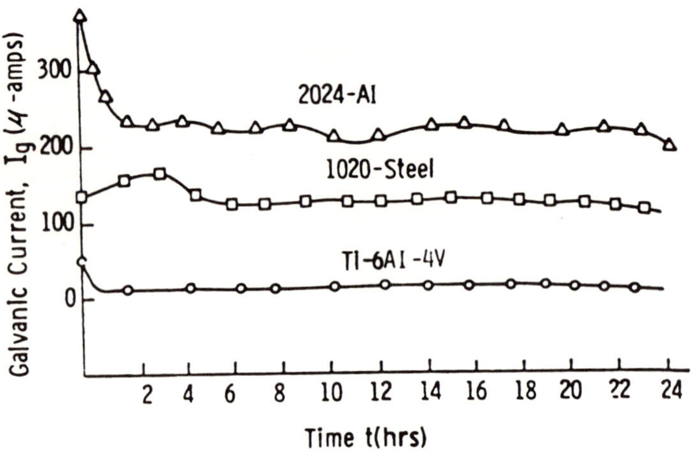

Fig. 8-6. Typical time versus I_g curves for the indicated alloys versus GECM.[14]

This method of comparing galvanic current was preferred over the electric potential differences.

Acceptable—less than 5 microamps/cm^2 (stainless steel and titanium alloys).

Borderline—5 to 15 microamps/cm^2 (aluminum alloys and some steels).

Unacceptable—over 15 microamps/cm^2 (steels: AF1410, 300M, and 4340; and aluminum alloys: 2020-T651).

Graphite behaves electrochemically like a noble metal such as gold or platinum, and exhibits a noble potential in most electrolytes, suggesting that it will

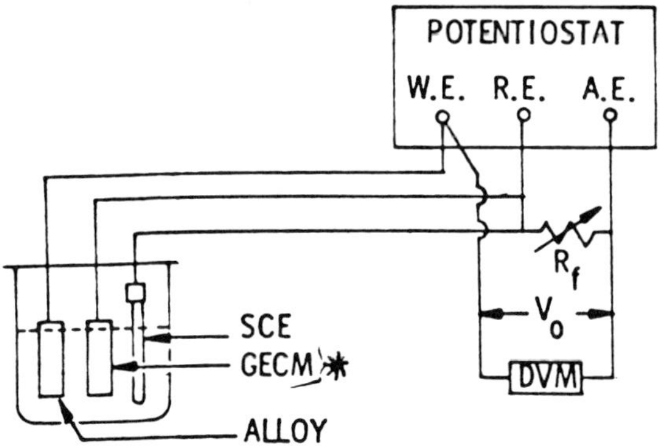

Fig. 8-7. Schematic of potentiostat ZRA used in measuring galvanic current.[14]

be cathodic to most metals upon galvanic coupling. Galvanic corrosion of aluminum alloys used on aircraft have been studied for some time, and this report draws attention to potential problems when assembled to graphite/epoxy composites. It is recommended that best protection on the aluminum alloys is achieved through the use of zinc chromate and strontium chromate epoxy primers, followed by an aliphatic polyurethane topcoat. Adhesively bonded, as compared to mechanically fastened (bolted) assemblies, they did not exhibit corrosion in either long-term or accelerated tests, even for aluminum or cadmium plated steel bonded to graphite composites.[15]

In the chapter on aerospace applications, the fabricated and proposed graphite/epoxy composite materials discussed for aerospace structures are impressive, and designers are reminded to take adequate precautions in uniting these structures to aluminum alloy members.

Antennas from Graphite/Epoxy Composites

Antennas for aerospace vehicles have received much attention in recent years, not only because of the substantial weight savings over their metal framework counterparts, but for even more compelling reasons (such as their dimensional stability under extremes of temperature). Several programs have been initiated in recent years, and some of the experiences gained from this work are reviewed here.

The reflector support truss for NASA's Application Technology Satellite, Models F and G, was required to maintain precise positioning of the reflector dish with respect to the earth moving module. The unidirectional thermal coefficient of expansion of graphite/epoxy composites (0.05 to -0.15×10^{-6} inches/inch/°F) minimized truss thermal growth or shrinkage, offering accurate positioning of the reflector with the expected temperature range of -265° to +200°F. Furthermore, the high unidirectional modulus/density ratio of the graphite/epoxy (450×10^6 inches versus 100×10^6 inches for aluminum) made possible significant weight savings. The graphite/epoxy prototype weighed 80 lb, one-half of the weight of an aluminum truss. The graphite/epoxy truss tubes were 170 inches long with a 2.50 inch I.D., and an 0.100 inch wall thickness. They were fabricated from 12 plies of prepreg of Hercules 2002M graphite/epoxy, according to the authors, Randolph *et al.*[16]

Graphite fiber ("Thornel" 75S) and ERLB-4617 (cycloaliphatic epoxies) advanced composites have undergone extensive evaluation and environmental tests in their application to millimeter wavelength antennas.[17] Specimens were subjected to vacuum exposure (10^{-6} torr), temperature cycling, ultraviolet light exposure, and radiation exposure. The 30 inch test model was a paraboloid of sandwich construction with 6 ply skins of "Thornel" 75 S/ERLB-4617 (0° ± 60°) bonded to an aluminum honeycomb core with FM-1000 nylon/epoxy film adhesive. After each temperature and humidity exposure, the model was visually inspected and contours were checked and found satisfactory. An overall film of graphite ("Grafoil" U.C.) was used on surface of the reflector. The overall thermal expansion coefficient of the graphite/epoxy panel remained substantially constant before and after exposure to 1800 megarads of gamma radiation (1 rad = 100 ergs/g of material of energy absorbed) and ultraviolet radiation equivalent to 986 hours of sun space exposure. Excellent reflective characteristics with "Grafoil" 0.004 inch surface or 0.008 inch aluminum were reported for the antenna at 60 and 94 GHz, both before and after radiation, with the focal wavelength variation plus or minus 0.014 inch, well within permitted tolerances. Penetration versus energy were given in this report for the graphite/epoxy panel (see Table 8-1).

Advanced concepts for large (12 ft diameter and larger), furlable space antennas have led to configurations utilizing conical main reflectors to simplify

Table 8-1. Penetration versus Energy.[17]

PARTICLE	ENERGY (KEV)	PENETRATION INTO GRAPHITE/ EPOXY COMPOSITE (mils)
Electron	1	Very slight
	100	3.6
	1000	108.0
Proton	100	0.04
	1000	0.70
Alpha	1,000	0.20
	10,000	3.3

unfurling in space.[18] High modulus graphite fibers, such as GY-70, "Thornel" 75, and Hercules HMS, gave superior performance when used in composites for the conical configuration. Important criteria were high thermal stability, high radio frequency reflectivity, and low thermal expansion. Radio frequency reflectivity losses in the resin binder above 16 GHz necessitated a metal reflector on the surface. The thermal expansion coefficient of the composite is very sensitive to the material constituents and processing parameters. The graphite/epoxy composite used on the magnetometer booms of the JPL "Mariner" spacecraft to Venus and Mercury were composed of Hercules 2002 epoxy and HMS graphite fiber. The thermal expansion coefficient was 0.3×10^{-6} inches/inch/°F.

A graphite/epoxy composite high gain antenna was developed for the "Viking" Orbiter 1975 spacecraft.[19] The low coefficient of thermal expansion of the graphite composite resulted in a dimensionally stable antenna structure for the thermal environment in the Mars mission (see the chapter on environmental exposure). A thin skin graphite composite sandwich structure was chosen because it offers many advantages over a conventional aluminum design. A cross-section of the high gain antenna used in the program is shown in Fig. 8-8. Not only stiffness requirements were met, but also the maximum allowable weight of

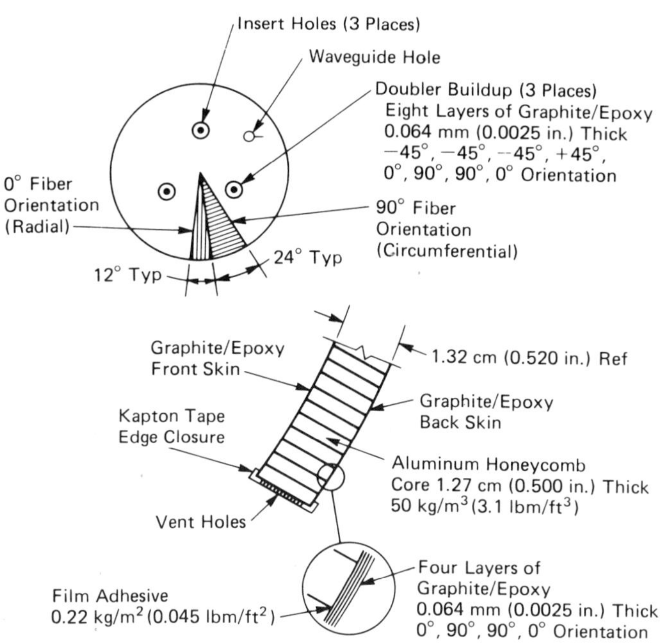

Fig. 8-8. HGA reflector cross-section.[19]

11.0 lb was not exceeded. The HGA reflector was basically a 58 inch diameter parabolic dish. Graphite/epoxy prepregs approved for the program were Ferro's C-5010 and Fiberite HyE 1330A.

Lightweight spacecraft antennas have been constructed by Hughes Aircraft Corporation for the Canadian government, utilizing high modulus graphite fibers in a matrix of a high temperature epoxy. The antenna reflector was designed as an offset parabolic shape with an aluminum honeycomb core and graphite fiber reinforced epoxy resin face sheets—a construction used in an earlier communications satellite. The lightweight, low thermal expansion composites combine strength and stiffness to form cost effective, thermally stable antennas for space vehicles.[20]

Haydostian *et al.*[20] reduced antenna weight significantly by a network of radial ribs and interconnected circular arches to support the gold plated metal reflective wire mesh. The open structure effectively reduced the "sail" effect of the 60 inch diameter parabolas, minimizing the effect of solar pressures. Spacing was analytically determined to minimize geometric surface distortion, while maintaining a maximum transparency to solar radiation pressure. The structures have withstood the maximum loads of the launch environment. In actual performance, the reflectors' dimensions yielded less than 0.1 db degradation from a perfect reflector. Typical structural configuration is shown in Fig. 8-9. Courtauld's HMS graphite fiber and E-42--1 G epoxy resin (Modified Bis-A-Novolac of U.S. Polymeric) system were selected for construction, according to the authors.

The application of graphite fiber reinforced metal composites has been investigated for large, deployable space antennas.[21] Ultra-high modulus (100 msi) pitch fibers from Union Carbide Corp. (VS0054) in aluminum and in magnesium matrices appeared to have the best properties in this study. It is anticipated that the graphite fiber volume in magnesium and aluminum will be 48%. It is expected that these spaceborn antennas will permit increased operational frequencies at a given diameter or allow larger antenna apertures for a particular radio frequency.

Applications to Radiological Equipment

In reviewing C/G fiber composites in X-ray equipment, Molyneux and Lyons[22] discussed the advantages of these materials. Attention was called to the transparency of C/G fibers to X-rays, making it possible to use carbon fibers to carry small electrical currents and monitor patients' vital signs during some X-ray diagnosis. Hence, X-ray photographs can be taken with the absence of wire definition in the photographs.

Important advantages were realized in the manufacture of therapeutic tables of C/G fiber composites, which not only possessed the necessary transparency to X-ray, but were also lightweight, accompanied by stiffness and strengths

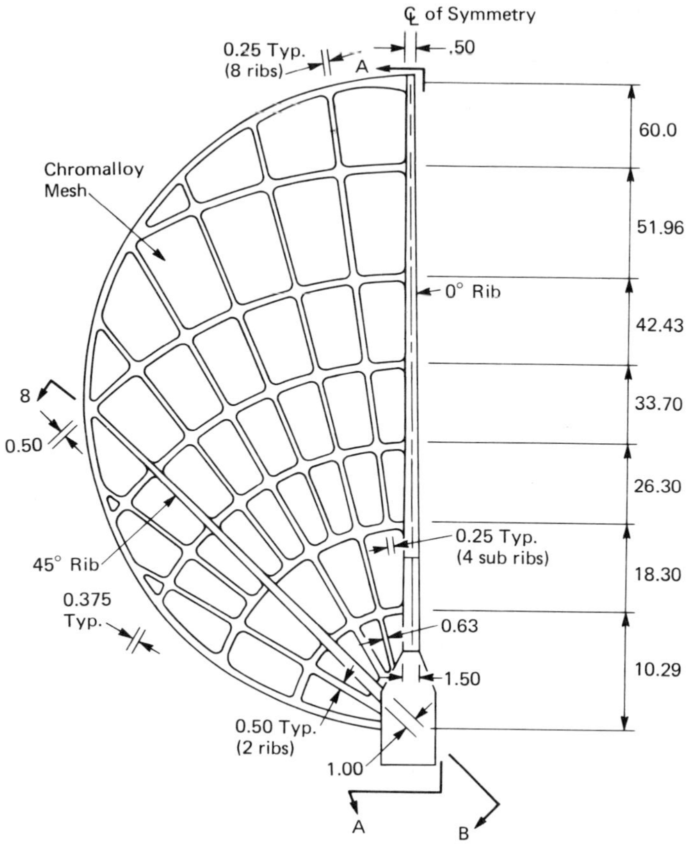

Fig. 8-9. Structural configuration graphite reflector.[20]

superior to aluminum and steel frames and tables. In some X-ray diagnosis, patients have to be moved and rotated on top of metal tables—an uncomfortable procedure during internal examinations. With a support couch of C/G reinforced materials, X-ray scanning may be accomplished by rotating the equipment about the patient lying on the table (see Fig. 8-10). Computerized tomography, based upon the fact that different anatomical materials absorb X-rays to different degrees, requires many X-rays as the patient or the equipment is rotated. A low and consistent absorption factor must be present not to interfere with the X-ray diagnostic beams. The C/G reinforced couches appear capable of fulfilling these requirements. The authors of this paper compared the X-ray transparency of several plastics to the X-rays used for diagnostic purposes (see Table 8-2). The data are given in millimeters of aluminum equivalents. Also, it will be recognized that carbon fiber reinforced plastics not only have the best transparency (least aluminum metal equivalents), but, as already noted for structural applications, the high modulus and low density of these materials make them ideal structural units for body scanner support couches.

Table 8-2. Comparison of X-Ray Transparency.[22]

	MM of Aluminum Equivalent	
	At 70 kv	At 100 kv
10 mm Radiographic phenolic resin	1.34	1.52
10 mm Acrylic	1.05	1.23
5 mm Radiographic phenolic resin	0.56	0.45
5 mm Acrylic	0.56	0.65
4 mm Polycarbonate	0.54	0.47
3-5 mm Carbon Fiber/reinforced plastic	0.40	–

Conductive Molding Compounds

It has been known for some time that the addition of finely divided carbon black (such as acetylene blacks) will improve the electrical conductivity of synthetic resins and rubbers in sheets and molding compounds. It has been suggested (and reduced to practice) that lesser amounts of carbon or graphite fibers may be used to achieve the same amount of conductivity. This is brought out in the Litant patent.[23] Chopped C/G fibers are dispersed into a synthetic resin using 0.05 to about 5% of randomly dispersed fibers which are 1/4 to 3/4 inch in length. The fine C/G fibers reduce electrical conductivity by establishing contacting relationships to provide paths for electrical current flow. In fact, Litant, as shown in Fig. 8-11, has a preference for the high conductivity graphite fibers produced at high temperatures of pyrolysis, which show better electrical conductivity as they approach graphitization temperatures. He also uses a polar plasticizer, orthonitroanisole, in preparing the semiconductive molding compounds.

Silicon Sheets for Solar Cells

As a randomly selected example of original thinking, attention is called to the experimental production of solar cell arrays, through the use of a grid-like sheet of graphite dipped into molten silicon. The grid could well be made of graphite woven fibers with large openings not more than 2 mm in diameter. As the graphite is withdrawn from the melt, silicon clings to it and solidifies as thin sheets in the holes. It was reasoned that the cost of producing solar cells by operating a continuous mode, using gauze-like graphite, was feasible.[24]

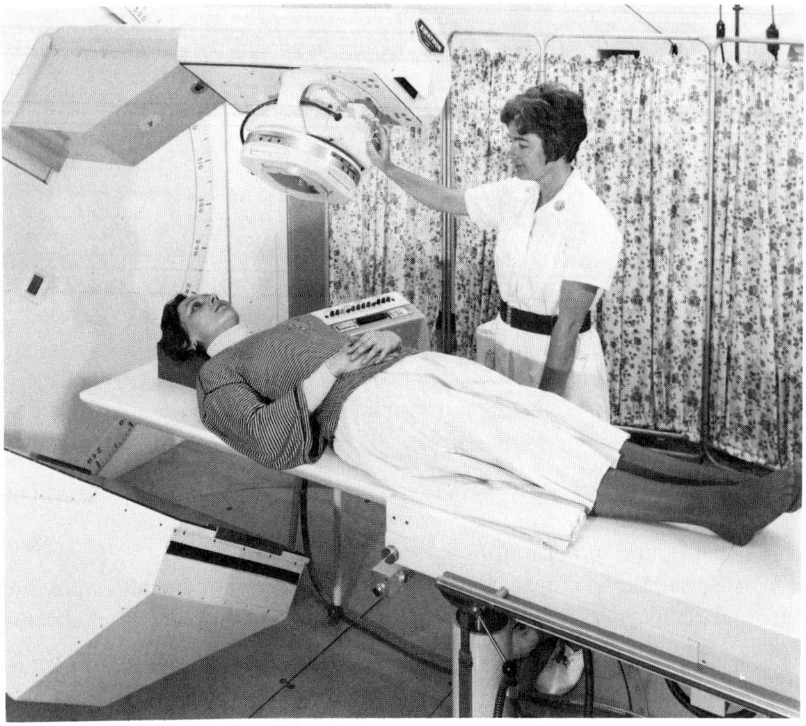

Fig. 8-10. Cantilevered carbon fiber table top molded by composites division of Fothergill & Harvey, Ltd., for Newcastle (England) General Hospital. (*Courtesy Fothergill & Harvey, Ltd.*)

In these and many other ways, attention is being focused upon the unique structural and electrically conductive properties of C/G fibers. It should also be noted, from one of the examples of deployable space antennas, that matrices for C/G fibers are not limited to organic materials, and that exciting opportunities are arising for structures involving fiber reinforced aluminum and magnesium.

Electric Storage Battery

In support of the current work on all-electric automobiles is the development effort on larger capacity electric storage batteries. The traditional lead-acid battery is more economical than more recent batteries, such as the zinc-nickel (which use materials in shorter supply). A recent development at Cal Tech in Pasadena involves the use of lightweight sheets consisting of graphite fibers molded into thin sheets of polyethylene used to conduct electricity from one

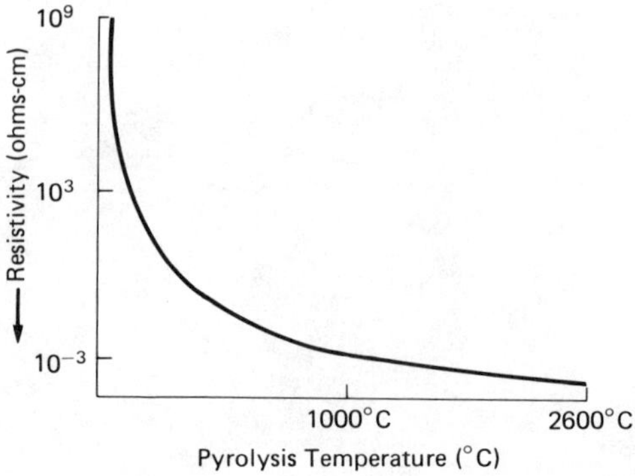

Fig. 8-11. Pyrolysis temperature versus resistivity.[23]

cell to another. The increased surface area leads to improved electrochemical performance of energy-storing lead materials. Experimenters claim four times the power per pound of conventional lead-acid batteries.[25]

The grids used in the conventional lead-acid storage batteries add considerable weight and restrict the battery's power performance, and the alternative approach, the "bipolar construction" does away with grids and uses thin conductive sheets (biplates) to conduct electricity from one cell to the next. Less overheating of the battery is reported in this construction.[25]

A further improvement in the performance of the lead and electrical storage battery was proposed by Clark *et al.*[26] They pointed out that graphite fiber reinforced lead plates for storage batteries would offer improved creep strength over regular lead. This would permit a reduction of plate thickness, an increase in the number of plates in a specific area, and improved battery efficiency.

A program has been initiated to achieve significant weight savings when applied to nickel-cadmium cells and batteries, resulting in a graphite fiber/epoxy cell case.[27] The objective was to manufacture a case resistant to the potassium hydroxide (KOH) electrolyte, resistant to gaseous diffusion through the case material, capable of withstanding internal pressures of 300 psi, and weighing significantly less than the current stainless steel variety. These cells have a power density of 12 to 15 watt-hours/lb. Construction evolved about Union Carbide Corp.'s T-300 graphite fiber used in a 2.5 mil prepreg tape with an epoxy matrix by Ferro Corp. (CE-9015). Processing of the cell cases was acceptable when the resin content was held at 43 to 47%. Porosity was manifested as a weeping of test fluid (water) through the cell walls during burst tests. The application of a 4 mil thick "Tedlar" (polyvinyl fluoride) film to the internal surfaces provided a satisfactory solution. Eight plies of graphite fiber/epoxy prepreg were laid up in contact over the polyvinyl fluoride film. Battery cell covers were also molded of one inch chopped graphite fiber reinforced epoxy compound. Data on 90 fabricated cell cases were presented. It was indicated that approx. 30% weight and cost savings could be achieved.

CARBON FIBER DISPOSAL

The behavior of atmospherically distributed, finely divided carbon or graphite fibers has aroused concern, particularly if such fibers settle upon electronic devices or semiconductors which have become integral parts of business and control equipment. As discussed in earlier chapters on resin matrices, some of the less oxidatively stable resin matrices may evaporate at high operating temperatures and release carbon fibers to the atmosphere or, in the event of an external, devastating fire, the destruction of structural C/G laminates would contribute to the release of carbon fibers to the surroundings. There are those who feel that this concern is of little consequence, though programs are underway to accommodate such exigencies, including disposal of C/G fiber waste products.

Waste disposal practices for carbon fibers were examined by McFerrin and Trulson,[28] who cautioned against the use of conventional incinerators and recommended disposal in certified landfills, considering their inert, non-leachable nature. Disposal of consumer goods which use carbon fibers is also being studied by EPA. While at the time of this study, large quantities of carbon fibers have not been used industrially, the strong trend in this direction has prompted studies to eliminate problems with airborne carbon fibers in the vicinity of electronic equipment. NASA data indicate that concentration levels in the range of 10^4 to 10^7 fibers/m^3 are where interference commences.[29]

Studies of carbon fibers release from Advanced Composites by Gilwee and Fish,[30] while evaluating fire resistant resins for aircraft applications, indicate that the higher the char yield, the more resistant the resin matrices are to ignition and burning—and less fibers are released into the atmosphere. Charred specimens were impacted in a special chamber and the percent of fiber weight loss was observed. Over a broad spectrum of char temperatures (427° to 750°C), polystyryl pyridine resins (PSP), followed by phenolic resins, had the highest yield of resin char and the least carbon fiber release. Woven carbon fabric composites exhibited less carbon fiber release after char than "quasi-isotropic" composites made from unidirectional tape.

Higher char yield resin matrices for advanced composites have been identified as reducing the carbon fiber release problem. Using bis(imide-amide) cure agents (produced from 2 to 1 adducts of the diamine/dianhydride reaction), Serafini *et al.*[31] cured the tetraglycidyl ethers of methylene dianiline to obtain a series of epoxy resins which demonstrated substantial improvements in char yield. A

200 to 300% gain in char yield was claimed under aerobic conditions, and a 100% improvement in anaerobic char yield. Bowles recognized the sensitivity of C-fiber release to fiber orientation.[32] He also noted the value of small additions of boron powder which stabilized the char and prevented fiber release from the burned material.

The electrical properties of C/G fiber composites differ significantly from the glass, boron, and aramid fiber composites. They are much more electrically conductive than the others, though not to the same degree as metal structures. The electrical conductive qualities have suggested applications related to electromagnetic shielding, heating, galvanic corrosion, and the effect of lightning strikes. These applications have suggested further marketing directions for the C/G fibers, though it appears that hybridization with metal fibers and/or metallic coatings will be necessary to attain optimum results as electrical conductors.

REFERENCES

1. Anon., "News Announcement," *Metal Progress,* page 115, May 1959.
2. Bandaruk, W., "Curing of Reinforced Plastics with Graphite Fibers as Heat Sources," *Technical Conference,* SPI, RP/C Institute, Section 15D, Chicago, Illinois, February 1966.
3. Carri, R., "The Effects of Lightning and Nuclear Electromagnetic Pulse on the Composite Aircraft," *SAMPE Tenth National Technical Conference,* Vol. 10, page 1, New York, October 1978.
4. Bacon, R. and Schalamon, W.A., *Applied Polymer Symposium,* Vol. 9, page 285, 1969.
5. McMahon, P., "Graphite Fiber Tensile Property Evaluation," ASTM-STP-521, page 376, 1973.
6. Woollam, J., "Review of Electrical and Thermal Properties of Carbon and Graphite," *ACS Symposium Series,* No. 21, "Petroleum Derived Carbons," Page 378, 1975.
7. Lin, R.Y., Smith, W.D., Coppol, A.D., and Economy, J., AFML-TR-74-113, 1974.
8. Bushman, E.F. and Bushman, G.R., "Carbon Fiber–Tooling," *Thirty-third Annual Technical Conference,* SPI, RP/C Institute, Section 21F, Washington, D.C., February 1978.
9. Kung, D.T. and Amason, J.P. (Douglas), "Lightning Conductivity Characteristics of Graphite Composite Structures," *SAMPE Twenty-third National Symposium,* page 1039, Anaheim, California, May 1978.
10. Skouby, C.D., "Relative Behavior of Graphite/Epoxy and Aluminum in a Lightning Environment," *SAMPE Quarterly,* Vol. 9, No. 3, page 20, April 1978.
11. Harvey, S.T. and Michaelson, G.L. (Boeing), NASA Contractor Report 145382-2, August 1978.
12. English, F.C., "Advanced Composite Aileron for L1011 Transport Aircraft," Contract NASI-15069, L.R. 29056, April 20, 1979.
13. Force, R. (Boeing), "Conductivity Measurements of Graphite/Epoxy Advanced Composites at Microwave Frequencies," *SAMPE Twenty-third National Symposium,* Vol. 23, page 1054, Anaheim, California, May 1978.
14. Miller, B.A. and Lee, S.G., "The Effects of Graphite-epoxy Composites on the

Galvanic Corrosion of Aerospace Alloys," NTIS Document AD A35029, September 1976.
15. Brown, A.G. and Coomber, R., "Resistance to Corrosion of Some Insulating Carbon Fiber Reinforced Plastic to Aluminum Alloy Joints," Royal Air Force Establishment, TR-72160, 1972.
16. Randolph, R.E., Jensen, L.C., and Martin, W., "Graphite Epoxy Reflector Support Truss," *SAMPE Symposium on Non-metallic Materials,* Vol. 4, page 209, Palo Alto, California, October 1972.
17. Goodyear Aerospace Corp., "Advanced Composite Material Study for Millimeter Wavelength Antennas–Environmental Tests," Technical Report AFML-TR-71-205, October 1971.
18. Robinson, E.Y., Stonier, R.A., and Lofgren, C.L. (JPL), "Development of a Unique Graphite/Epoxy Antenna Subreflector," *Third ASTM Conference on Composite Materials,* STP 546, pages 632–652, 1974.
19. Stonier, R.A. and Hillesland, H.L., "Fabrication of a Graphite/Epoxy Antenna for the Viking Orbiter Spacecraft," *SAMPE National Symposium,* Vol. 19, page 596, April 1974.
20. Haydostian, H., Robinson, S., Taormina, F., and Walker, G. (Hughes), "Lightweight Composite Spacecraft Antennas," *SAMPE Journal,* Vol. 13, page 14, September/October 1977.
21. Wade, W.D. and Ellison, A.M., "Application of Metal-matrix Composites to Spaceborne Parabolic Antennas," *SAMPE Twenty-fourth National Symposium,* Vol. 24, page 1265, May 1979.
22. Molyneux, M. and Lyons, B.R. (Fothergill and Harvey), "Carbon Fiber Reinforced Plastics Parts in Radiological Equipment," *Thirty-third Annual Technical Conference,* SPI, RP/C Institute, Section 12E, Washington, D.C., February 1978.
23. Litant, I., "Conductive Synthetic Resin Composition Containing Carbon Filaments," U.S. Patent 3,406,126, October 15, 1968.
24. Cal Tech/JPL, *NASA Technical Brief,* Vol. 2, No. 4, page 432, Winter 1977.
25. Rippel, W., "Battery Power," *Engineering and Science* (Cal Tech), Vol. 43, page 16, December 1979.
26. Clark, G., Paproki, S., Meyerer, W., and Kizer, D., "Recent Trends in the Primary Fabrication of Graphite Fiber-metal Matrix Composites," *SAMPE National Technical Conference,* Vol. 11, page 341, Boston, Massachusetts, November 1979.
27. Bauer, J.L., "Manufacturing Methods of a Composite Cell Case for a Ni-Cd Battery," JPL Publication 79-101, NASA Contract No. NAS7-100, December 15, 1979.
28. McFerrin, J.H., and Trulson, O.C. (U.C.C.), "Electrical Equipment Protection and Waste Disposal Practices–Carbon Fiber Manufacture," SAE Technical Paper 790034, Detroit, Michigan, February 1979.
29. NASA, Technical Memo 78562, "Observed Effects on Electrical Systems of Airborne Carbon/Graphite Fibers," January 1978.
30. Gilwee, W.J. and Fish, R.H. (NASA–Ames) "Test for Fiber Release from Carbon Composites," *Conference on Advanced Composites,* Technology Conference Associates, El Segundo, California, page 45, December 1979.
31. Serafini, T., Delvigs, P., and Vannucci, R. (NASA–Lewis), "High Char Imide-modified Epoxy Matrix Resins," *SAMPE National Technical Conference,* Vol. 11, page 571, Boston, Massachusetts, November 1979.
32. Bowles, K.J., "Burning Characteristics and Fiber Retention of Graphite/Resin Composites," Thirty-fifth Annual Technical Conference SPI, RP/C Institute, Section 11B, February 1980.

9 Environmental Influences on Carbon/Graphite Fiber Composites

The influence of environmental extremes upon the performance of C/G fiber composites is of concern to those responsible for the selection and adaptation of these materials. Aerospace requirements are the most demanding and C/G composites behavior will be examined closely. This is certainly prudent because, should the materials prove acceptable to aerospace needs, they are quite likely to satisfy the needs of other industries, such as the marine, antennas, energy storage devices, pressure tanks, and automotive and recreational equipment. The chemical industries require materials of construction to withstand the attack of many chemicals. These requirements are so diverse that, as the need arises for chemical applications, specialized attention and testing will have to be conducted in determining the suitability of the C/G composite.

Environmental influences are numerous and include, in addition to moisture and temperature, the effects of radiation of all types, including ultraviolet light, electrons, protons, charged particles of outer space, and the thermal and fast neutrons of nuclear reactors; chemical exposure, as from fuels, gases, and hydraulic fluids; as well as the effect of the high vacuum of outer space. Combinations of these exposures with mechanical stresses and strains provide the environment under which advanced composites must operate. There has not yet been sufficient time to develop adequate data in all of these categories, other than exposures of advanced composites to extremes of moisture and temperature, accompanied by cyclic stresses in some instances. For this reason, emphasis in this chapter is placed upon the moisture influences.

The evaluation of the environmental influences upon C/G fiber composites generally commence with determination of the effect of water and water vapor upon salient properties of the materials. For structural applications, the concomitant application of stress at time of moisture exposure is important. Resin matrices and the interface between the carbon and graphite fibers and the resin have received considerable attention. Surface preparation of the fibers and bonding to the fiber surface are important to successful in-service use of the composite in the presence of water. The early history of composites of all types is replete with physical tests upon the wet and the dry state of the composites. These basic tests become more involved when questions are raised as to the influences of thermal and humidity cycling, as well as stress cycling—all of which exacerbate the effects of dimensional differences and property changes in the C/G fiber composites. More advanced studies of the effects of environmental changes are those tests concerned with fracture mechanics and fatigue behaviors of the composites. "Real-life" exposure will usually introduce variables which are not anticipated by prior routine laboratory exposures, for example, the thermal spike test alluded to when an aircraft experiences large temperature differences in the space of a few minutes. The U.S. government agencies have wisely programmed service testing on advanced components prepared from C/G fiber composites. Some of these aerospace applications are reviewed further in this book. Emphasis in this chapter will be upon the effect of moisture upon the behavior of C/G products, though, where necessary, the experiences developed upon glass fiber laminates and other reinforcements will be drawn upon.

Hertz[1] called attention to the fact that in some circumstances high temperature strength properties of graphite fiber and boron fiber reinforced epoxy composites fell significantly after aging at room temperature in a standard laboratory environment. Among the tests reported were results on longitudinal flexural strength of HT-S graphite/epoxy composites which suffered losses in strength at 350°F of approximately 20% from the initial 350°F after aging for about 3 months at room temperature under ambient conditions. This was also reported for several of the neat epoxy resin systems tested under the same conditions. Further post curing the composites at elevated temperatures before aging permitted the composites to retain better 350°F performance after aging at room temperature or accelerated aging at high humidity. The loss in 350°F properties after room temperature aging was attributed to moisture absorption qualities of the resin matrix. It was also pointed out that high temperature properties were recovered after redrying of the specimen after humidity aging.

Continuing tests were run under NASA sponsorship to ascertain the influence of moisture upon the high temperature performance of C/G fiber/epoxy composites.[2] It was evident that the water absorption properties of the resin matrix had an important influence on results. Hence fiber dominated properties, such as tensile tests parallel to fibers on unidirectional composites, were not as useful as longitudinal flexural data on quasi-isotropic laminates, which would reflect, to a greater degree, changes in strength than tensile tests during environmental aging.

Browning[3] developed a comparative report on the effects of moisture on the properties of high performance structural resins and composites in 1972. Graphite and boron fiber reinforced composites were evaluated to determine the effects of moisture and/or high humidity on their room temperature and elevated temperature physical properties. Associated with moisture absorption is a loss of elevated temperature tensile strength, as demonstrated by composites prepared with the cycloaliphatic epoxy ERLA-4617/mPD system, which undergoes a substantial loss at 350°F. All of the composite systems show weight gains and thickness increase when subject to a high humidity environment. Standard epoxy resins based on bisphenol A and cycloaliphatics were used in the study as well as P-13N polyimide (TRW). The graphite fiber reinforced plastics of these tests contained Hercules' HT-S graphite fibers.

Environmental exposure cycling used by Browning in Figs. 9-1, 9-2, and 9-3 followed these steps:

1. Relative humidity 95 to 100% at 120°F. Time—22.5 hours.
2. Specimens removed and placed under normal room conditions for 15 minutes.
3. Temperature at -65°F. Time—1 hour.
4. Repeat of step 2.
5. Temperature 250°F. Time—30 minutes.
6. Repeat of step 2.

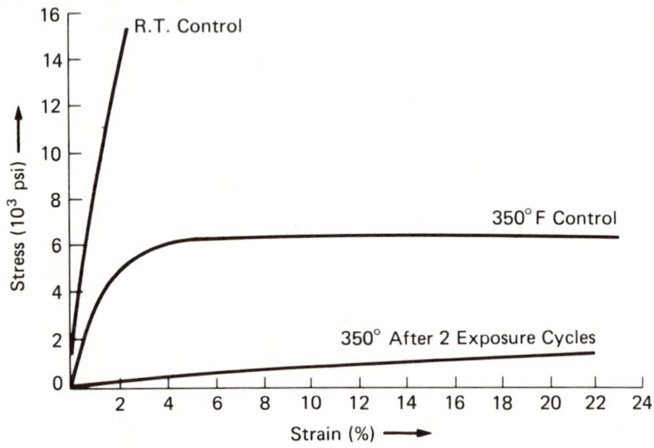

Fig. 9-1. Effect of exposure cycles on the stress-strain behavior of ERLA-4617 cast epoxy specimens.[3]

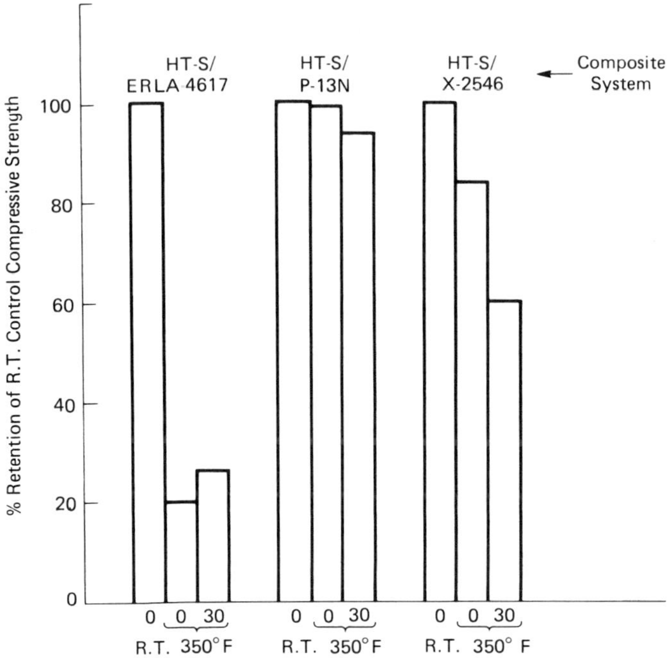

Fig. 9-2. Effect of exposure cycles on compressive strength of quasi-isotropic graphite/resin composites.[3]

The effect of the number of exposure cycles on the properties of graphite fiber composites, illustrated in these figures, are reproduced in part from Browning's paper together with test data on the neat epoxy resin system. It should be noted that properties are undoubtedly higher due to the redrying interval of step 5 before testing. Quasi-isotropic laminates were prepared from prepregs supplied by Narmco Corp. and Fiberite Corp. The tensile and compressive data were obtained on these laminates. Shear data (not shown) were also obtained on unidirectional laminates.

It is also interesting to observe the better performance of the graphite fiber/polyimide (P-13N) composites after environmental exposure cycling. These data are typical of information being developed for advanced composites at that time, though it should be noted that more recent epoxy and polyimide resins, as well as improved C/G fibers, have replaced many of the earlier products which were tested. Continued tests revealed the presence of ionic impurities on the surface of some carbon and graphite fibers, a situation which contributed to

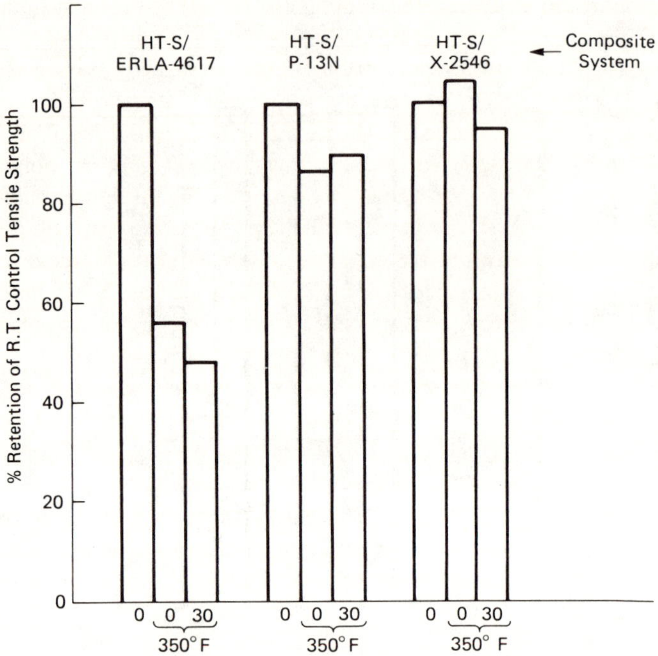

Fig. 9-3. Effect of exposure cycles on tensile strength of quasi-isotropic graphite/resin composites.[3]

higher water absorption and loss of strength when tested at 350°F.[5] Corrective measures by the suppliers of the fibers by washing removed contaminants to improve the resistance of the composite to moisture penetration, as well as the high temperature stability of carbon and graphite fibers. However, the problems caused by water absorption of the resin matrix had to be countered by post-curing of the resin and desorption procedures for the composite.

Further tests on the effects of moisture on high temperature strengths of C/G fiber reinforced epoxies failed to establish a correlation of flexural modulus retention at elevated temperature with water absorption, equilibrium water content, or volume change.[4] The authors of this report used cured/epoxy resin matrices, pulverized them, and determined equilibrium moisture content in 10 to 20 days.

The epoxy resins were prepared into composite panels using Courtaulds' HT-S graphite fiber. Flexural and shear specimens were cut from cured unidirectional panels and tested at 75°F and 350°F, both before and after exposure to boiling water for 24 hours. At 350°F, property retention varied from 40 to

75% of original room temperature strengths for the several resin systems tested.

Some of the epoxies were selected for further tests and then compared with polyimide resins, the latter demonstrating better strength retention after boiling water tests. The HT-S/RS 6234, graphite/polyimide system was deemed the best in these series of tests. It was noted that significant degradation of all specimens occurred in 350°F testing when there was 1% weight gain in moisture in the resin–irrespective of how this moisture was absorbed (high humidity versus boiling water). Early blistering of composite samples occurred during high humidity and water boiling tests. This was traced to some ionic contamination on the fibers–a situation corrected by rewashing of the fibers before lay-up into composites. When fiber water extracts were below 5 micro-ohm/cm conductivity blistering did not occur.

As part of an effort aimed at developing high temperature, high performance laminating resins for use with C/G fibers, Eckstein reported on the weight gains in epoxy systems after exposure to boiling water up to 168 hours for 3 mm thick castings of neat resins. Equilibrium was not reached in this time, though data were more significant than the water gain measurements after 72 hours in boiling water. Best water resistance was displayed by Shell's experimental ERX-67 resin (N,N-diglycidyl ether of 2,4,6-tribromoaniline), though the resin system did not have a high heat distortion temperature.[6] Data in Table 9-1 are reproduced from this paper for aromatic epoxy and cycloaliphatic epoxies which have been used in advanced composites of C/G fibers, though the data shown is for the neat resin system only. Refer to the chapter on resin matrices for further information on the resins and their curing agents.

As the author points out, there is no relationship between the heat distortion temperature and the water pick-up for the systems in Table 9-1. However, when an epoxy resin is fully cured and optimum heat distortion is obtained, better water resistance and chemical resistance are realized.

The effect of absorbed moisture on the properties of neat epoxy resins and some of their composites have been explored by DeIasi and Whiteside.[7] Excellent agreement among the diffusion coefficients, activation energy, and entropy was observed for the majority of the resins. Good correlations are reported between neat resins and their composites. In addition, the effect of absorbed water on the glass transition temperature of epoxy resins was shown (see Fig. 9-4). Also see, for example, the curve of the effect of moisture concentration upon a tetraglycidyl ether of methylene dianiline ("Fiberite" 934), using thermomechanical analysis. Figure 9-5 (from this paper) demonstrates the plasticizing action of moisture of several epoxy resins by examining the glass transition temperature versus the equilibrium moisture content.

Browning and Hartness[8] discussed the effects of moisture on high performance structural resins and composites. The flexural strength of epoxy composites

Table 9-1. Moisture Absorption by Epoxy Resin Matrices[6]

SYSTEM AND RATIO OF RESIN/Hn.	HEAT DISTORTION TEMPERATURE (ASTM-D 648)	PERCENT WEIGHT CHANGE/ IMMERSION IN WATER AT 100°C 24 hr	72 hr.	180 hr	REDRY 6 days/150°C
Standard Bis A Epoxy/mPDA 100/14.5	161°C	1.62	2.41	2.92	0
Standard Bis A Epoxy/MDA 100/27	167°C	1.49	2.08	2.38	-0.07
Standard Bis A Epoxy/ MNA/BDMA 100/80/1	143°C	1.01	1.21	1.48	0.21
Epoxy Novolac Resin/MDA 100/28	185°C	1.48	2.24	2.74	0.04
Epoxy Novolac Resin/MNA/BDMA 100/87/1.5	183°C	1.15	1.56	1.86	0.30
ERX-67/MDA 100/20.5	132°C	0.47	0.81	1.14	0.11
ERLB-4617/mPDA 100/27.5	195°C	5.23	8.48	10.31	1.27
ERLB-4617/MDA 100/50	180°C	3.12	5.08	6.59	0.32
ERLB-4617/Hn9245/ /BF_3MEA 100/22/1.5	242°C	5.97	9.86	10.83	0.51

(X-2546 cycloaliphatic) resin and several structural fibers are shown as a percent retention of unexposed room temperature flexural strength after 350°F and 15 cycles of humidity aging as described earlier. Composites were laid up with several unidirectional graphite fibers and a cycloaliphatic epoxy comparing the decrease in flexural strength with humidity cycling (see Fig. 9-6). The characteristics of the fiber filaments used in these composites were reported as follows.

FIBER IDENTIFICATION	MODULUS ($\times 10^6$ psi)	TENSILE STRENGTH ($\times 10^3$ psi)	DENSITY (g/cc)
A-S–graphite	32	400	1.79
HT-S–graphite	38	380	1.75
HM-S–graphite	50	320	1.90
Boron	60	500	2.63
PRD 49-111 (du Pont aramid)	19	400	1.47
S-glass	12.5	500	2.49

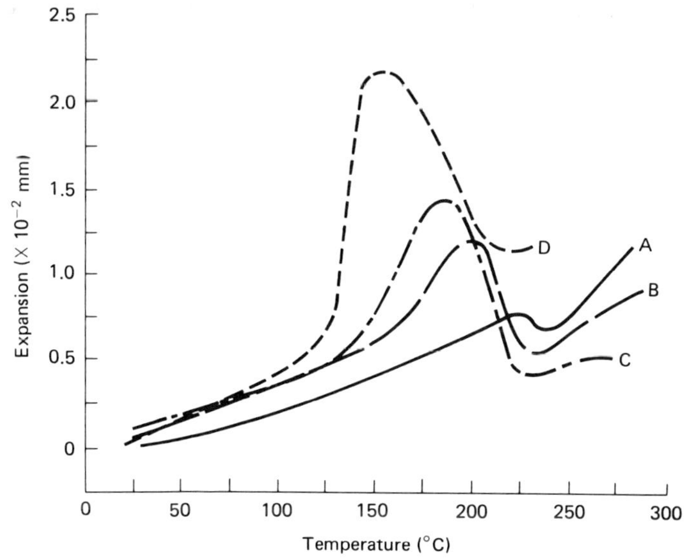

Fig. 9-4. Effect of moisture concentration on T_g of Fiberite 934 epoxy resin: (A) 0.0% H_2O, (B) 2.2% H_2O, (C) 4.1% H_2O, (D) 6.6% H_2O.[7]

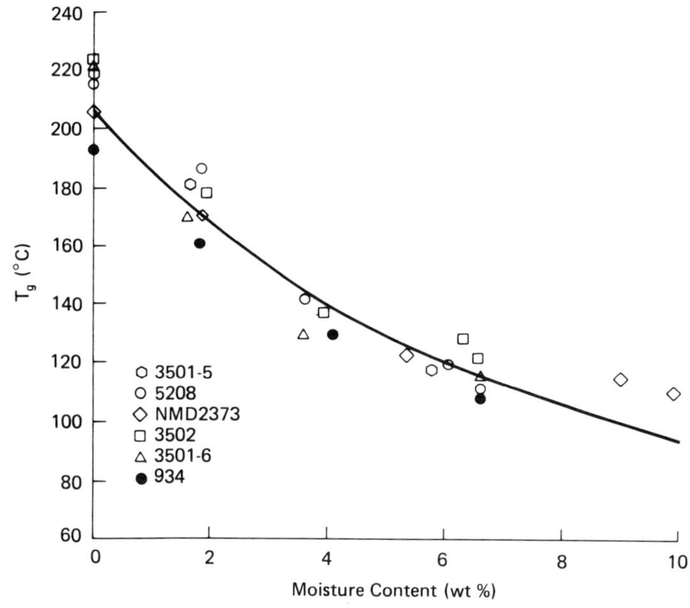

Fig. 9-5. Moisture content versus T_g.[7]

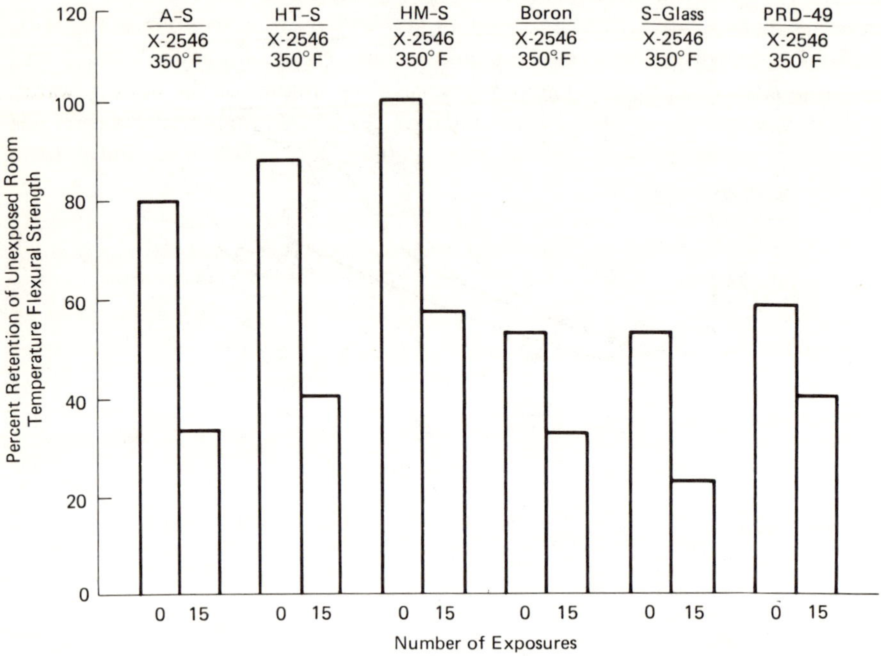

Fig. 9-6. Effect of cyclic exposures on the flexural strengths of X-2546 epoxy composites reinforced with various fibers.[8]

Further data indicated better percent strength retention at 350°F for the unidirectional composites over the quasi-isotropic lay-ups. The absorption of water by the composites caused different failure modes—the dry 350°F control exhibiting a sharp tensile failure with graphite fiber breakage and the wet 350°F specimens (as from water boil tests) showing a plastic deformation with no filament breakage. Chung and Crognola[9] developed equations relating the weight changes in advanced composites due to water absorption to the loss of mechanical properties. Moisture absorption was studied with the aid of activation energies derived from time-temperature superposition. The data developed by Hertz and by Browning (see earlier references) was used in analysis.

Such analyses become more complicated when it is considered that the moisture distribution within the thickness of a graphite fiber epoxy is not uniform. Sandorff and Tajima demonstrated the moisture distribution in a T300/5209 laminate after conditioning at 98% relative humidity for 148 hours to an average weight gain of 0.49% weight.[10] Their technique involved micro-toming of 0.25 mm slabs and then placing them in a drying environment to determine absorbed water at different distances from center plane. See typical results for moisture distribution within a section of graphite/epoxy composite in Fig. 9-7.

Judd[11] presented data on interlaminar shear (ILS) of standard epoxy and cycloaliphatic (ERLB-4617) epoxy resins, after immersion in boiling water. Though this method is useful for comparing systems and materials, it does not necessarily predict service performance. Among the conclusions drawn from this work are the following:

1. Low initial ILS in graphite fiber reinforced epoxies is associated with high void content.
2. Resin C/G fiber bonds are not permanently degraded after moisture exposure.

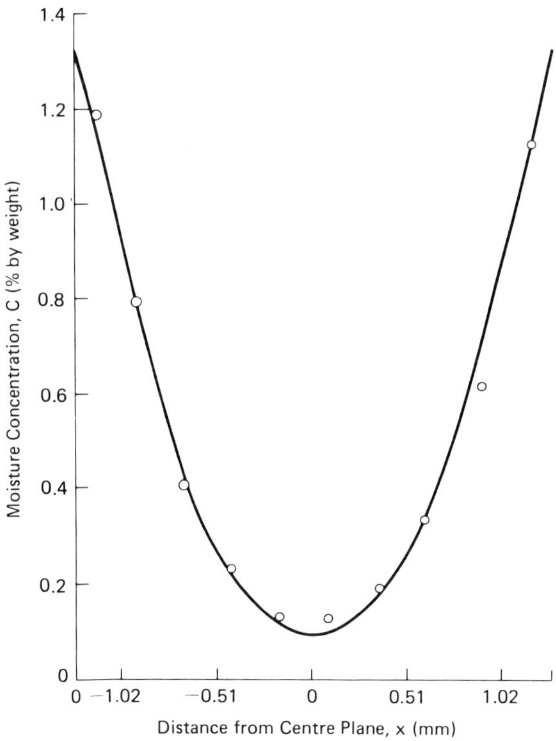

Fig. 9-7. Diffusion solution (solid line) fitted to moisture distribution obtained after conditioning a predried T300/5209 laminate at 66°C/95% RH for 148.5 hours to a weight gain of 0.49% (C_S = 1.31 wt %, $D = 3.26 \times 10^{-7}$ mm^2/s and the average variance is 1.7×10^{-3} (wt %)^2)[10] *(Courtesy of IPC Science and Technology Press–England)*

3. Surface treatments of C/G fibers produce marked improvements in ILS and resistance to loss of strength in long time water immersion.
4. Standard bisphenol A resin graphite fiber composites showed a higher loss in ILS during boiling water periods up to 168 hours, though the cycloaliphatics that were employed in these tests had higher values of weight changes and lower loss of ILS.

Natural weathering of selected graphite/epoxy composite materials has been studied by measuring changes in specific mechanical properties.[12] Panels of several commercial composite materials from England and the United States were exposed at Panama, Warminster, Pennsylvania, and in North Carolina sites. Six ply panels with 0° and ±45° fiber orientations were prepared. Half were coated with an epoxy-polyamide primer (MIL-C-23377C) followed by a polyurethane topcoat (MIL-C-81773B). All panels were exposed at the various sites in a state of flexural stress. The tensile strain on the panel side facing the exposure was calculated to be 4000 microinches per inch. The Panama exposure effected the largest losses in mechanical properties. Natural weather exposure for 18 months showed virtually no change in ILS properties of panels of "Modulite" 5206/Modmor II graphite fibers, while T400/2544 panels showed the greatest degradation on exposure. The pigmented coated specimens uniformly showed

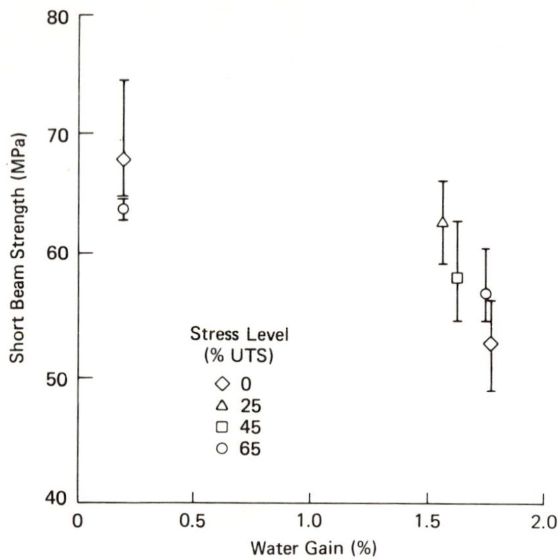

Fig. 9-8. Current short-beam strength of SP-313 cross-ply laminate as a function of its moisture content.[13]

enhanced shear strength due to prevention of ultraviolet degradation and reduced rate of moisture penetration.

The effect of external stress on moisture diffusion and mechanical properties degradation was investigated upon a commercial epoxy/graphite prepreg "Scotchply" SP-313 (T-300 graphite fibers).[13] The external stress applied to the composite during exposure to water affected the degradation process by accelerating the moisture absorption. In Fig. 9-8 the short beam strength of a cross-ply laminate is shown at two different moisture contents and at several stress levels. Significant increases occur above certain combinations of stress and moisture. In this study the critical external stress moisture combination is 45% ultimate tensile strength (UTS) and 1% moisture content. Only static stress tests were used in this program.

As indicated below, moisture pickup by the prepreg will affect results. Aging studies were carried out to examine the effects of out-time and humidity at room temperature on the properties of a polyimide/graphite prepreg (Hexcel F-178/T300). As aging continued at 95% R.H., the prepreg lost its tackiness and became dry and brittle.[14] Mechanical properties of laminates fabricated with polyimide/graphite prepreg at 95% R.H. showed the influence of high humidity aging before laminating, as follows:

FLEXURAL STRENGTH (ksi)	PREPREG AGING TIME (weeks)	SHEAR STRENGTH (ksi)
265.	0	10.5
235.	4	14.0
200.	24	17.0

THERMAL SPIKES

Tests were conducted by McKague *et al.* to determine whether aircraft flight conditions change moisture absorption in graphite/epoxy composites.[15] Union Carbide Corp.'s "Thornel" 300 and Narmco's epoxy resin (5208) system were used to prepare the composites for test. Panels were cured at 355°F and 86 psi, followed by cooling under pressure, then post-cured for 4 hours at 400°F. A "real-life" environmental cycle was established based on aircraft performance. This entailed a simulated supersonic dash with rapid heating of the composites to 300°F followed by rapid cooling. This was described as a thermal spike resulting in a permanent change to subsequent moisture diffusion behavior of graphite/epoxy panels with the result that moisture absorption was double that experienced under constant humidity at constant temperature.

More recent investigations by McKague[16] suggest that the lower glass transition temperature from absorbed moisture of epoxy resin in the graphite fiber/epoxy composite limits the supersonic service conditions of military aircraft. Moisture has caused greater reductions in glass transition values than previously believed, and hence this curtails flight patterns that can be used without triggering a thermal spike effect. He suggests the relationship of T_g to relative humidity forms a service envelope that indicates the maximum flight temperature that can be tolerated.

Temperature and humidity criteria for design of advanced composite structures were under examination by Wollner.[17] Complete environmental history of the composite experienced during service life will determine the moisture characteristics. Ground storages, particularly under humid conditions, will build up moisture in the composites. The rapid changes in flight profile are shown in Fig. 9-9. Changes in moisture distribution commence near exposed edges of the

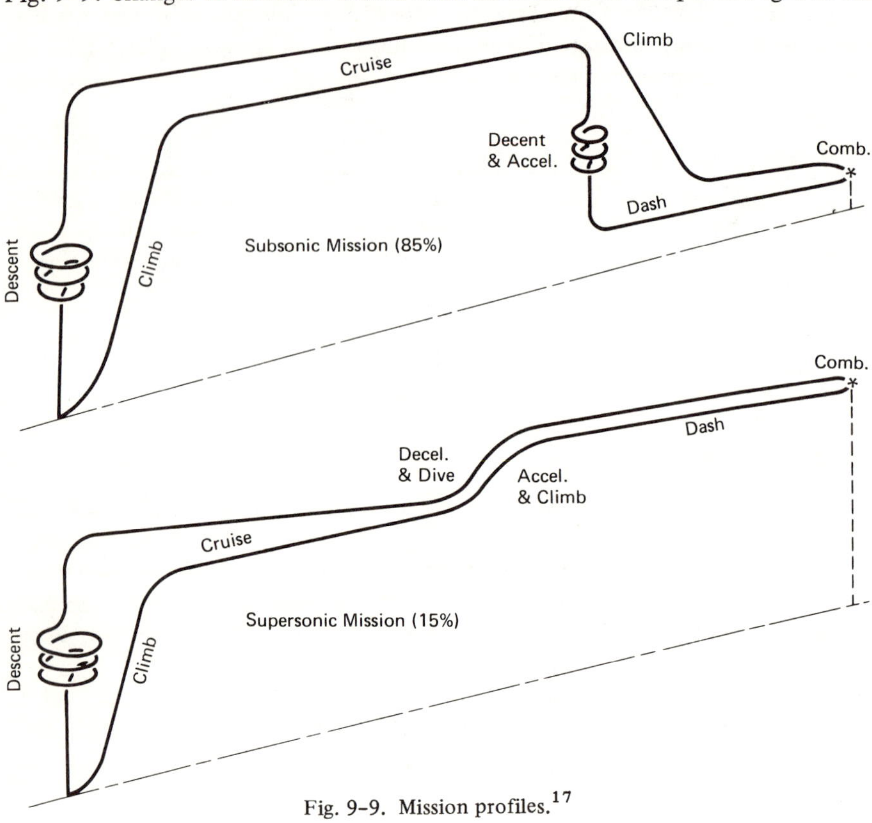

Fig. 9-9. Mission profiles.[17]

composite and, as suggested by earlier studies, the thermal spikes encountered in supersonic flight will increase the diffusion rates of moisture into composites. Moisture absorption increases even more with repeated thermal spikes, probably due to progressive micro-cracking.

Data have been published on the effect of thermal spikes on moisture gain in epoxy resin matrices.[18] Data in Fig. 9-10 were obtained at 100% relative humidity and 160°F. The 300°F thermal spike, as noted above simulates the activity of a high speed aircraft. The condition is even more aggravated in a composite where micro-cracking can be traced to differences in thermal expansion between the fiber and resin matrix. In this current work, moisture gains greater than equilibrium amounts were observed without T_g changes. Moisture gains are attributed to moisture entrapment during micro-cracking in the resin. Epoxy resin used was the tetraglycidyl ether of methylene dianiline, cured with 4, 4′-diaminophenyl sulfone.

Comparative absorption and desorption data were developed for several epoxy resin systems in cross-plied collimated tape laminates containing "Thornel" 300 graphite fibers and Fiberite-934 epoxy matrix. Moisturization was effected at 140°F and 95% relative humidity and drying was accomplished at 250°F. Moisture levels of 0.6%, 0.8%, and 1.0% were developed and specimens thermally spiked up to 100 times.[19]

	STEADY STATE MOISTURE
Unspiked moisture	1.4 to 1.5%
Specimens with 0.6% moisture and spiked	1.75%
Specimens with 0.8% moisture and spiked	1.75%
Specimens with 1.0% moisture and spiked	2.0%

A study has been made by Sykes *et al.* of the effect of moisture absorption on the dynamic thermo-mechanical properties of a graphite/epoxy composite being considered for aircraft structure. T-300 graphite fibers and Narmco's 5209 epoxy were used in preparing the composite specimens. Torsional braid analysis and thermo-mechanical analysis techniques were used to measure changes in the glass transition temperature of composites exposed to soaking in distilled water. Moisture induced changes produced greater changes in the ASTM heat distortion test than observed in the glass transition temperature. A 1% moisture absorption resulted in a 12°C lower T_g and a 49°C lower heat distortion temperature.[20] This could be attributed to the greater possibilities of creep in the ASTM test. Figure 9-11, which is from this report, summarizes some of the findings on a graphite/epoxy composite.

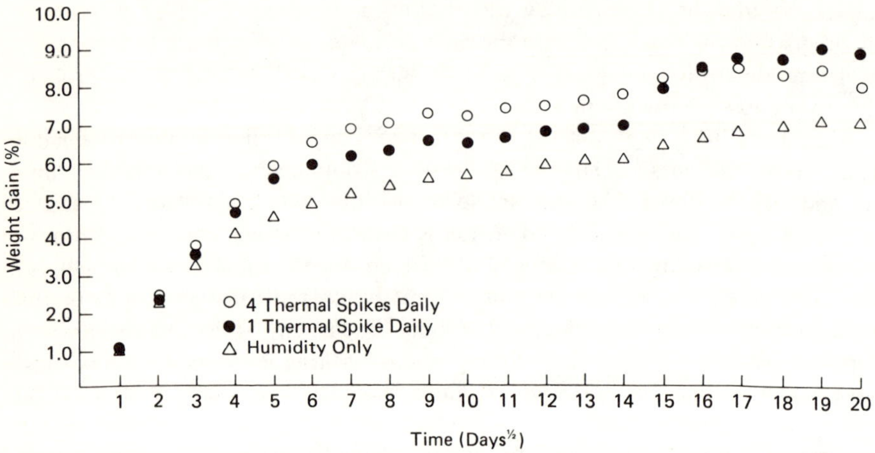

Fig. 9-10. The effect of thermal-spikes on epoxy moisture absorption (160°F/100% relative humidity.[18]

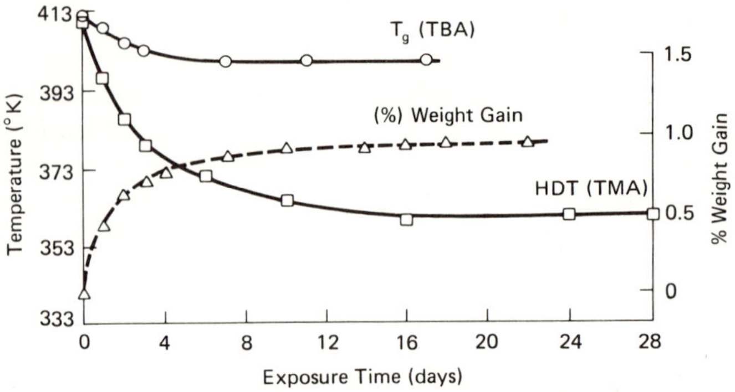

Fig. 9-11. Effect of soak time on the HDT, T_g, and moisture pick-up of a two-ply unidirectional "Thornel" 300/5209 composite cured at 400°K.[20]

Reports on moisture absorption on graphite/epoxies appear to have proliferated in recent years, with most of them supporting these general conclusions.

1. Lowered T_g and lowered heat distortion temperature with increased moisture pick-up.
2. In general, the reversibility of absorption and desorption of moisture.
3. The greater moisture after application of thermal spikes.

4. Pronounced fall-off in high temperature performance at 350°F after moisture pick-up.
5. Variables in moisture pick-up depending upon the laminate construction.
6. The increased moisture pick-up after high temperature post-cure.
7. Edge to surface area ratio will influence absorption and desorption rates of the composite construction.
8. Contributions to increased moisture gain from micro-cracking.

Lacking in most cases are data on surface treatments of carbon fibers and lot identities to assist identification of process and impurity variables. Similarly, the generic nature of the resin matrix system and its cure cycle have not always been disclosed. The multiplicity of resins and C/G fibers used have made it difficult to make direct comparisons in some instances. The problems of resin matrices on environmental exposure have been appropriately highlighted and positive design and manufacturing methods now require detailing to insure good environmental performance.

The deleterious effects of moisture at elevated temperatures were noted recently in tests upon graphite/epoxies and boron/epoxies.[21] Some unusual test results noted, showed a two-stage increase in moisture gain, in which there was an initial levelling off of absorption, and after a continued exposure, the water gain commenced again. The formation of surface cracks on rapid cooling and subsurface microvoids were noted under some conditions, contributory to increased water absorption.

Welhart[22] noted better results with the polyimides over the epoxies in environmental exposure tests. In his report, environmental phenomena included water immersion, high temperature aging, lightning strikes (see the chapter on electrical applications), galvanic action, electromagnetic interferences, sand and rain erosion, and thermal/vacuum outgassing.

Considerable effort has been expended to show that moisture affects the performance of composites and sandwich constructions. Naval aircraft will encounter considerable moisture during service lives and typical saturation moisture levels for the composite skins fall in the range of 0.8 to 2.5% of the original dry composite skin weights.[23] A typical aircraft component utilizing a composite sandwich construction would be a wing flap. Construction variables in this study included: 1) thicknesses of the face skin of epoxy/graphite (9 versus 14 plies), 2) presence of glass interlayer between the composite face skin and honeycomb core, 3) painted and unpainted surfaces, and 4) holes penetrating the skin. Environmental conditioning entailed: 1) high humidity, 2) elevated temperatures, 3) stress cycling, and 4) stack gases and sea water—simulating major load excursions in a 1000 hour flight segment, followed by 336 hours exposure to sea water and stack gases. Residual strength testing after exposure, generated compressive rupture of the composite structure (4 point beam loading test) and failure of the beam due to disbonding of the adhesive. Results at 75°F showed increased resistance to environmental degradation with glass interlayers present between the epoxy composite and the honeycomb core, while at 218°F shear type failures predominated. In Fig. 9-12, triangles show data with glass interlayers versus circles–without glass interlayers. The improvement in resistance of the composite sandwich to environment with the glass interlayer between the face skin and the honeycomb core is evident.

Time, temperature, and moisture dependent dimensional changes of GY 70/CE 339 advanced composite exposed to variations in temperature and moisture were calculated by Crossman and Flaggs.[24] Experimental verification procedures were employed to compare predicted and measured changes. Good correlations were established between viscoelastic analysis and test results. The measured and predicted curvature of a non-symmetric $(0_4/90_4)_T$ GY70/339 laminate were shown as follows.

EXPOSURE CONDITION			LAMINATE CURVATURE AT 75°F AFTER EXPOSURE (% OF INITIAL DRY VALUE)	
TEMPERATURE (°F)	RELATIVE HUMIDITY (%)	EQUILIBRIUM MOISTURE CONTENT (%)	MEASURED	VISCOELASTIC ANALYSIS
75°	95	0.65	20	21.4
125°	95	1.28	35	5.5
160°	95	2.20	7	4.0

Particular attention was given to the swelling of graphite fiber/epoxy composites with "zero" coefficients of thermal expansion, when they were exposed to high humidities. While no expansion was noted below 0.4%, absorbed moisture stress and dimensional changes were calculated for hygrothermal cycling. These data are particularly important to telescopic and other optical mounts, and antenna supports fabricated of advanced composites of graphite fibers.

Wolff and Eselun[25] have derived coefficients of moisture expansion (CME), analagous to the coefficient of thermal expansion, defining CME as a change in length per unit length per weight percent of water at constant temperature and pressure. Dimensional changes caused by loss of water in space can be either negative or positive in a multi-directional angle-ply laminate.

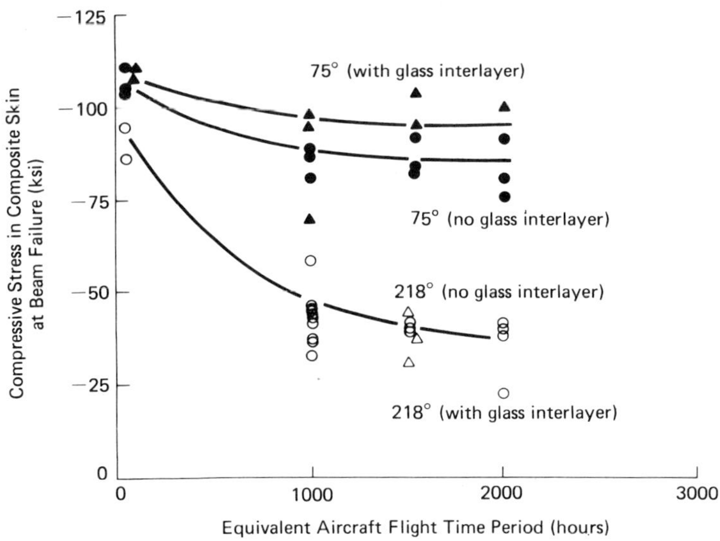

Fig. 9-12. Sandwich beam strength degradation versus exposure period comparison of sandwich beam strength with a glass interlayer and without a glass interlayer.[23]

MARINE EXPOSURE

Marine uses of advanced composites are described in a later chapter on miscellaneous industrial applications which includes several examples in the boating industry. There have been reports[26] on long-time exposure of carbon and graphite fiber composites immersed in sea water and distilled water.

Mazor *et al.* reviewed weight changes and mechanical properties of Naval Ordnance Laboratory (NOL) ring composites fabricated by Union Carbide Corp. in 1965 from rayon based carbon yarn and a graphitized (at 2400°C) rayon yarn–bonded with ERL 2256 cycloaliphatic epoxy and cure agent MPDA (metaphenylene diamine). Specimens were tested after 11 years. Maximum weight loss of NOL ring specimens upon drying were reported as follows.

ENVIRONMENTAL HISTORY (11 YEARS)	MATERIAL	MAXIMUM WEIGHT LOSS (%)	DRYING TIME (DAYS)
Distilled water	Carbon/epoxy	18.0	45
Sea water	Carbon/epoxy	14.0	55
Distilled water	Graphite/epoxy	2.3	30
Sea water	Graphite/epoxy	2.0	30

The unexpected high water absorption of the carbon fiber specimens was attributed to water absorption by the fibers themselves, through pores associated with the fibers (measured density of carbon fibers was 1.57 g/cc). The long term water immersion of 11 years and accelerated drying affected the diffusion constants upon re-exposure to humid environment. A considerable shear strength reduction (short beam test) was found for carbon/epoxy specimens having long-time water environment exposure. It was presumed that micro-damage and irreversible degradation of the matrix and fiber interfaces occurred during the 11 years of water immersion. Re-immersion after drying out of the composites yielded high wet strength in the C/G epoxy laminates, suggesting internal stress relaxation during the long prior immersion period.

Carbon/graphite composites are not alone in being adversely affected by marine exposure for long immersion periods. Glass composites have been known to rapidly deteriorate under stress in water—particularly if stress is cyclically applied. Fried *et al.*[27] outlines results upon the deep submergence in water of filament wound glass composites, though further work emphasized the importance of the glass to resin interface and the right coupling agents. Fried[28] reported on the physical properties of glass fiber reinforced epoxy and polyester composites after three years immersion at a depth of 5700 feet in water.

Romans *et al.*[29] commented upon the durability of the glass to resin bond as a major determinant of the composite to withstand repeated stress and strain under water. Microscopic examination of NOL rings (5.75 in I.D. and 6.00 in O.D.), fatigued in water, indicated that failure is initiated in the area of maximum fiber stress. Test specimens were of S-glass and anhydride cured epoxies. Minute cracks occur in the resin, followed by failure of the glass-resin bond. Cyclic stressing in water may be as much as 100 times more damaging than if the stress level is static, as observed in their tests upon glass/epoxy composites.

The effect of the marine environment on stressed and unstressed graphite/epoxy composites has been observed on C/G fiber composites prepared with high tensile strength (HTS) graphite fibers (void content reported as less than 1%) and epoxy resin matrix. There was 97% strength retention after extended immersion in water, compared with 92% retention for composites involving hybrids of HTS graphite fibers and ultra-high modulus fibers. Under steady load stress, the HTS/epoxy laminates retained 86% of their initial strength, as compared to 90% when tested in the dry state. No discernible correlation was reported for the amount of water absorbed and strength retention.[30] Further references will show influence of cyclic stress on C/G composites in water.

Morley[31] examined the water sensitivity of composites used in Europe which contained epoxy resins with Courtaulds' fiber HT-S. A proprietary surface treatment was applied to the graphite fibers. Epoxy resin matrices were present to 40 to 45% by weight. Three systems are compared in Table 9–2 are:

A. Standard bisphenol A type epoxy resin, cured with dicyandiamide,
B. Precondensate standard epoxy resin cured with BF_3MEA, and
C. Experimental resin DX-284 (Shell) cured with 4,4′-diaminophenyl sulfone (DADS).

Table 9-2. "Grafil" HT-S Fiber/Epoxy Composites—After Four Weeks Immersion in Deionized Water at 23°C—Unstressed[31]

	RESIN A		RESIN B		RESIN C	
PROPERTY	INITIAL	AFTER 4 WEEKS	INITIAL	AFTER 4 WEEKS	INITIAL	AFTER 4 WEEKS
Flexural strength (MN/m^2)	1480	1450	1130	1130	1430	1300
Modulus (GN/m^2)	131	122	151	153	124	113
Interlaminar Shear Strength (MN/m^2)	72	60	73	74	100	100

Data were also presented on the composites after thermal aging for 28 months at 150°C. It was interesting to observe that resin A had very little change in flexural strength and flexural modulus.

Scola[32] reported very little effect of room temperature or 600°F shear strength of flexural strength of S-glass in PMR-11 or "Thornel" 300 graphite fiber in PMR-11 polyimide resin matrices. These observations were made after 24 to 72 hours in boiling water or 8 weeks at 95% relative humidity at 120°F.

The moisture effects upon the tetraglycidyl ether of 4,4′-diamino diphenyl methane epoxy (TGDDM) resin cured with 4,4′-diaminophenyl sulfone (DADS) —see the chapter on epoxy resin matrices—have been investigated, and the effects of curing agent concentration and water absorption upon the viscoelastic properties have been identified.[33] The alpha glass transition (T_g) did not show any sensitivity in these tests to moisture, due to the fact that thin films of 60% C/G fiber composites were used in the examination for determining dynamic mechanical properties and moisture content was suspect. The weaker beta transition (at -50°C and omega transition at 100°C) did exhibit some dependence upon moisture content. These data appear at variance with other reports on the influence of moisture on T_g of advanced composites examined earlier.

Various effects of moisture on graphite/epoxy composites have been described by Hertz, with particular emphasis upon changes in dimension.[34] Ultra-high modulus GY-70 graphite fibers were used in the composites, intended for spacecraft structures, where optimum dimensional stability was being sought. As for optical mountings and antennas, see Table 9-3 reproduced from this report:

Table 9-3. Graphite Fiber Selection for Dimensionally Stable Composites.[34]

COMPOSITE PROPERTY	HIGH STRENGTH GRAPHITE FIBERS (T-300, A-S, AND "CELION" 6000)	HIGH MODULUS GRAPHITE FIBERS (HM-S, PAN-50, 5-T)	ULTRA-HIGH MODULUS GRAPHITE FIBER (GY-70)
Tensile Modulus (msi)	7.4	9.3	15.0
Density (g/cc)	1.59	1.66	1.75
Coefficient of Thermal Expansion (micro-in./in./°F)	1.04	0.610	0.038

Moisture absorption is described as a function of laminate thickness, ply orientation, relative humidity, and temperature. The weight gain (ΔW) as a function of time is correlated to a change in length ($\Delta L/L$). Figure 9-13 from Hertz's paper, shows dimensional changes in GY-70/X-30 composites for different thicknesses, at various changes in moisture content. The author suggested use of aluminum foil with the composite to minimize water absorption. The ultra-high

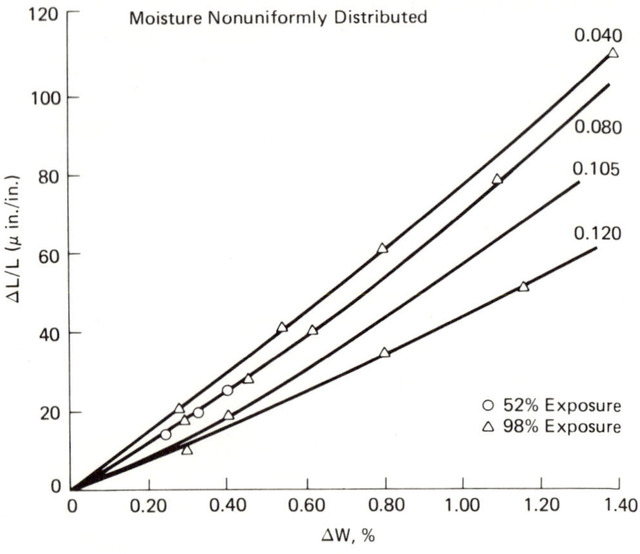

Fig. 9-13. $\Delta L/L$ versus ΔW for $[(0/45/90/135)_S]_2$ GY-70/X-30 exposed to humidity.[34]

modulus composite led to the most dimensionally stable configuration by reason of low expansion coefficient and higher modulus of the composite.

Aerospace structures have concern with corrosion caused by the coupling of graphite fiber composites and aluminum. Some of the galvanic studies in this area are reviewed in the chapter on electrical applications of advanced composites. In examinations of the mean strength of composite to metal adhesively bonded joints, Wolff[35] developed data on the bond strength of glass composites to aluminum with AF-147 adhesive. With 2% moisture content in the composite, tests at various temperatures showed the following results.

At room temperature, adhesive bond strength tested at 82% of dry strength.
At 175°F, adhesive bond strength tested at 60% of initial.
At 250°F, adhesive bond strength tested at 22% of initial.

Because of the adhesive bonding, no corrosion of the aluminum alloy was reported for bonds to C/G composites. As part of F-16 composite certification, composites for the aircraft were adhesively bonded to aluminum alloys. Outer skins of graphite/epoxy and glass/epoxy were involved on the horizontal tail and rudder.

A study on the influence of moisture on the impact behavior of hybrid glass fiber/graphite fiber/epoxy composites revealed some interesting synergistic results.[36] S-993 glass rovings/"Narmco" 5208 epoxy and "Thornel" 300 graphite fibers/"Narmco" 5208 resin were used for preparing specimens. Two types of laminates were prepared, unidirectional and quasi-isotropic (a ply-wise blend of 0°, 90°, and ±45° layers), which place the high modulus graphite fibers in the 0° direction in the outermost layers. The presence of moisture did not degrade impact resistance of either single phase of hybrid composites and frequently improved the impact energy to fracture, as the moisture plasticized the resin matrix. As observed from Hofer and Porte's paper,[36] the impact resistance of hybrid/glass/graphite/epoxy composites were improved over the all-glass or all-graphite composites.

EFFECTS OF THERMAL CYCLING

Some of the effects of abrupt changes in temperature, after exposure of the graphite/epoxy composites to moisture, have been reported earlier in this chapter—particularly under the aegis of thermal spikes. There is continuing interest because of the pronounced difference in the coefficients of thermal expansions between the resin and the high modulus fibers which have been assembled into a composite structure. In most cases the cracks that are formed run parallel to fiber axis, on the outermost plies of the laminate. This situation has been examined in a recent study[37] which reported on interfiber cracks and interlaminar cracks. A serious problem arose in fabricating carbon fiber/reinforced plastics (CFRP) with thick unidirectional doublers, as cracks developed during cooling from cure to room temperature. The most effective preventative measure was the distribution of unidirectional plies uniformly between the angle plies. Slow cooling from cure or interspersing of flexible interlayers were not as effective. Corrective measures after cracking were followed by vacuum impregnation with a liquid epoxy resin and then curing the assembly.

Post-curing and annealing of C/G epoxy composites lowered residual stresses in the matrix induced by prior hygrothermal histories.[38] Data indicated that thermal fatigue resistance was enhanced, dimensional stability improved, and outgassing minimized.

The effects of a thermal cycling environment on the dimensional stability and microstructural integrity of C/G epoxy composites were evaluated by Camahort *et al.*[39] Severe micro-cracking was found in several high modulus graphite fibers (HM-S) composites after exposure to 25 thermal cycles between liquid nitrogen (-320°F) and boiling water (212°F). A hybrid system combining "Thornel" 300 fabric and HM-S tape exhibits fewer micro-cracks with the fabric serving as a "crack-stopper." Gold plating (by sputtering) of the polished metallographic specimens makes the micro-cracks clearly visible. The HM-S graphite fibers and the ultra-high modulus fibers (GY–70) were provided in tape form. Fiberite Corp., Hercules, and U.S. Polymeric provide tapes with epoxy systems that cured at 350°F—all of them yielding a negative thermal coefficient of expansion (range -0.25 to 0.45 microinches/inch/°F). An elastomeric modified epoxy yielded better results in resisting micro-cracking. The differences in longitudinal C.T.E. as cured and after thermal cycling were relatively small. Lower temperature curing resins and thinner plies of prepreg will also reduce micro-cracking.

Composite durability characteristics and ultrasonic scanning, coupled with direct measurements of bulk moisture content, provided a nondestructive testing method for the detection of hydrothermal aging effects on large composite, graphite reinforced structures according to Kaeble.[40] Exposure of the composite to high moisture and high temperatures produced irreversible deterioration in shear strengths and modified the Weibull distribution of survival probability. Analysis of water diffusion kinetics showed that strength degradation was dominated by bulk matrix properties. Ultrasonic 2.25 MHz wave velocity transverse to the graphite fiber axis was sensitive to moisture and acoustic attenuation was correlated with prior moisture history and strength degradation. Effects of hydrothermal damage was determined by ultrasonic methods and dynamic mechanical spectroscopy.

ENVIRONMENTAL EFFECTS ON FATIGUE PROPERTIES OF C/G COMPOSITES

Several references have reported on the effects of water and high humidity environments on the fatigue properties of C/G fiber composites. It has been observed that the application of full reversed stress cycling under high humidity environmental conditions is more severe in its effect upon structural integrity of advanced composites than the application of stress in one direction. Thus, for example, when in repetitive flexural loading the fibers are alternately under tensile and compression stresses, which approximates "real-life" conditions, the fatigue of the material is accompanied by decrease in the modulus of elasticity.[41] As a nondestructive test, measurements of moduli are useful on the same specimens after different intervals of cylic reversal of stressing—especially in presence of water. Comparisons were run between glass laminates and graphite/epoxy laminates, with the latter demonstrating better endurance by retaining flexural moduli for a larger number of cycles before fall-off of values of the moduli of elasticity.

Sumsion[42] reported on the effects on torsional and flexural fatigue properties of graphite/epoxy composites. He clearly points out that the effect of environment is greater in the fully reversed cycling. Torsional fatigue testing was conducted at a frequency of one Hertz under conditions of controlled torsional strain. The specimens were prepared from Hercules' AS/3501 resin system and prepreg tape. Torsional fatigue data of ±45° fiber oriented graphite/epoxy composites is shown in Fig. 9-14. Changes occurred much more rapidly in a moist environment. Specimen stiffness, $\gamma\theta$, is shown as a function of the number of applied stress cycles, N, where γ is the applied shear stress and θ is the torsional deflection angle in radians per unit gage length. Reproducibility of data for the cross-plied materials was ±5%, and for woven materials, ±10%. The amount of fatigue damage from torsional cycling in water at 74°C compared to that in water at 24°C or air at 74°C is significantly greater. In the example cited, the maximum torsional stress applied was at 41% of the static torsional stress limit, and a reversed torsion cycle was used. After an apparent damage free period, the cross-plied (±45°) graphite epoxy was found to undergo a decrease in torsional stiffness with increasing stress cycles. The limiting stiffness in cross-plied specimens that were torsion tested at 24°C appears to correspond to a limiting torsional stress level equal to about 15% of the torsional failure strength. This stress may be looked upon as a fatigue limit for these composite materials.[43]

These environmental influences are recognized, and in view of the important designs to be fulfilled by C/G composites, reliance upon "real life" or service oriented data are essential. To draw attention to a specific example, the adhesion

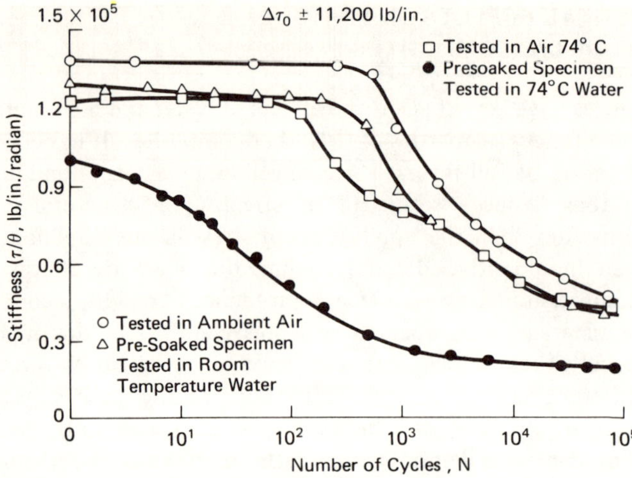

Fig. 9-14. Torsion fatigue of ±45° fiber-oriented graphite-epoxy composite. Decrease in stiffness (τ/θ) with number of cycles (log N) as a function of environment.[42] *(Reprinted with permission American Institute of Aeronautics and Astronautics (A.J.A.A.))*

of epoxy matrices may prove less under cyclic loading than under the same stress in sustained loading—particularly under high humidity conditions.

As noted in the beginning of the chapter, there are many aspects to environmental influences and emphasis has been placed upon the effects of moisture, principally because much of the recent research on environmental influences has focused on these aspects. While advanced composites of C/G fibers have demonstrated very low, controllable thermal coefficients of expansion, this chapter has pointed out problems related to dimensional changes in composites with humidity changes. As mentioned, there are applications in optical devices and in radio antennas, where changes in moisture content of the structure will affect the dimensional accuracy of the structure. Even in space the desorption taking place could influence the dimensional accuracy.

Further space applications of C/G fiber composites are reviewed in a later chapter, though to exemplify the hostile environments, a photograph of the magnetometer boom used on the Viking spacecraft, which landed on the planet Mars, is shown in Fig. 9-15. The C/G fiber advanced composites employed in the construction of the magnetometer boom were expected to maintain good functionality after takeoff from Earth and after the subsequent landing on Mars. Continued exposure to the radiation of space environment and the relative unknown hostile environment of Mars pose problems to all materials and all equipment. The high specific stiffness and specific strength of C/G advanced composites, accompanied by low order of thermal expansion, make this class of materials a good candidate for space applications.

Fig. 9-15. C/G fiber advanced composite magnetometer boom used on Viking space vehicle for Mars. (*Courtesy Jet Propulsion Laboratory.*)

The radiations of space pose a complex spectrum of conditions which are not fully understood. While the penetration of electrons, protons, and alpha particles into C/G fiber composites has been examined in the chapter on electrical applications, the effects of intense thermal radiation on C/G composites–ultraviolet light degradation (particularly of more recent candidate resin matrices), the high energy particles of Van Allen belts, and solar winds and their consequences–are part of the hostile environments in which advanced composites have not been fully evaluated. Studies on earlier structural plastic laminates exposed to high intensity thermal radiation have been reported,[44] though such studies have not appeared on more recent advanced composites. This report covered failure incident energy versus tensile stress and the spectral transmission and spectral reflectivity for certain polyesters, heat resistant phenolics, and silicone resins.

Studies are in progress on the effects of space environment on mechanical properties of graphite/epoxy composites.[45] Candidate materials will be placed in orbiting vehicles for 2 to 10 year space environment exposure. In time much more basic information will become available on the subject of environmental influences. When the subject of exposure to chemicals arises—as it will for applications of advanced composites to chemical industries—attention will be focused on the resin matrix, the most likely material to be attacked by chemicals. In fact, in the chapter on industrial applications, the success of valves and fittings of heat and chemical resistant polyphenylene sulfides in conjunction with graphite fibers has been noted in connection with a wide variety of industrial chemicals. Advanced composites may be expected to find important uses in the chemical industry environments.

Methods have been outlined for predicting performance of newly developed materials, so these may be used to construct vehicles suitable for long duration missions in known but variable space environments.[46] Accelerated aging by intensifying space environment components is one procedure, though the author utilizes the concept that like materials perform in a similar manner and uses real time performance of proven materials to predict performance of a new material containing like functional groups. It is probable that both procedures will be mandatory in selecting materials.

REFERENCES

1. Hertz, J., "High Temperature Strength Degradation of Advanced Composites," *SAMPE Symposium,* Vol. 3, Huntsville, Alabama, October 1971.
2. Hertz *et al.,* "Investigation into High Temperature Strength Degradation of Fiber Reinforced Resin Composite During Ambient Aging," Report GD CA-DBG-73-005, Contract NASB-27435, June 1973.
3. Browning, C.E., "Effects of Moisture on the Properties of High Performance Structural Resins and Composites," AFML–NTIS, AD 917821, September 1972.
4. Hertz, J., Christian, J.L., and Varias, M., "Advanced Composites Applications for Spacecraft and Missiles," Tech Report AFML TRMI-186, Vol. 2, page 269, March 1972.
5. Hertz, J., "Moisture Effects on High Temperature Strength of Fiber Reinforced Resin Composites," *SAMPE Symposium,* Palo Alto, California, October 1972.
6. Eckstein, B.H., "Moisture Absorption by Epoxy Laminating Resins," UCC Paper, Parma, Ohio, 1977.
7. DeIasi, R. and Whiteside, J.B., "Effect of Moisture on Epoxy Resins and Composites," *Advanced Composite Materials–Environmental Effects,* ASTM, STP 658, pages 2-20, 1978.
8. Browning, C.E. and Hartness, J.T., "Effects of Moisture on Properties of High Performance Structural Resins and Composites," ASTM, STP 546, pages 284-302, 1974.
9. Chung, E. and Crognola, A., "Time-temperature Moisture Studies on Graphite Fiber Reinforced Epoxy Composites," *Thirtieth Annual Technical Conference,* SPI, RP/C Institute, Section 9A, Washington, D.C., February 1975.
10. Sandorff, P.E. and Tajima, Y.A. (Lockheed), "Experimental Determination of Moisture Distribution in Carbon/Epoxy Laminates," *Composites,* Vol. 10, No. 1, page 37, January 1979.

11. Judd, N.C. (Royal Aircraft Establishment), "Effects of Water on Carbon Fiber Composites," *Thirtieth Annual Technical Conference,* SPI, RP/C Institute, Section 9A, Washington, D.C., February 1973.
12. Trabocco, R.E. and Standor, M., "Effect of Natural Weathering on the Mechanical Properties of Graphite/Epoxy Composite Materials," ASTM, STP 602, pages 67–84, 1976.
13. Gillat, O. and Broutman, L.J., "Effect of an External Stress on Moisture Diffusion and Degradation in a Graphite-reinforced Epoxy Laminate," ASTM, STP 658, pages 61–83, 1978.
14. Nash, H.C. et al "Effect of Prepreg Out-time and Humidity on Polyimide/Graphite Prepregs," Composites Technology Review, vol. 1, no. 4, p. 16, Summer, 1979.
15. McKague, E.L., Haklias, J.E., and Reynolds, J.D., "Moisture in Composites," *Journal of Composite Materials,* Vol. 9, page 2, January 1975.
16. McKague, L., "Environmental Synergism and Simulation in Resin Matrix Composites," ASTM, STP 658, pages 193–204, 1978.
17. Wollner, B. (Northrop), "Temperature-humidity Criteria for Advanced Composite Structures," *SAMPE Tenth National Conference,* page 674, New York, October 1978.
18. Browning, C.E., "Mechanisms of Elevated Temperature Property Losses in High Performance Structural Epoxy Matrix Materials After Exposure to High Humidity Environments," *SAMPE Twenty-second National Symposium,* Vol. 22, page 365, San Diego, California, April 1977.
19. Powell, J.H. and Zigrang, D.J., "Moisture Absorption and Desorption of Three Graphite/Epoxy Systems," *SAMPE Eighth National Conference,* Seattle, Washington, October 1976.
20. Sykes, G.F., Burks, H.D., and Nelson, J.B. (NASA-Langley), "Effects of Moisture on Dynamic Thermomechanical Properties of a Graphite/Epoxy Composite," *SAMPE Twenty-second National Symposium,* page 350, San Diego, California, April 1977.
21. Shirell, C., "Moisture Sorption and Desorption in Epoxy Resin Matrix," *SAMPE Twenty-third National Symposium,* Vol. 23, page 175, Anaheim, California, May 1978.
22. Welhart, E.K., "Environmental Effects on Selected Resin Matrix Materials," NASA, CR–150938, March 1976.
23. Waring, G., Hofer, K.E., Vadala, E., and Trobocco, R., "Failure Mechanisms for Advanced Composite Sandwich Construction in Hostile Environments," Conference on Advanced Composites, page 83, Technology Conference Associates, El Segundo, California, December 1979.
24. Crossman, F.W. and Flaggs, D.L., "Dimensional Stability of Composite Laminates During Environmental Exposure," *SAMPE Journal,* Vol. 15, No. 4, page 15, July/August 1979.
25. Wolff, E.G. and Eselun, S.A., "Coefficient Moisture Expansion," Conference on Advanced Composites, page 55, Technology Conference Associates, El Segundo, California, December 1979.
26. Mazor, A., Broutman, L.J., and Eckstein, B.H., "Effect of Longtime Water Exposure on Properties of Carbon Fiber Reinforced Epoxies," *Polymer Engineering and Science,* Vol. 18, No. 5, page 341, April 1978.
27. Fried, N., Kaminetsky, J., and Silvergleit, M., "The Effect of Deep Submergence Operational Conditions on Filament Wound Plastics," *Twenty-first Annual Technical Conference,* SPI, RP/C Institute, February 1966.

28. Fried, N., "Effect of Water on Glass Reinforced Plastics," *Fifth Symposium of Naval Structures, Mechanics* (1967), pages 813-837, 1970; and *Chemical Abstracts,* 76, 100429j, 1972.
29. Romans, J.B., Sands, A.G., and Cowling, J.E., "Fatigue Behaviors of Glass Filament Wound Epoxy Composites in Water," *Industrial Engineering Chemistry Product Regulatory Division,* Vol. 11, No. 3, page 261, 1972.
30. Macander, A. and Silvergleit, M., "Effect of Marine Environment on Stressed and Unstressed Graphite/Epoxy Composites," *Naval Engineering Journal,* Vol. 89, No. 4, page 65, 1977 (CA 87-152800V-1977).
31. Morley, J.M., "Role of Matrix in Preparation and Properties of Carbon Fiber Composites," *ACS Reprints, Organic Coatings and Plastics Chemistry,* Vol. 38, page 666, Anaheim, California, March 1978.
32. Scola, D., "Thermo-oxidative Stability and Moisture Absorption Behavior of Glass and Graphite Fiber Reinforced PMR-11 Polyimide Composites," *SAMPE Twenty-second National Symposium,* Vol. 22, page 238, San Diego, California, April 1977.
33. Keenan, J.D., Seferis, J.C., and Quinlivan, J.T., "Effects of Moisture and Stoichiometry on Dynamic Mechanical Properties of Carbon Reinforced Composites," *ACS Organic Coatings and Plastics Chemistry,* Vol. 40, page 700, April 1979.
34. Hertz, J., "Moisture Effects in Spacecraft Structures," *SAMPE Twenty-fourth National Symposium,* Vol. 24, page 965, San Francisco, California, May 1979.
35. Wolff, R.V., "Effects of Moisture Upon Mean Strength of Composite to Metal Adhesively Bonded Joint Elements," *SAMPE Twenty-second National Symposium,* Vol. 22, page 183, San Diego, California, April 1977.
36. Hofer, E.E. and Porte, R., "Influence of Moisture on the Impact Behavior of Hybrid Composites," *SPE, Thirty-sixth ANTEC Conference,* page 286, Washington, D.C., April 1978.
37. Mocho, A. and Ishai, O., "Thermal Cracking of CF/RP Laminates," *SAMPE National Meeting,* Vol. 10, page 255, New York, October 1978.
38. Crossman, W.F., Warren, W.J., and Pinoli, P.C., "Time and Temperature Dependent Dimensional Stability of Graphite-epoxy Composites," *SAMPE National Symposium,* Vol. 21, page 424, Palo Alto, California, 1976.
39. Camahort, J.L., Rennhack, E.H., and Coons, W.C., "Effects of Thermal Cycling Environment on Graphite/Epoxy Composites," ASTM, STP-602, pages 37-49, 1976.
40. Kaeble, D.H. and Dynes, P., "Moisture Degradation in Graphite/Epoxy Composites," *Materials Evaluation,* Vol. 35, No. 4, page 103, 1977.
41. Almon, M., McCrory, L., and Delmonte, J., "Behavior of Composites Under Cyclic Stress in Presence of Water," *SPE, Retec,* Las Vegas, Nevada, September 1975.
42. Sumsion, H.T., "Environmental Effects on the Torsional and Flexure Fatigue Properties of Graphite Epoxy," *American Institute of Aeronautics and Astronautics,* Paper 75-771, Denver, Colorado, May 1975.
43. Sumsion, H.T. and Williams, D.P., "Effects of Environment on the Fatigue of Graphite-epoxy Composites," ASTM, STP-569, pages 226-247, 1975.
44. Schwartz, H.S. and Lisle, B.J., "Effects of High Intensity Thermal Radiation on Structural Plastic Laminates," WADC Technical Report 57-638, December 1957.
45. Felbeck, D.H., U.S./R.D. Publication, page 3952, Washington, D.C., April 1978.
46. Long, R., "Materials Evaluation for Use in Long Duration Space Missions," *SAMPE Quarterly,* Vol. 11, page 23, October 1979.

10 Test Methods for Advanced Fiber Composites

Test methods for advanced fiber composites are reviewed with particular emphasis on carbon and graphite fibers and their applications, particularly by the aerospace industry. The basic guidelines for testing arise from the American Society for Testing Materials (ASTM) and its counterparts in other countries, though new techniques have been born of necessity to accommodate the vagaries of newer materials, products, and processing measures. The larger suppliers of C/G fibers (see Chapter 2 for the American, English, and Japanese manufacturers) have also instituted test methods appropriate to advanced fiber composites. Among the United States government agencies most active in fostering the use of advanced composites and establishing test procedures are the Air Force Material Laboratory, the Naval Research Laboratory, and NASA research centers. The larger aerospace corporations and the resin material and prepreg suppliers, who have basic interests in advanced fiber composites, have supported tests and programs to standardize the quality and performance of the ultimate products. The media by which testing information is disseminated are numerous and the professional scientific societies play a leading role through their symposia and conferences—such as the Society of the Plastics Industry (SPI); the Society for the Advancement of Materials and Process Engineering (SAMPE); the American Chemical Society (ACS); the Society of Plastics Engineers (SPE); ASTM; the American Carbon Society; the Society of Automotive Engineers (SAE); government agencies, which release publications through the National Technical Information Service (NTIS); Federal Standard Specs.; and the Royal Aircraft Establishment (RAE). There are many other organizations and publications concerned with testing information, and the test methods established by manufacturers of glass fibers have also been invaluable as source material.

The fact that contributions come from so many sources underscores the multiple disciplines involved in testing advanced fiber composites. This situation also accounts for the empiricism which will be observed in many instances, though this is being reduced as the intrinsic qualities of the materials become better known and established. As pointed out in previous chapters, C/G fibers and their technology are developing, and test methods applicable several years ago are not always relevant in the light of current practices. However, a start must be made, and what better test method to begin with than basic tensile strength of C/G fiber?

TENSILE STRENGTH

In the tensile testing of C/G fibers, considerable study has been undertaken on testing tensile properties of the single filamentary fiber to methods for tensile testing of larger laminated composites used for structural purposes. For the engineered applications, the physical properties and mechanical behavior of the final composites are most relevant, though for comparison of fibers and resin matrices, intermediate tensile tests are necessary for quality control purposes. Hence, there are test methods which are intended to compare C/G fibers from different sources, without the variables introduced by the sequences of ply stacking and the processing methods used to prepare the final advanced fiber composite. There are several good papers which correlate the tensile strength properties of the filaments with the properties of the final composite, and some of these will be referred to in this chapter.

In drawing comparisons between the tensile strength of various graphite filaments, it is important for one to understand the test methods involved and the test specimens used. Ultimate tensile strengths of C/G fibers are reported as described below.

The tensile strength of single monofilaments is a laborious test because of the fragile nature of the fine filament (typically 6 to 12 microns in diameter). The selected fiber is mounted across the cut-out area of a firm piece of cardboard and held in place with sealing wax. The cardboard is placed in the jaws of a testing machine after axial alignment with the testing device, and the edges of the cardboard are cut out by a heated wire or a laser beam, to avoid disturbing the fine filament. The breaking loads of multiple specimens are determined (see Fig. 10-1).

There is a standard ASTM specification for the preparation, mounting, and testing of high modulus, single filament materials over 21 GPa (over 3×10^6 psi).[1] The method is limited to single filaments utilizing a fixed gage length at least 2000 times longer than the nominal filament diameter. Typical mounting of a test specimen is illustrated in Fig. 10-1. Applied loads for the C/G fibers are

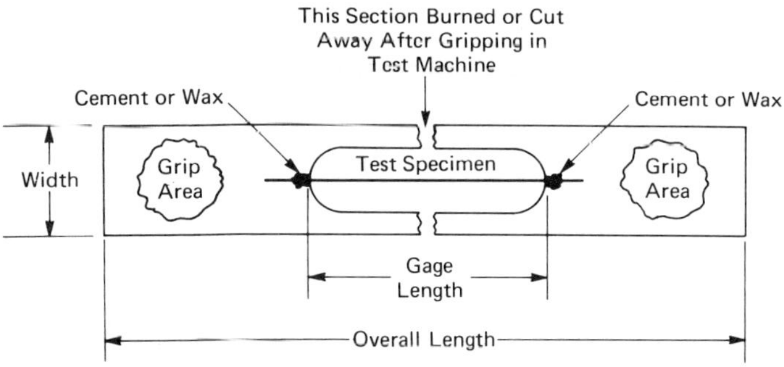

Fig. 10-1. Tab showing typical specimen mounting method.[1]

usually in the order of grams. That portion of the indicated elongation contributed by the load strain system and specimen gripping system (compliance) may be first determined experimentally, though it is usually found to be insignificant. Obtaining accurate readings becomes increasingly difficult with the higher modulus fibers (some of the new high modulus graphite fibers show moduli over 100×10^6 psi), because of the very low elongation.

In tensile testing of yarns and tows comprising many hundreds or thousands of filaments, a portion of the tow is selected for tensile testing. There may be as many as 720 filaments for rayon fibers and up to 160,000 filaments for some polyacrylonitrile (PAN) precursors (GY-70 contains 400 filaments). The filaments are preferably handled by polyethylene gloves and then impregnated with a resin matrix to maintain their integrity and reduce breakage and misalignment. A portion of the tow is passed through a resin bath, dried, and cured. The tows or yarns which are impregnated and then tested in tension should be uniformly collimated and cured. Excess resin is removed by a sizing groove or a loop formed from a coiled wire, which provides external pressure to smooth out external protrusions or catenaries. Specimens showing broken filaments in the gage area, extreme fuzziness, or imperfections which would mitigate against meaningful results should be discarded. Two alternatives are available for attachment to the testing machines: 1) a cast-on, large diameter collar of liquid resin over the exposed fiber ends, fitting within a similar receptacle on the testing machine, or 2) a rubber faced gripping device with a threaded clamp to ensure tight grasping of specimens. It is desirable that during drying and cure there should be some tension maintained upon the impregnated fibers to ensure axial alignment of the individual filaments.

Bell[2] advanced the idea that broken filaments are locally ineffective at the fracture point and that the resin will redistribute the broken fibers' tensile load into the strong adjoining fibers. It was felt that as more tows are laminated together, weaknesses in some filaments can be reinforced by adjacent filaments.

He attributed scatter in the data to the large distribution in filament strength. (See also the data supplied by Union Carbide in Fig. 10-2 on filament tensile strength of "Thornel" 50, a C/G fiber prepared from regenerated cellulose.)

This problem was treated effectively by McMahon,[3] who drew attention to the fact that, as the gage length of the test specimens was increased, when measuring the tensile strength of high modulus graphite fibers, lower tensile strengths were obtained. This indicated that the tensile strengths were flaw related: when more of the specimen length was exposed for test, the greater was the probability of seeing more flaws. When the tensile strengths are plotted as a function of the logarithm of the gage length (as seen in Fig. 10-3), the relationship is nearly linear. Several carbon and graphite filaments are shown in this chart of logarithm of gage length versus the ultimate tensile strength. Boron fibers and S glass fibers show a similar relationship. To obtain these data requires tensile measurement at a minimum of several gage lengths, permitting extrapolation to very short gage lengths—with minimum fiber defects. Reports are published, in some instances, specifying the gage length used for tensile measurements.

The tensile testing of resin impregnated portions of the filaments of the tow or yarn appears to be the most popular method for checking of the tensile strength of C/G fibers. Some manufacturers and suppliers of prepreg have prepared detailed process sheets for preparing tensile test specimens, and much may

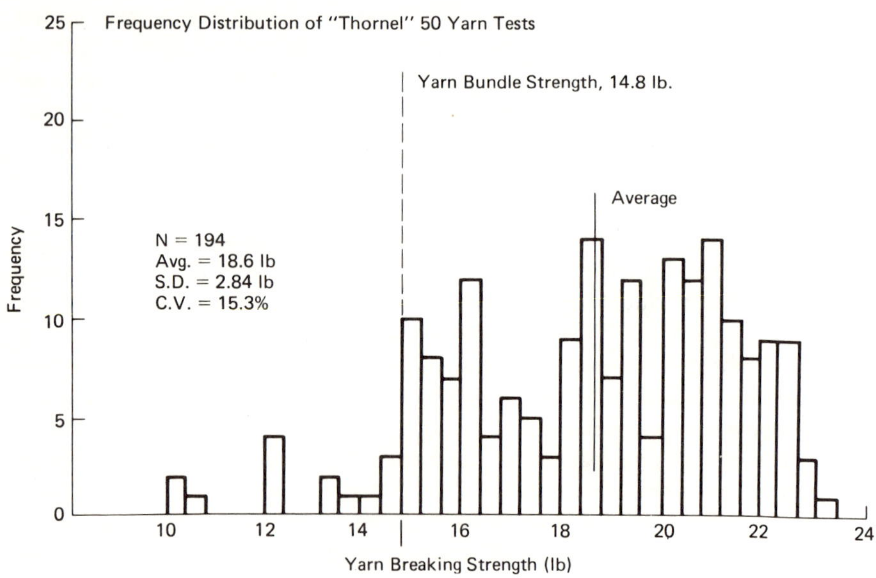

Fig. 10-2. "Thornel" 50 yarn data.[2]

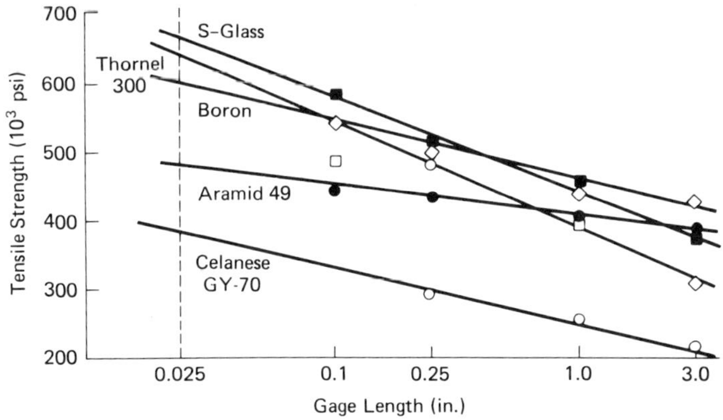

Fig. 10–3. Tensile strength versus gage length for various high modulus reinforcement fibers.[3]

be learned from an examination of such reports. As an example, Hercules' specifications are cited in the References.[4] The contribution of the resin content to support of the tensile load is usually very low because of the large differences between the moduli of the resin matrix and that of the fiber (0.5 msi typical for the resin matrix and 20 to 70 msi for the C/G fiber). Because the elongation of the fiber at the time of breakage is of the order of magnitude of 1% or less, the resin contribution to strength is ignored, and the tow or yarn strength is determined as the maximum load divided by the cross-sectional area of the fibers alone (see procedure below). For modulus of elasticity determination, the schematic illustration in Fig. 10–4 depicts the mounting of the extensometer or the strain gage.

Dry strand testing of fibers was pursued by McMahon,[5] who compared the test data of the impregnated strands of carbon fibers with dry strand techniques and vibroscopic denier measurements. Good correlation of test results were reported for the tensile strengths of C/G fibers. The dry strand method of testing of fibers requires more care in handling and positioning fibers than does the impregnated method described above, and in general dry strands are mounted across a cut-out, as in Fig. 10–1 for the single strands, and are held in place by bonding with an epoxy adhesive to provide greater strength and avoid slippage.

The cross-sectional area of unidirectional C/G fibers selected from a tow or yarn is generally determined in the manner described herein. Rather than laboriously counting the number of fine filaments that are present, the denier of the unimpregnated fiber, which is the weight (g) for a length of 9000 m, is determined. The cross-sectional area is calculated in square inches by multiplying the denier of the fibers (g/9000 m) by the factor of 1.72×10^{-7}, divided by the

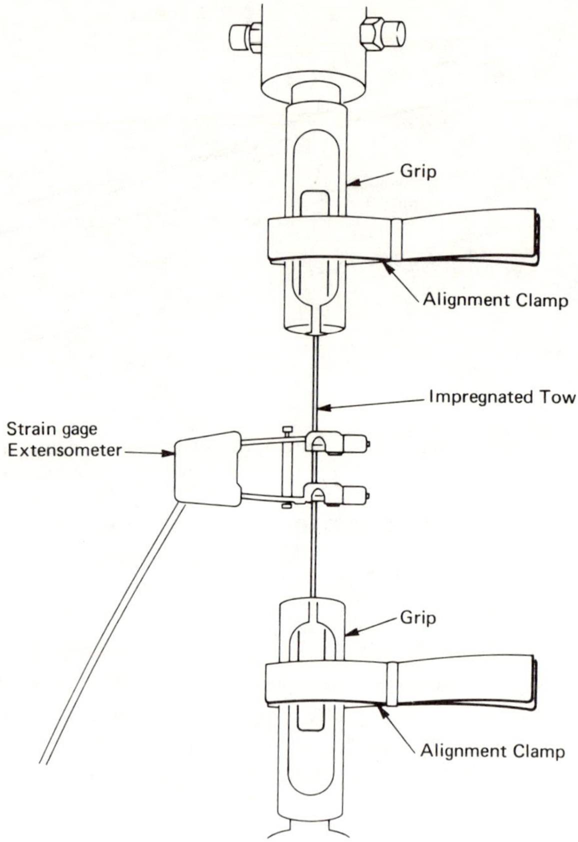

Fig. 10-4. Tensile specimen mounted in test equipment.[4]

density (g/cm^3) or converted to area (m^2) with the factor 1.11×10^{-7} divided by the density in kg/m^3.

Tensile strength and modulus expressed in grams per denier (gpd) units can be converted to psi units by multiplication of the former with the factor of 12,800 × density (g/cc) or to MN/m^2 with the factor 88.2 × density (g/cc). The tensile strength in gpd units is referred to as the tenacity.

While in 1979 complete standardization of testing of carbon and graphite fibers and their composites was still in committee, references to other fibers may be cited. Testing of glass fibers has been standardized for some time.[6] The schematic outline for impregnating rovings, shown in Fig. 10-5, is applicable to other types of fibers. Cure of resin impregnants may follow, rather than wrap-up about the bobbin, when testing of carbon fiber tows takes place, because of the

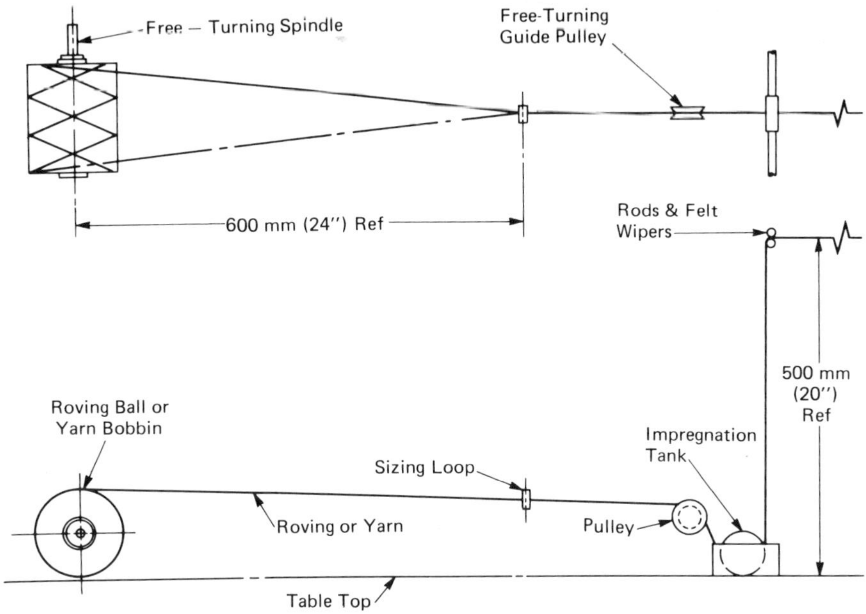

Fig. 10-5. Typical impregnating apparatus for rovings and yarns.[6]

rigidity and stiffness of the test specimens. Glass fibers are stronger in some instances than C/G fibers, though do not have the stiffness of C/G fiber specimens.

Measurements of ultimate strain or elongation are very important to the design of advanced composite structures. While conventional stress analysis procedures provide significant reference data, studies and analyses are dependent upon strain or deformation data. The anisotropic nature of advanced fiber composite structures lends itself to more useful interpretation, when examining allowable strains.

C/G fiber composites require sophisticated measuring techniques for determining strain. These techniques must be performed with precision by strain gages or extensometers. For extremely low strain, optical interferometers are used. The usual care of using tensile testing machines with a constant rate-of-crosshead movement, and care in providing proper axial alignment of test specimens, should be taken. Details within ASTM specifications are pertinent.

When testing tensile properties of cured resin matrix composites reinforced by oriented continuous or discontinuous high modulus fibers (greater than 20 GPa or 3.0×10^6 psi moduli), there is an applicable test specification.[7] Laminates of unidirectional reinforcing fibers, as well as symmetric orthotropic construction, follow these test procedures. The typical tensile test specimens are shown in

Fig. 10-6, taken from ASTM D3039, together with tables on recommended thicknesses for glass, boron, and graphite fibers. Exact and precise positioning of reinforcement tabs are important. There are no recommendations for the hybrid constructions, though one should be governed by the principal fibers employed. Standard strain rates between 0.01 and 0.02 inches/inch/minute are recommended for testing by ASTM D3039. An alternate method uses constant crosshead speed, determined by multiplying the strain rate by the distance between tabs, in millimeters or in inches. While the specification indicates that variation in specimen width must be maintained within ±1% (nominal width is 0.5 inches), testing laboratories are frequently called upon to maintain much closer tolerances for evaluation of high performance fiber composites from C/G fibers (usually ±0.25%). Furthermore, minute inspection to ensure proper axial alignment of the unidirectional fibers in the composite test specimens is an important measure, as well as care in machining to avoid fraying and disorientation of high strength, high modulus fibers in preparing specimens. High speed, diamond gritted grinding wheels have proven necessary in preparing specimens for accurate tensile tests. It has been found of advantage to use a backing plate such as an acrylic undersheet under the composite while cutting, to avoid fraying out of fibers.

Type of Reinforcement	Continuous Fibers		Discontinuous Fibers	
	mm	in.	mm	in.
Boron	0.508 to 2.54	0.020 to 0.100	1.016 to 5.08	0.040 to 0.200
Graphite	0.508 to 2.54	0.020 to 0.100	1.016 to 5.08	0.040 to 0.200
Glass	0.762 to 3.275	0.030 to 0.125	1.524 to 6.55	0.060 to 0.250

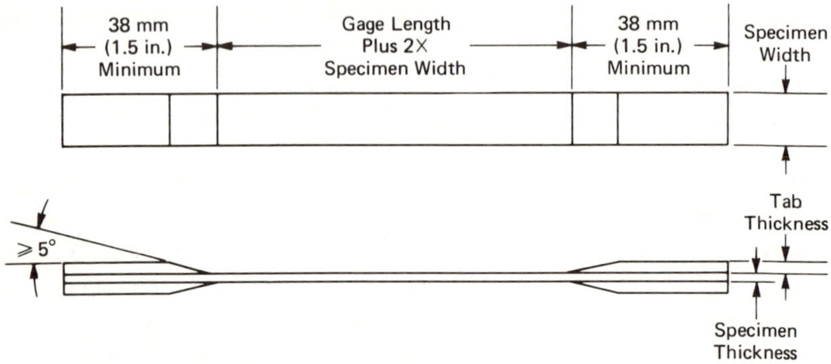

Note – Bond laminate tabs on two sides and both ends. Tabs are applied to the end of the test composite with a suitable adhesive. Each tab is a minimum of 38 mm (1.5 in.) long by the width of the laminate and a thickness of approximately 1.5 to 4 times the thickness of the test composite.

Fig. 10-6. Composite tensile test specimens and recommended thicknesses for various reinforcements.[7]

For those familiar with the testing of plastics materials, particular attention should be given to the use of tabbed tensile specimens in lieu of the reduced cross-section radiused members prepared from lower strength laminates. Precise tab alignment, mounting, and bonding of the tabs, which are usually thicker than the test specimens, with suitable adhesives, are important testing details. The ASTM D3039 also outlines the conventional procedure for determination of Poisson's ratio (which is measured in the linear region of the stress-strain curve)–a ratio of the strain measured perpendicular to the applied load to the strain parallel to the applied load.

The utilization of flat tensile specimens with tabs added at the ends for insertion into tension testing machines appears to be widely accepted, particularly in aerospace industries concerned with the high modulus, high strength C/G fiber composites. There have also been other approaches which recommended reduced thicknesses and reduced widths at the center of the test specimens to promote failure in the gage section located at the center.[8] The use of tabs was motivated by the difficulty which appears when unidirectionally reinforced composites are tested in compression or tension with the test direction parallel to the fiber direction, resulting in splitting of test specimen when reduced sections are employed. Data are also sought when the test loads are applied transverse or across the fiber direction–as this failure mode is strongly dependent upon the properties of the resin matrix.

The considerable work of Brown[9] in measuring elastic modulus and tensile strength of high modulus graphite was focused upon the C/G fibers made from rayon precursors. Cross-sectional areas were examined optically and photomicrographically recorded. The crenulated or dumb-bell cross-section of the fibers from rayon made area measurements more difficult. Frequency distribution of cross-sectional area was widespread. A minimum sampling recommended was 30 filaments. The author stressed the value of testing complete strands, with an effective strand cross-section calculated from the weight per unit length and density. High modulus fiber strands are hard to grip and are easily frayed in the testing fixture, a problem minimized by impregnating the strands with a resin and pulling it as a composite. In using the strand test for both quality and process control for more than a year on over 500 samples, the following data were typical of C/G fibers from rayon precursors:

Elastic modulus = 50.7×10^6 psi; standard deviation = 2.4×10^6 psi.
Tensile strength = 287,000 psi; standard deviation = 24,000 psi.

An effort was made to establish a relationship between electrical resistivity and elastic modulus for the C/G fibers prepared from rayon precursors. The wide scatter of data is shown for these particular tests (see Fig. 10-7). For the ultra-

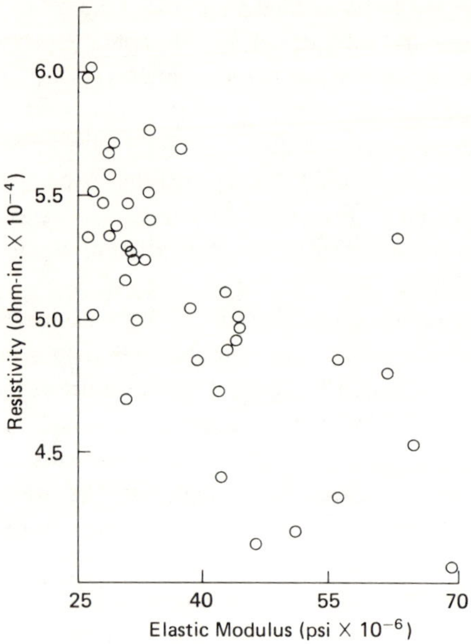

Fig. 10-7. Relationship of electrical resistivity and elastic modulus on carbon fiber from rayon.[9]

high modulus graphite fibers, such as GY-70 of Celanese, better correlation of elastic moduli with electrical resistivity was shown by McMahon.[5]

The tensile strengths of notched composites, studied by Phillips and Beaumont,[10] required the preparation of special test specimens. Tensile strengths of different aligned fiber/epoxy composites containing cracks perpendicular to the fibers have been measured as a function of crack size. Surface treated carbon fiber composites showed themselves to be notch sensitive, whereas the untreated surface carbon fiber composites were notch insensitive.

Considerable data on testing of advanced composites appeared in a report authored by the Convair division of General Dynamics.[11] Standard practices for tensile testing, as already described, were employed in developing information, though on other physical properties (such as compressive strength and shear), standardization by industry had not been achieved at that time. There were valuable comments made which reflect aerospace industry concern for good test specimen preparation. For example, the careful inspection prior to testing to eliminate specimens with misaligned fibers, accurate dimensional measurements to the nearest 0.0001 inch, and the polishing of edges after machining are typical of the details emphasized. Two main types of C/G fiber (GY-70 and HM-S) epoxy composites were compared in the report and individual tensile curves reproduced at temperatures ranging from -320° to 350°F.

In Figs. 10-8 and 10-9, two graphite fiber constructions are reported upon—GY-70/HM-S/X-904, emphasizing high modulus fibers, and HT-S/X-904, with the principal fiber a high tensile strength material. Unidirectional tensile data are shown for the composites, and the stress-strain curves reflect the fiber and the dominating physical properties. Performance at three temperatures is illustrated: -320°F, room temperature, and +350°F. Elastic moduli are quite similar for each material at the three test temperatures (HT-S, 20 msi; GY-70/HM-S composites, 40 msi). At -320°F, failure mode was more brittle than at room temperature and at 350°F. Each plot consisted of several hundred data points from six strain-gaged specimens at each condition. Standard ASTM specimen details were essentially followed in preparing composite test specimens.

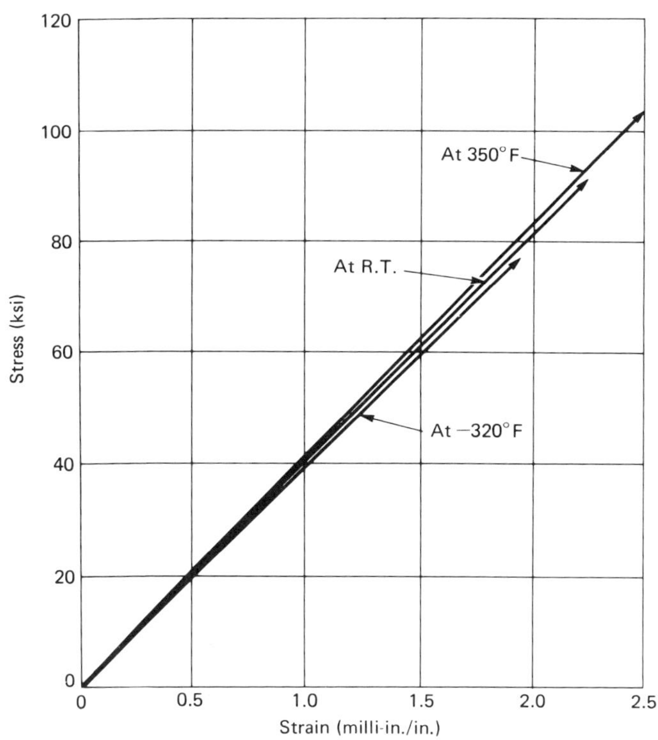

Fig. 10-8. Longitudinal tension (Graphite/Epoxy) (GY-70/HM-S/X-904).[11]

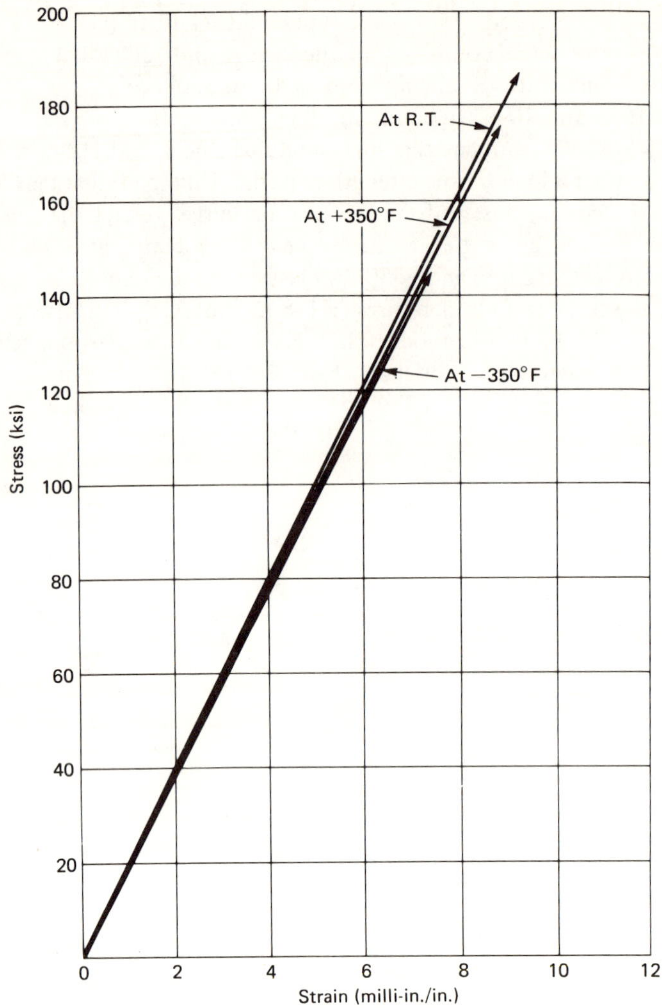

Fig. 10–9. Longitudinal tension (Graphite/Epoxy) (HT-S/X904).[11]

The cure cycle for preparing test specimen panels followed this procedure:

1. Heat prepreg to 190° to 200°F at 3° to 5°F/minute under vacuum.
2. Hold at 190° to 200°F for 15 minutes and apply 100 psi autoclave pressure.
3. Hold at 190° to 250°F for an additional 60 minutes.
4. Raise temperature to 290° to 310°F at 3° to 5°F/minute and hold for 60 minutes.

5. Raise temperature to 365° to 385°F at a rate of 3° to 5°F/minute and hold for 4 hours ± 15 minutes.
6. Cool under pressure below 150°F.
7. Post cure in oven at 360° to 390°F for 20 hours ± 15 minutes under vacuum bag pressure.

Tensile testing continues to be a valuable means for examining advanced fiber composites, and the methods and care for preparing, machining, and conditioning test specimens are applicable to other testing procedures, which are reviewed in the pages to follow. The reader is urged to become familiar with ASTM standards and federal specifications applicable to non-metallics and to C/G fiber structural composites, as well as specifications developed by material and component suppliers. For comparative purposes, tensile stress-strain curves are illustrated in Figs. 10-10 and 10-11 for several grades of C/G fibers, glass fibers, boron fibers, and aramid fibers. Also shown are the stress-strain curves of three types of epoxy resin materials. When compared with stress-strain curves of advanced composites (Figs. 10-8 and 10-9), the composites reflect the overriding influence of the fiber reinforcement in the shape of their curves.

COMPRESSIVE STRENGTH

Compressive strength determinations of fibrous structural composites require special testing techniques and fixtures which accommodate the anisotropic character of the composites. For thin laminates, compressive loads are applied and evaluated upon relatively short sections to forestall buckling of the columnar fibers under load. The compressive tests have been developed principally for evaluation of materials parallel to the fiber direction, though there are bearing applications where compressive loads are applied perpendicular to the axis of the fibers. Close machining tolerances are essential, as with tensile test specimens, and, on specimens being loaded directly, parallel machined faces are important for applying compressive loads.

Standardization of test fixtures has not yet been accomplished for compression testing of advanced fiber composites, though one will find trade references to fixtures such as the "Northrup" compressive fixture, which has been used by some laboratories to obtain meaningful test data. Sandwich beam tests, using unidirectional face plies, have provided data on tensile and compressive properties during bending in a four point flexure fixture, though there is sensitivity to the honeycomb core density. Evaluations at General Dynamics[11] drew attention to several compressive fixtures, although no one method was singled out over the others. The standard ASTM method of determining the compressive properties of oriented fiber composites offers procedures for the testing of high modulus

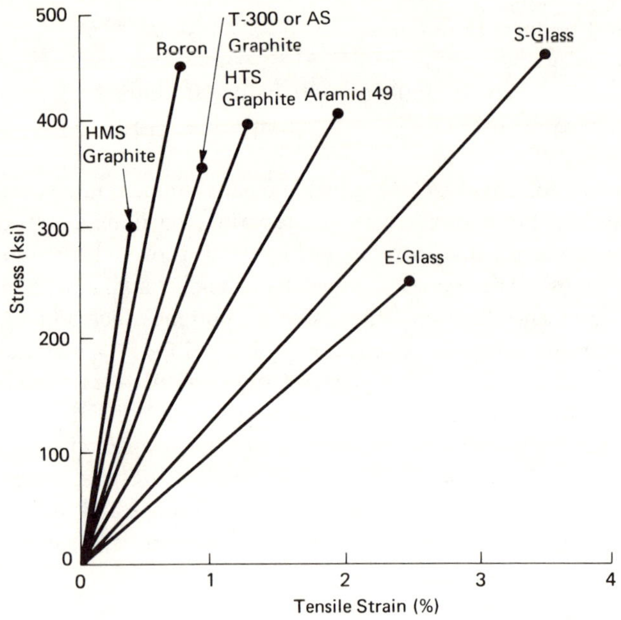

Fig. 10–10. Stress-strain diagrams of various fibers. (E.I. du Pont technical data.)

fibers over 3×10^6 psi (21 GPa).[12] Reinforcing tabs are used on compressive, as well as on tensile specimens. Recommended composite specimen thicknesses for different fiber types are as follows.

FIBERS	SPECIMEN THICKNESS
Boron	0.060 to .080 inches
Graphite	0.060 to 0.120 inches
Glass	0.125 to 0.155 inches

Specimen thicknesses are greater than those recommended for the tensile testing of composites, primarily to reduce specimen buckling. In Fig. 10–12 is illustrated the recommended fixture for compression testing and the dimensions of the test specimen and bonded tabs. The fixture is placed between the platen of the testing machine, at least one of which is coupled to the testing machine with a hemispherical ball and swivel joint to assure alignment. The positioning of the strain gage for deformation under load appears in Fig. 10–12.

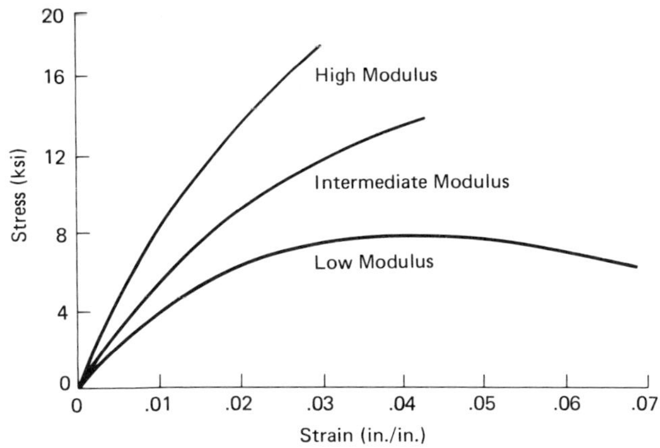

Fig. 10–11. Stress-strain diagrams of epoxy matrix resins.

Experimental and theoretical studies were presented on the failure modes of unidirectional fiber composites subjected to compressive loading parallel to the fiber direction by Greszczuk.[13] Compression test data were obtained on composites with carbon fibers imbedded in resin matrices ranging in elastic moduli from 2 to 460 ksi. For anisotropic composites reinforced with graphite fibers, there exist three different compression failure modes: 1) micro-buckling in the shear mode, in the case of composites made with low modulus resin; 2) transverse tension failure, including fiber "splitting," in the case of composites made with intermediate modulus resin; and 3) compression failure of the reinforcement, in the case of composites made with high modulus resin. Micro-buckling initiates transverse deformation of individual fibers, and when deformation occurs, bending stresses are induced in the test specimens. Misaligned and broken fibers within the test specimens will decrease compressive strength. Compressive strength was found to increase with higher shear modulus of the resin.

SHEAR STRENGTH

The advanced fiber composites have required special test methods for developing shear strength data useful for engineering design and analysis. Difficulties arise from the anisotropic character of the laminated structures that are produced and the environmental and temperature influences upon the materials. Attempts have been made to simplify tests so that the data generated are applicable to a variety of composite configurations and load distributions. Because they represent the efforts of many qualified individuals, the ASTM standards represent a good starting point in examining the available procedures for the determination of

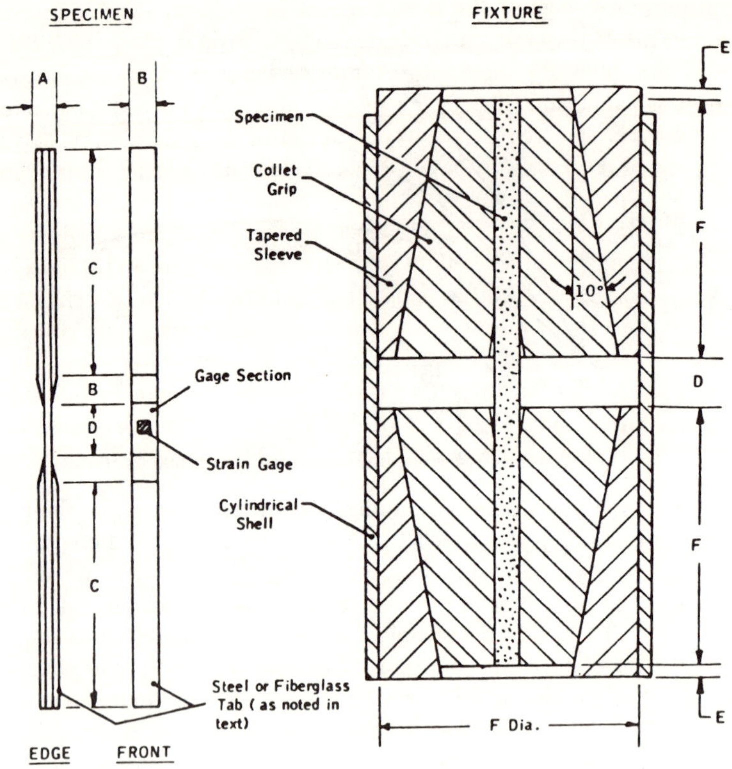

Dimensions	m	in.	Dimensions	m	in.
A	0.00399	0.157	D	0.01270	0.500
B	0.00635	0.250	E	0.00318	0.125
C	0.05715	2.250	F	0.06350	2.500

Fig. 10-12. Compression test specimen and fixture.[12]

shear strength. ASTM D3518-76 recommends procedures for the determination of the ultimate shear strength and the in-plane shear stress-strain response of unidirectional reinforced plastics—such as laminates of C/G composites.[14] A tension test is performed upon a ±45° symmetric laminate prepared in accordance with ASTM D3039.[7] Loads and longitudinal and transverse strains are recorded to approximate the complete shear response of lamina. The uniaxial stress-strain curves of the ±45° tensile coupon are used to tabulate values of longitudinal strain and transverse strain, corresponding to particular values of load.

There are other alternative methods for the determination of in-plane (intralaminar) shear of advanced composite materials. During the loading process there is a continuous change of the lamina stress distribution. The major contribution to the non-linear response of composites stems from in-plane shear behavior. At higher values of stress, specimens may fail because of transverse stress components.[15] Chamis and Sinclair[16] suggested the use of a 10° off-axis tensile test specimen in order to reduce the contribution of the normal in-plane stress components. They were able to obtain higher values of in-plane shear than possible with the 45° specimen.

The preparation of precise 45° angle-ply laminates from high strength unidirectional, high modulus graphite fiber tape requires care and skill in lay-up and cure of the composite. This procedure has its drawbacks in that the compression loading of the specimen may give substantially different values of mean failure stress.

Frequently, in current literature (and in this text), there are references to the short beam interlaminar shear strength of advanced fiber composites (see ASTM D2344-72[17]). Using a shortened span on a flexural test fixture, with a span to specimen thickness ratio of 4 to 1, testing is performed as shown in Fig. 10-13. Modes of specimen failure also accompany this sketch. The load nose is a dowel pin with a diameter of 0.250 inches. The supports have diameters of 0.125 inches. The short beam shear strength (sometimes called interlaminar shear strength—ILS or ILSS) at failure is calculated as follows:

$$\text{ILS} = 3P/4\,bd$$

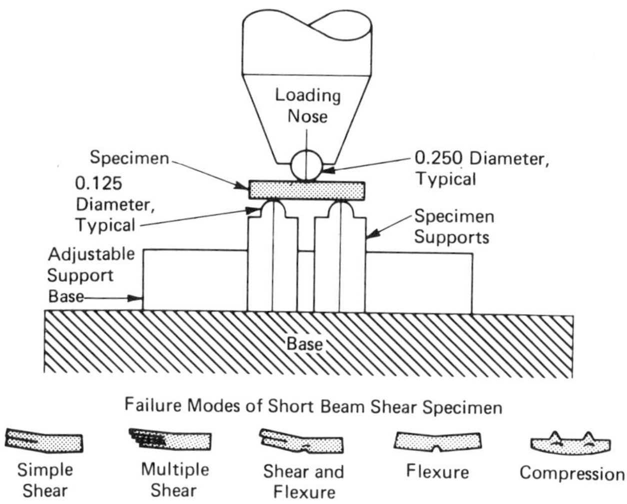

Fig. 10-13. Short beam shear testing fixture.

ILS = shear strength, N/m^2 (or psi)
P = breaking load, N (or lb)
b = width of specimen, m (or inches)
d = thickness of specimen, m (or inches).

Flaws and defects in test specimens would disqualify the specimen from the test. As apparent from earlier chapters, the nature of the surface treatment on the C/G fibers has an important bearing upon the interlaminar shear strength and should be noted when reporting test results. The ultra-high modulus graphite fiber composites have not shown as good ILS data in advanced composites as the general purpose C/G fibers.

Among the flat coupon type shear tests, the rail shear test is one of the most widely used methods in the aerospace industry.[18] A rail shear test specimen is shown in Fig. 10-14. Part A shows the fixture, Part B is the idealized pure shear, and Part C reveals that a bending moment must exist. The method consists of clamping the long sides of a laminated composite plate to steel rails by bolts. Shear stresses are induced in the test specimen by a displacement of one rail parallel to the other. Though the method is simple, inexpensive, and easily adapted to high and low temperatures, it has not won favor with stress analysts because of the presence of large combined stresses that may initiate premature failure in unknown modes. In particular, the existence of free edges causes the stresses in the laminate to differ from pure shear.

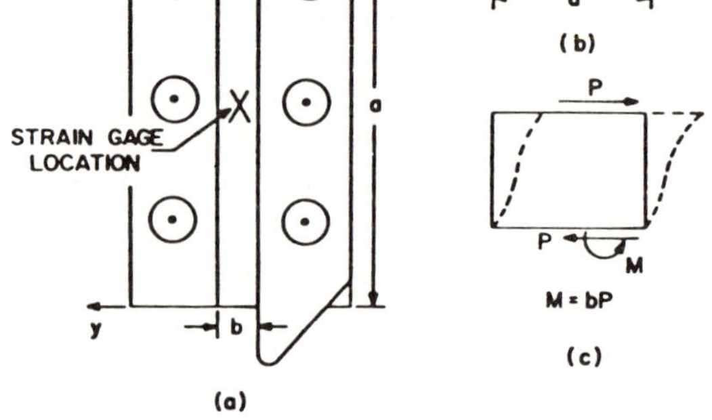

Fig. 10-14. Rail shear specimen.[18]

Duggan[19] offered a modified lap shear test, wherein a unidirectional laminate is bonded between two metal bars, with the fibers oriented parallel to the specimen loading axis. This method may cause failure of the bond before the high strength composite. Also investigated was a slotted tension test, whereby a conventional tensile coupon of C/G composite is loaded axially in tension, and simultaneously in compression, along a portion of its length through a rectangular test section in a transverse direction. Good shear data were developed and reported by this method. The analysis predicts that the shear stress in the center of the test zone is within 2% of the stress calculated by simple statics.

In reviewing the shear stress-strain response of unidirectional composites, Sims[20] placed emphasis on results obtained with the rail shear test fixture. The fixture used in these experiments is illustrated in Fig. 10-15, with the load being applied through the center part of the fixture. In fastening the test specimen into position, a torque of 40 ft-lb was applied to each of the nine bolts, and the compressive load applied at a rate of 20 lb/second. It was felt that a better approximation to a state of pure shear was obtained by this method. Furthermore, two independent calculations of the shear stress-strain response can be made from one specimen.

Chiao *et al.*[21] have evaluated the measurement of shear properties of fiber composites. They have recommended the ±45° angle ply laminate tensile shear test, as described by the ASTM standard given earlier. However, this method has its limitations, in that, while in-plane shear data are provided, the out-of-plane shear modulus is not determined.

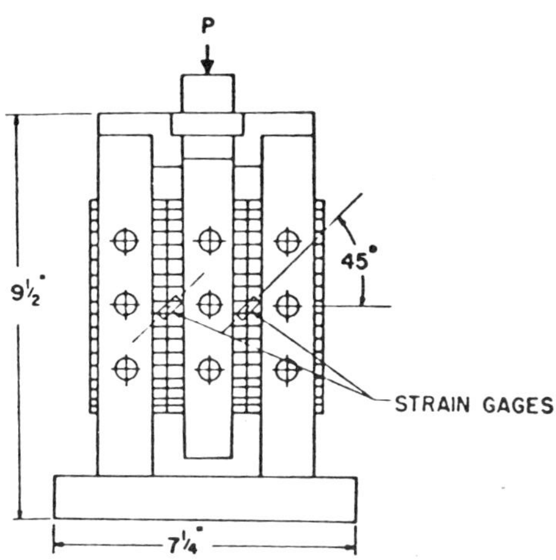

Fig. 10-15. Rail shear test on 0/90 laminate.[20]

While the application of pure torque to a tubular specimen has been used as a technique for the determination of the in-plane shear stress-strain response of unidirectional composite materials, the tube test has the disadvantage of special specimen fabrication and testing equipment. Sumsion *et al.*[22] have proposed and reported on torsion tests of flat specimens of rectangular cross-sections. The effective in-plane and out-of-plane shear moduli were determined on flat sheets 2 mm thick, and of different widths, of A-S/3501 and T-300/5209 graphite fiber epoxy composites. Torsion tests were run under controlled deflection, constant angle of twist, and the shear moduli were calculated from an equation derived from the theory of elasticity, which relates applied torque, torsional angle of twist, and specimen geometry. One of the reasons the authors recommended this test method was because of its value to studies of environmental degradation of composites. Another factor was the opportunity of developing accelerated test methods on moisture-induced property changes. Material properties at or near the surface (where moisture concentration is at the highest) are given more weight than material properties in the bulk specimen.

FLEXURAL STRENGTH

The flexural strength of reinforced plastics and advanced fiber composites is arrived at by standard test methods, entailing the use of a three point or a four point test fixture. For the high modulus C/G fiber composites, the three point fixture (see Fig. 10-16) is preferred and is the most direct and simplest technique to use. (Refer also to ASTM D-790 procedure.) Flat specimens machined with care and precision, as described earlier for tensile testing, are selected for measurements. The usual practice is to prepare a test panel from prepregs, complete the cure, and plan a specimen cut-out procedure—certain areas retained for tensile specimens, for compressive specimens, for resin content determination, for flexural strength, etc. For the ultimate flexural strength—stress in outer fiber at failure (FS)—calculations in psi and modulus of elasticity in flexure (E_B) are as follows.

For three point flexure test:

$$\text{FS} = \frac{3PS}{2bt^2} \qquad E_B = \frac{S^3}{4bt^3} \times \frac{\Delta P}{\Delta d}.$$

For four point flexure test:

$$\text{FS} = \frac{3PS}{4bt^2} \qquad E_B = 0.21 \times \frac{S^3}{bt^3} \times \frac{\Delta P}{\Delta d}.$$

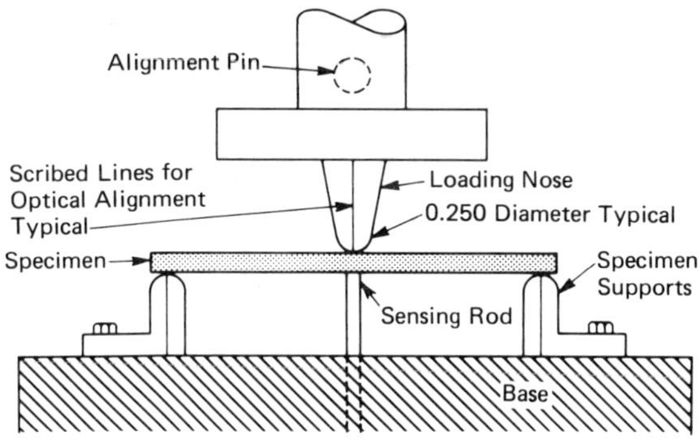

Fig. 10-16. Three point flexure test fixture.

where

P = maximum load carried by test specimen (lb)
S = span (inches)
b = width (inches)
t = thickness (inches)
$\Delta P/\Delta d$ = slope of initial straight line portion of load deflection (lb/inch).

The span to thickness ratio is usually about 32 to 1—considerably larger than the span used in the short beam test for interlaminar shear, described earlier. While the longitudinal flexure specimen does not provide data that can be used directly for design, because compression, tension, and shear loads are all present, flexure tests are valuable for performing comparisons between materials or for evaluating the effects of various exposures, as well as for receiving inspection and quality control purposes.

The four point flexural loading procedure, as illustrated in Fig. 10-17, is often used in testing composites, especially at elevated temperature. This fixture minimizes the local crushing beneath the central loading nose of the three point loading procedure. Loading at one-quarter of the span distance from each support is recommended. Flexure testing offers a simple testing method for measuring the modulus of elasticity without having to resort to the application of strain gages usually required for high modulus, advanced composites.

RESIN CONTENT AND VOID CONTENT

It is frequently necessary to determine the resin content of cured advanced fiber composites and also to calculate the voids that were present from density

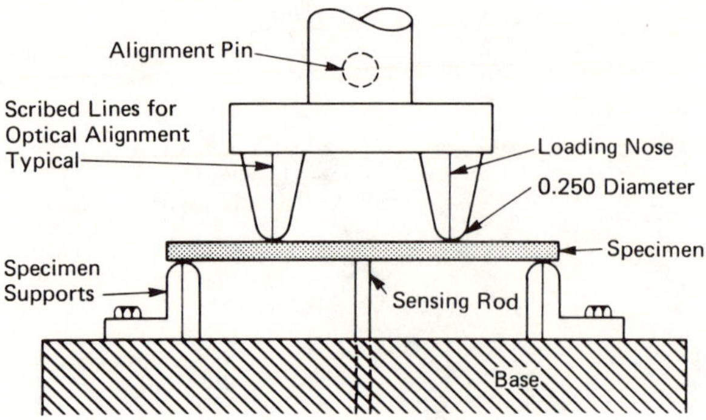

Fig. 10-17. Four point flexure test fixture.

measurements. When handling glass fiber composites, techniques are available for "burning" off the resin content in a muffler furnace, relying upon the stability of weight of the glass fiber at those temperatures which decompose the organic matrices. In the case of the C/G composites, temperatures of resin decomposition may approach those of the active oxidation of carbon and graphite fibers, where they, too, will lose weight. Hence, chemical means are available for removing cured resin matrices from the C/G fibers. Problems will be compounded in the event of hybrids which involve synthetic organic fibers and/or glass fibers with C/G fibers, though procedures may be developed to accommodate these situations.

There are ASTM methods available for determination of the C/G fiber content of resin matrix composites, based on the digestion of the resin matrix by chemicals which will not attack the fiber excessively.[23] The digestive media for epoxy resins is concentrated nitric acid (greater than 70%), and for polyimides and phenolics a mixture of concentrated sulfuric acid and 50% aqueous hydrogen peroxide is recommended. These are hazardous chemicals, and proper safety measures and handling under the supervision of trained chemists are necessary. Small specimens are digested at 75°C for several hours, for the C/G/epoxy composite; the fibers are separated in a filtering crucible and are washed per the ASTM specifications. For the polyimides and phenolic-based specimens, initial oxidation with the sulfuric acid is followed by completion of the oxidation with hydrogen peroxide—observing the detailed procedures of the referenced specification. If correction for fiber weight change by the digesting medium is necessary, a blank is generally run in digestion without the presence of the matrix resin binder to provide correction factors. The resin content of the test specimen is determined from the following equation.

$$\text{Percent resin content } (W_R) = \frac{W_1 - (W_3 - W_2)}{W_1} \times 100$$

where

W_1 = initial sample weight
W_2 = tare weight of filtering crucible
W_3 = final weight of separated fiber and crucible.

$$\text{Percent Fiber Volume} = \frac{W_F/\rho_F}{W_C/\rho_C} \times 100$$

The void volume percent =

$$100\left[1 - \frac{\rho_C}{W_C}\left(\frac{W_R}{\rho_R} + \frac{W_F}{\rho_F}\right)\right]$$

where

ρ_C = composite density (g/cc.).
ρ_F = fiber density (g/cc.).
ρ_R = resin density (g/cc.).
W_C = weight of composite.
W_F = weight of fiber.
W_R = weight of resin.

Typical values for the above equations are:

Cured Epoxy Resin Density
- Standard Epoxy - 1.22 g/cc
- MY720 System - 1.24 g/cc

C/G Fiber densities - 1.7 to 2.0g/cc

There are other alternate methods for determing the fiber fraction of cured epoxy composites (reviewed by Jackson[24]). Two methods were singled out as alternatives to the use of concentrated nitric acid: 1) pyrolysis of the test sample in molten lead (at 400°C), followed by stirring in concentrated sulfuric acid after the specimen has cooled down from its initial decomposition period, and 2) the use of concentrated sulfuric acid at 220°C, which has been equally effective in removing the cured epoxy resin from the graphite fibers.

IMPACT TESTING

Charpy and Izod impact tests (ASTM D256-72) are available for the testing of impact strength of advanced fiber composites, though they require the sophistication of instrumentation and recording to ascertain the elements of fracture toughness. The impact or fracture toughness tests are quite varied, and the selection of test methods is influenced by the makeup of the composite structure. Toughness is commonly defined as the work or energy per unit volume to cause fracture, and can usually be approximated by the area under the stress-strain curve for an isotropic material.

In an interesting review of impact test methods upon composites, Pike and Novak[25] compared Charpy impact tests upon standard thicknesses of 0.25 and 0.5 inch with tests upon thin specimens, because the influence of shear failure on total impact is decreased with the thinner specimens. The importance of shear and flexural stress interactions demonstrates the importance of the span to depth ratio on test specimens. These tests included A-S and HM-S graphite/epoxy composites and hybrids with aramid and S-glass fibers. An instrumented pendulum was used in the test. There was an attempt to correlate ballistic impact tests and pendulum impact tests upon thin specimens. In the ballistic impact study, high pressure air was used to fire gelatin spheres at velocities of 30 to 270 m/second at composite specimens 9 inches long, 12 inches wide, and 0.100 inch thick.

Rhodes[26] described the impact testing of high strength graphite/epoxy and aramid/epoxy composites in sandwich structures. Impact was initiated by low velocity impacting aluminum spheres 1.27 cm in diameter fired from an air gun at velocities of 16 to 67 m/second. The purpose of the study was to examine the load-carrying capability of composite sandwiches subject to impact damage. Four to six plies (0.006 in thickness per ply, average), were bonded as face plies to 1.0 inch "Nomex" honeycomb cores. One set of panels was examined after low velocity impact and another set was subjected to a four point beam bending, imposing a substantially uniform stress field on the test area, and then impacted. Aramid/epoxy composites sustained higher impact energy than graphite/epoxy before exhibiting visible damage. Above the visible damage threshold, percentage reduction of ultimate tensile stress is about the same for the two reinforcements. Improvements through added face plies were negligible. Figures 10-18 and 10-19 illustrate the relationship between load and impact energy for a typical tension loaded and unloaded graphite/epoxy sandwich composite and impact damage length of four ply graphite/epoxy sandwich face sheets. The threshold for visible damage was about 0.6 joules.

According to Greszczuk,[27] resistance to impact damage in graphite fiber reinforced composites increases by reason of:

1. Modulus of graphite fiber decrease;
2. Young's modulus of matrix decrease;
3. Strength of matrix increase;
4. Fiber strength increase; or
5. Bidirectional ply lay-up being more effective than unidirectional lay-up.

Of course, it is not easy to achieve an increase in strength of the resin matrix without increasing the modulus of elasticity, and the spread available in resin matrices for effective impact contribution to composites is limited. Multidimensional advanced composites offer a solution to inadequate impact strength,

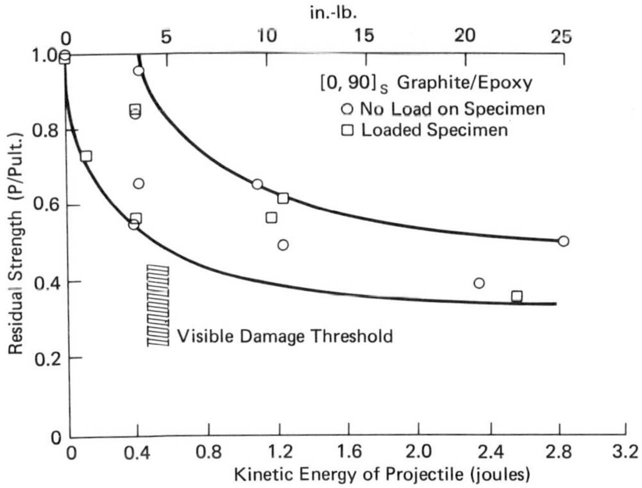

Fig. 10–18. Residual strength of graphite/epoxy sandwich structure after impact by 1.27 cm diameter aluminum projectile.[26] (*Reprinted with permission of AIAA*)

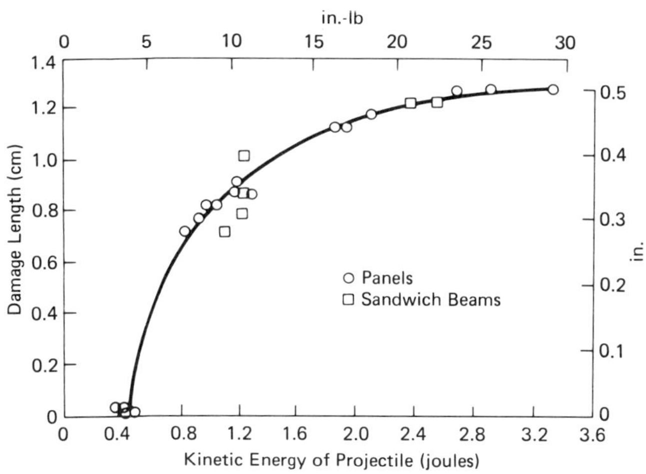

Fig. 10–19. Effect of kinetic energy on impact damage of four-ply graphite/epoxy sandwich face sheets impacted by 1.27 cm diameter aluminum ball.[26] (*Reprinted with permission from AIAA*)

according to Herrick.[28] Under low velocity impact, he reported, there were significant improvements of three dimensional graphite/epoxy woven goods over the more conventional two dimensional fabrics, which resulted in increased interlaminar shear strength. Tests were conducted on short beam flexure specimens.

Impact resistances of epoxy/graphite composites have received much attention. Internal damage from impact—arising from many sources—is not always visible. The damage may arise from random impact of ground equipment or high velocity particles stirred up from the ground by exhausts of engines. Whatever the causes, repairs are essential on the composites to ensure that the structure has the capability of operating within anticipated stresses. Laboratory tests on impact damage (by falling weights) indicate that notable improvements may be expected in graphite/epoxy composites from the following measures.[29]

1. Significant increase in energy absorbed by the laminate when one ply of S-glass is used in a 16 ply composite. Best results are realized when the glass ply occupies a position at the bottom of the stack, away from the point of impact.
2. Woven prepregs offer better impact resistance than non-woven structures. Particularly good improvements in impact resistance and the onset of incipient damage take place when a hybrid woven fabric of glass (every tenth tow) and graphite fibers is used. The woven materials clearly contain any damage to the material.

To these improvements in impact resistance may be added the reminder that the matrix epoxy resin has an important influence, and a trade-off between improved impact and decrease of modulus of the matrix is entirely feasible.

Among the manifestations of impact damage are: 1) delamination, which absorbs much energy; 2) broken fibers, which may be a few or many, depending upon the amount of impact; and 3) in the case of honeycomb sandwiches, crushed cores. It is prudent to expect that such damages will be further aggravated by environmental exposures of temperature and humidity extremes.

PREPREG TESTING

Many carbon and graphite fiber products are supplied as resin impregnated prepregs, ready for fabrication and lamination into structural components. As part of the acceptance of these prepregs, there are test procedures followed to affirm the quality of the materials before they are assembled into molds or tools for fabricating the final components. These tests are typical of those performed by the prepregger upon impregnated carbon and graphite fibrous products as they are being produced and made ready for shipment or cold storage. Many details, such as labeling and packaging, are not covered herein, as they are usually a matter of procurement understanding between the supplier and the buyer. Resin content, volatiles content, and resin flow from the prepreg of C/G fibers and resin matrices are probably the most important qualities examined, and usually accompanied by additional observations on gel time of the resin system at specified temperatures, tackiness, and drape qualities—all of which are important. The latter two qualities are subject to empirical observations, where tack describes the self adhesion of semi-cured prepreg laminates to themselves, and drape refers to their ability to conform to the shape of the tools during lay-up of prepreg layers, before heat and pressure are applied to cure the prepreg.

Resin Solids Content

These procedures employ chemical methods to separate the resin matrix from the C/G fibers, though with less drastic chemical agents than those employed on the cured sections of composite, which have been described earlier. There is an established procedure for resin content of uncured prepregs of carbon fiber epoxy tapes and sheets.[30] Well known solvents for the uncured epoxy resin include methyl ethyl ketone, acetone, and dimethylformamide, and procedures outlined in ASTM D3529-76 are recommended. Some manufacturers of prepreg cover many more solvent extraction procedures for different resin matrices, though techniques follow the ASTM practices.[31]

Volatiles Content

In some prepreg systems, the application of a high molecular weight matrix resin and a cure agent system into the C/G fiber tows or yarns requires the presence of a minor amount of low viscosity solvents during processing. While the bulk of these solvents are removed in ovens before the tape or fabric are wound upon a roll, there is a small amount of solvent retained in the fibrous structure. This small amount of volatile solvent is difficult to remove, because the temperatures and time required would advance the cure of the resin matrix in the event of thermosetting products, and mitigate against the required flow during molding. On the other hand, the small amount present can facilitate the forming and shaping of the prepreg, and also contribute to the necessary tack and drape essential to good fabrication of structural parts. Some prepreggers minimize volatiles through the use of hot melt systems for the epoxy matrices, with volaties usually under 1%.

During high temperature forming and shaping, followed by a post-cure (for thermoset materials) or a post-bake for thermoplastics, all traces of volatiles should be removed to permit optimum strength properties in the final composite component. For the determination of volatile content in the prepreg furnished to the fabricator, there is an applicable test method.[32] Test specimens measuring 5 inches by 5 inches are cut from the prepreg and are accurately weighed and suspended in a heated air circulation oven for 10 minutes either at 250°F or at

350°F, depending upon the volatiles that were used. The percent volatiles present is reported. Exceptions to the above procedure exist when hot melt resin systems are used or when low viscosity monomeric reagents are employed, as in the case of the polyimides. If volatiles are formed during cure, this procedure would provide useful information.

Resin Flow

The resin flow of carbon fiber/epoxy prepreg is also covered by an ASTM standard.[33] Two plies of 2.0 inch (50 mm) square tape of C/G fiber prepreg are cross-plied (0° to 90°), sandwiched between a release film and an unimpregnated fibrous bleeder material, and placed in a platen heated press. The temperature of the platen is held either at 250°F or at 350°F, depending upon the processing temperature of the prepreg. Under a pressure of 690 kPa (100 psi), the plies are pressed until the resin gels. The cooled sandwich assembly is removed, the resin that has flowed to the edges of the specimen and beyond is removed, and the specimen is reweighed. The change in weight is expressed as a percent of the original weight, and reported as the percent flow of the resin matrix. As volatile content is lost on storage, and as resin cure advances with time, this test is useful in estimating the useful remaining life of the prepreg tape or sheet.

In the chapter on high temperature resin matrices, chemical methods for following the aging of the polyimide monomers used in C/G composites were reported. Although these test methods appear mundane, the experienced fabricator of advanced composites knows what resin flow is necessary in the prepregs to provide satisfactory parts and obtain the desired properties in the composite. Fabrication inadequacies or insufficient quality control measures could well contribute to unsatisfactory parts of fiber/resin composites.

Overage indicators purport to monitor the age of the prepreg and correlate the aging progress with the loss of some relevant, critical properties. Sanjana's paper[34] presents data on the aging of Hercules 3501-6/A-S, a solventless epoxy/graphite prepreg using principally dielectric analysis and dynamic mechanical analysis. The curing agent is 4,4′-diaminodiphenyl sulfone (DADS) and the catalyst a BF_3 complex. Tack and drapability were lost before any other property deteriorated, establishing an apparent limiting factor to the useful life of the prepreg, as this indicates continued polymerization in the material.

CHEMICAL QUALITY ASSURANCE

With the development of sophisticated and experimental resin systems for advanced fiber composites, the need for chemical quality assurance of the matrix resin formulations becomes necessary. It is generally accepted that the sensitivity of mechanical tests to resin variables in graphite/epoxy and graphite/polyimides is not adequate. Lobb[35] specifically examined short beam shear, 10° tensile, and Celanese compression tests on composites which were impregnated with variations in the curative used in the epoxy system, without significant property changes, indicating the necessity for chemical quality assurance. This was also pointed out by Hadad,[36] who observed that chemical changes in the composition cannot be monitored by changes in mechanical tests. These chemical changes may cause problems in the manufacture of parts or in their performance, and hence she outlined in her paper current procedures at Lockheed in examining the resin matrices used with C/G fiber composites. The techniques of applying gel permeation studies and liquid and thin layer chromatography to evaluating resin variables have proven useful. These testing techniques have not only been valuable to the user of prepreg, but also to the supplier of prepregs, who analyze the raw materials being supplied by chemical manufacturers for the impregnation of C/G fibers. Commercial laboratories experienced in this work, such as the Delsen Testing Laboratory, perform these services on a routine basis.

Chen[37] described procedures for the quality assurance of incoming composite material at Boeing. The chemical characterization of high temperature (350°F) graphite/epoxy composites included: 1) reverse phase liquid chromatography (LC) for component polarity separation, 2) gel permeation chromatography (GPC) for the molecular weight distribution, 3) differential infra-red spectroscopy (IR) for the hardener concentration, 4) differential scanning calorimetry (DSC) for thermoproperties, and 5) thermomechanical analysis (TMA) for glass transition temperatures. Examples were shown in this report to illustrate the test methods.

May[38] presented an overview of the history of the evolution of chemical characterization as a technique for the assurance of the quality of advanced composite prepreg matrices. His chemical characterization scheme, reproduced in Fig. 10-20, illustrates the techniques currently available to a quality assurance group. There are other sub-groups which may be included, though from a functional overview, this chart is representative.

NON-DESTRUCTIVE TESTS

There are several non-destructive tests that are used to examine the preparation and the quality of advanced fiber composites. These test methods have not been standardized and are not in general use, though they have been developed to improve quality assurance and the resulting performance of advanced composites, particularly in the aerospace field.

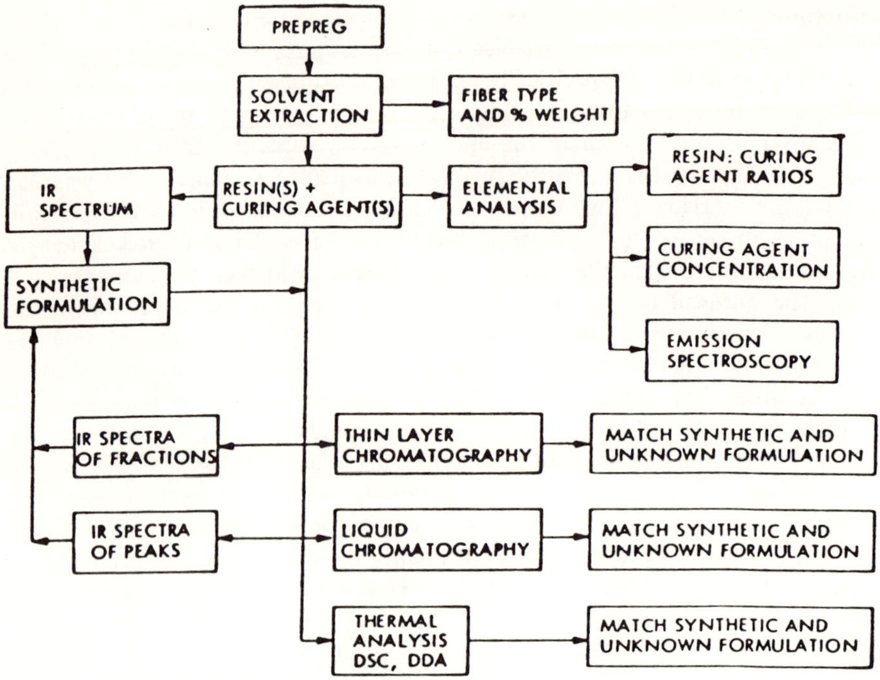

Fig. 10-20. Chemical characterization scheme.[38]

Dielectrometry

The use of dielectrometry for examing the molecular changes in thermoset epoxy resin during cure was first reported by Delmonte.[39] Currently, it is used in examining the progress of cure of the resin during fabrication. In the case of C/G composites, resin flow from outside the composite is monitored because the lower electrical resistivity of C/G fibers affects the sensitivity of measurements. Dielectric measurements will also detect variables in resin distribution in laminated structures.

Neutron Gaging of Resin Content

Martin[40] reported on the determination of resin contents in selected composite panels ranging from 25 to 38% using neutron radiographs. The source was 252_{CF} for thermal neutrons and the neutron film density was ploted against the resin content.

Ultrasonic Techniques

Ultrasonic techniques are used for detecting gross flaws such as delamination and high void content in a qualitative fashion. They will detect the presence of flaws unobserved by other methods. Measurements of the attenuation of ultrasonic waves to evaluate strength-related factors in various materials have been described. Tests made with graphite fiber composite panel specimens have produced strong correlations between the stress energy measurements and the interlaminar shear strength. Figure 10-21 shows the attenuation of a pulsed signal traveling through the test specimen. Variation in output count is related to a change in the propagation properties (and strength).[41]

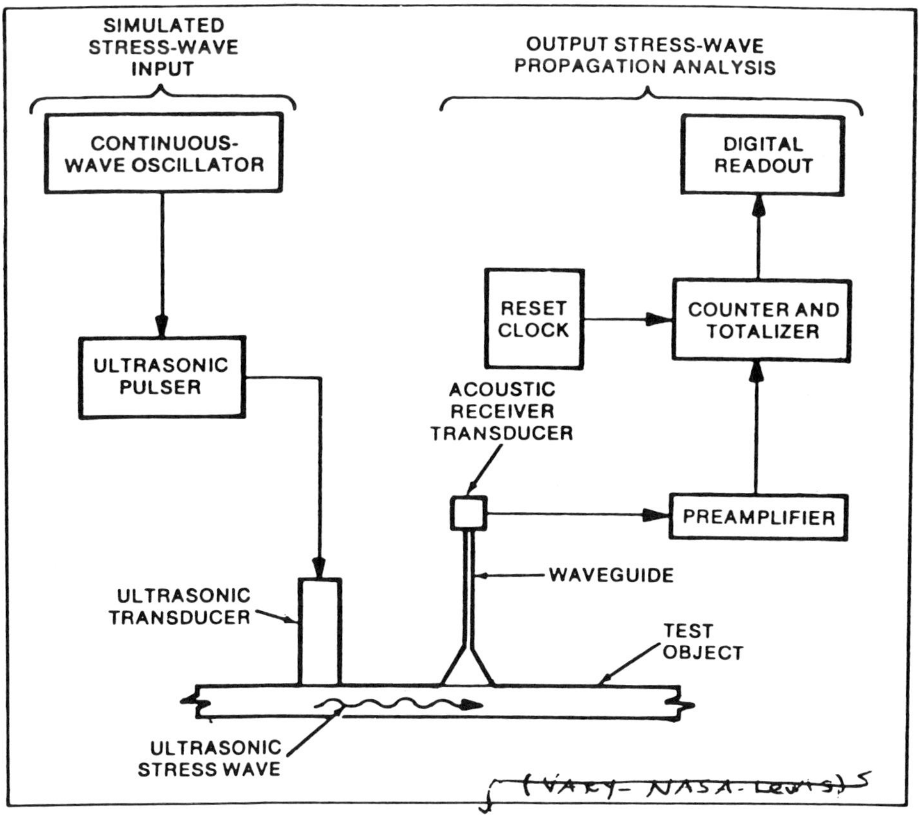

Fig. 10-21. Ultrasonic test of composite strength.[41]

Ultrasonic techniques were used to monitor flaw growth in graphite/epoxy laminates containing four types of flaws and subjected to fully reversed spectrum fatigue loading. The flaws included circular holes, embedded film patches, internal ply gaps, and surface scratches. It was observed that environmental fluctuations in addition to load spectrum, induced much more pronounced flaw growth. Flaw growth maps were developed for many specimens by ultrasonic scanning during the fatigue testing.[42] Accelerated fatigue testing at a cycling frequency of 3 Hz covered 12 hours during which 127,500 cycles of tension-compression loading simulated a lifetime of exposure. Among conclusions reached are the following.

1. Greater flaw growth occurred during fatigue in lay-ups of graphite/epoxy (AS3501-5A) of $[(0°/\pm45°/90°)_S]_2$ than in those of $[0°_2/\pm45]_{2S}$.
2. Environmental fluctuations induced more pronounced flaw growth.
3. Specimens with internal ply gaps show the least amount of flaw growth.
4. Specimens with surface scratches develop extensive delaminations near surface plies.

Acoustic Emission

One of the newer techniques adaptable to the testing of advanced composites is acoustic emission. While it might be argued that this is not strictly non-destructive testing, the fact is that through acoustic emission one is able to record the breaking of internal fiber lengths as test specimens are stressed. Such acoustic observations are made even during the thermal cycling, as from a cure temperature down to room temperature. Results of field application indicate that acoustic emission does provide early warning of structural problems. The ASTM definition of acoustic emission (AE) is "a transient elastic stress wave generated by the rapid release of energy from a localized source within the material." An AE transducer attached to the surface will convert the stress wave to an electrical signal. The most appropriate resonant frequency for a particular application is chosen by evaluating the background noise. Large structures can be monitored and weak areas located by mounting several transducers at defined spacings.[43]

Green and Landy[44] reviewed AE and its application to non-destructive evalution (NDE) of composite structures. Deformation and fracture events at a microscopic level are detected as AE signals at very low levels of deformation. Sensors operating in the 50 to 500 KHz ultrasonic frequency range enable ambient noises to be discounted. Low noise, high gain amplification, following the AE sensors and additional second stage amplification, allow the microscopic signals to be handled by conventional data processing methods. Voltage measuring circuits permit following matrix failures, observing total AE signal level, while counting threshold crossings are employed to follow the high amplitude signals from filament failures or delamination in the composites.

In most materials there is a short-term irreversibility of the AE data, known as the Kaiser effect, whereby the signal observed in loading the material is not observed again until the loading reaches a level beyond that originally reached. A new technique involving the use of an ultrasonic sensor to inject a stress pulse into the composite specimen, and a lower frequency AE sensor to receive the transformed wave, permits studies of the flaw population as well as fiber-resin bonding variables. A stress wave factor based upon the rate of ultrasonic pulses, a predetermined time interval, and a counter evaluating the number of oscillations in each burst, exceeding a fixed threshold of voltage, has been correlated to material and process variations in the specimen. Vary and Lark[45] felt that the stress wave factor may assist in predicting potential failure locations in their composite laminates.

Ryder and Wadin[46] reported on the use of AE to follow the progressive failure of a quasi-isotropic graphite/epoxy specimen under tension in fatigue

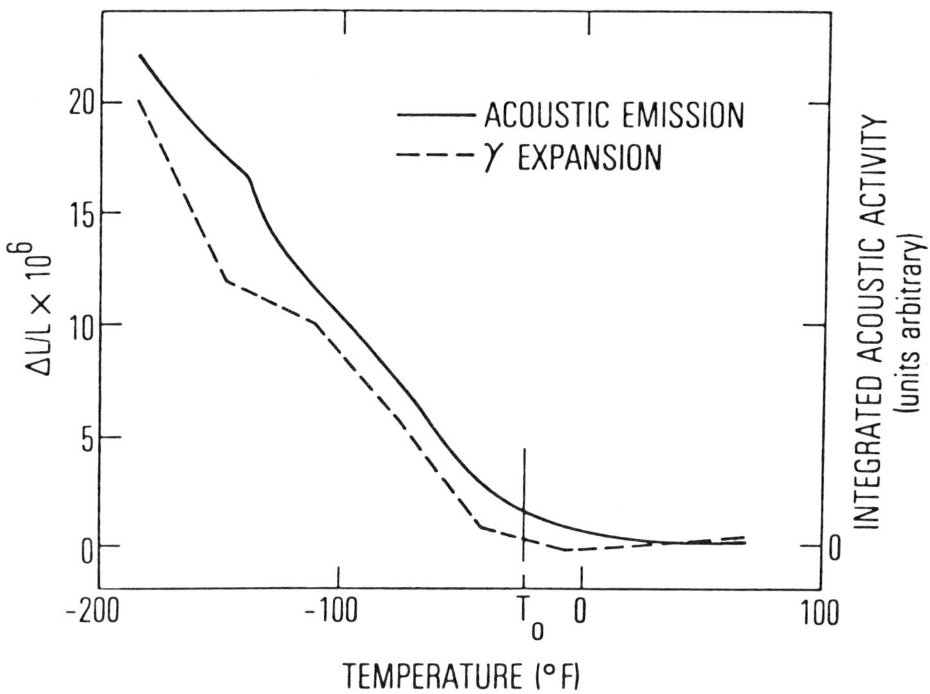

Fig. 10-22. Acoustic emission and γ expansion for the 0/±60/0 tube.[47]

loading. Relationships were established between acoustic emission events and fatigue life of composites. Wolff[47] measured acoustic emission and observed the changes in length of a 0°/±60°/0° lay-up graphite/epoxy tubing (GY-70/934) during the initial thermal cycling. The AE emissions data is plotted against the change in length, showing strain differences between thermal cycles one and two (see Fig. 10-22). There was a net length change of 20×10^{-6} inches/inch following the first cycle—attributed to micro-cracking. It was estimated that about 96% of cracking occurred on the first cycle.

Ultrasonic, radiographic, and acoustic methods are described by Hamstad and Chiao.[48] Cited was an early Grumman patent on acoustic techniques (U.S. Patent 3,985,024–1970). Defect depth determination, detection of radii or edge defects, and capabilities of predicting ultimate or residual strength of a structure were reviewed.

The acoustic emission from the testing of filament wound graphite fiber bottles (4.0 inches in diameter) has been reported.[48] A typical response is shown in Fig. 10-23. Acoustic emission was studied for the effects of different winding patterns, artificial flaws, winding-induced fiber fraying, different resins, and different fibers. It was observed that lower cure temperature epoxy systems increased by a factor between two and four, the pressure at which bottles first produced significant acoustic emissions. S-glass and aramid fibers were also examined in this series of tests.

Guidance in non-destructive inspection (NDI) policies has been suggested for assuring eqality of primary structures of advanced composites of boron/epoxy.[49]

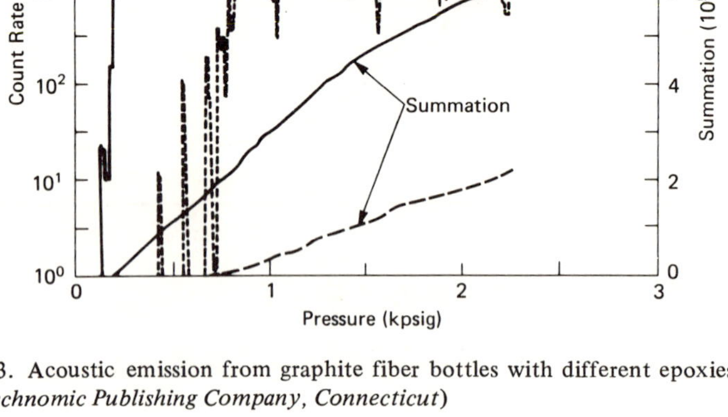

Fig. 10-23. Acoustic emission from graphite fiber bottles with different epoxies.[48] (*Courtesy of Technomic Publishing Company, Connecticut*)

Vary[50] also emphasizes the need for advanced non-destructive evaluation (NDE) for quantitative assessment of the mechanical strength and integrity of fiber composites during manufacturer and service, and following repair operations.

Thermal Field Techniques

Thermography was employed as a tool for the non-destructive testing of composites to locate below the surface flaws, sub-surface laminar cracks, and delaminations. In the presence of a flaw or damaged region, the discontinuous thermal conductivity would cause a perturbation in the heat flux and temperature field.[51] Temperature gradients were induced in specimens by applying heat externally and employing an infra-red camera with videoisotherm readout. Thermography has potential for being able to examine large surfaces rapidly without the need for sophisticated data reduction. This technique has been applied to advanced fiber composites, including graphite/epoxy systems. Delaminations, blind surface impact damage, and surface cracks can be detected qualitatively. Stress generated thermal fields were also investigated to examine internal frictional heat developed at flaws.

Test methods for evaluation of C/G fiber composites have been receiving considerable attention, and while there are various physical, electrical, and thermal techniques which were not covered in this chapter because conventional ASTM or federal specifications were available for evaluation, emphasis has been placed upon reviewing those methods unique to the testing of advanced composites.

REFERENCES

1. ASTM D3379-75, "Tensile Strength and Young's Modulus for High Modulus Single Filaments," *Annual Book of ASTM Standards,* Part 36, page 847, 1978.
2. Bell, J.E., *SAMPE Symposium,* Vol. 15, page 219, April 1969.
3. McMahon, P.E., "Relationship Between High Modulus Fibers and Unidirectional Composite Strength," *SAMPE Symposium,* Vol. 14, page 214, 1977.
4. Hercules, Inc., "Test Methods for Determining Physical Properties of Carbon and Graphite Tows," HD-SG-2-6001B, September 1974.
5. McMahon, P.E., "Graphite Fiber Tensile Property Evaluation," *Symposium of ASTM D30 Commitee,* STP-521, pages 367-389, ASTM, 1973.
6. ANSI/ASTM D2343, "Standard Test Methods for Tensile Properties of Glass Fiber Strands, Yarns and Rovings Used in Reinforced Plastics," *Annual Book of ASTM Standards,* Part 36, page 352, 1978.
7. ANSI/ASTM D3039-76, "Tensile Properties of Oriented Fiber Composites," *Annual Book of ASTM Standards,* Part 36, page 721, 1978.
8. Weidner, J.C., "New Tensile and Compressive Test Specimens for Unidirectional Reinforced Graphite Epoxy Composites," Technical Report AFML-TR-70-264, February 1971.

9. Brown, H.J., "Measuring Elastic Modulus and Tensile Strength of High Modulus Graphite," *SAMPE Symposium,* Vol. 15, page 239, April 1969.
10. Beaumont, P.W.R. and Phillips, D.C., "Tensile Strengths of Notched Composites," in *Carbon Reinforced Epoxy Systems,* Hilado, C. (Editor), page 5. Technomic, Westport, Connecticut, 1974.
11. Hertz, J., Christian, J.L., Varias, M. *et al.* Convair Division of General Dynamics), "Advanced Composite Applications for Spacecraft and Missiles," Vol. 2, "Material Development," Technical Report AFML–TR–71-186, 379 pages, March 1972.
12. ANSI/ASTM–D3410-75, "Standard Test Method for Compressive Properties of Oriented Fiber Composites," *Annual Book of ASTM Standards,* Part 36, page 853, 1978.
13. Greszczuk, L.B., "Compressive Strength and Failure Modes of Unidirectional Composites," *Analysis of Test Methods for High Modulus Fibers and Composites,* ASTM STP 521, Part 30, pages 192–217, 1973.
14. ANSI/ASTM D3518–76, "Inplane Shear Stress-strain Response of Unidirectional Reinforced Plastics," *Annual Book of ASTM Standards,* Part 36, page 875, 1978.
15. Lifshitz, J.M. and Gilat, A., "Experimental Determination of Non-linear Shear Behavior of Composites Under Impact," *SESA Meeting,* San Francisco, California, May 1979.
16. Chamis, C.C. and Sinclair, J.H., *Experimental Mechanics,* Vol. 17, No. 9, page 399, September 1977.
17. ANSI/ASTM D2344–76, "Apparent Horizontal Shear Strength of Reinforced Plastics by Short Beam Method," *Annual Book of ASTM Standards,* Part 36, page 361, 1978.
18. Whitney, J.M., "Free-edge Effects in the Characterization of Composite Materials," ASTM–STP 521, pages 167–180, 1973.
19. Duggan, M.F., "Experimental Evaluation of the Slotted Tension Shear Test for Composite Materials," Paper R79–117, *Society for Experimental Stress Analysis,* San Francisco, California, May 1979.
20. Sims, D.F., "Inplane Shear Stress-strain Response of Unidirectional Composite Materials," in *Carbon Fiber Reinforced Epoxy Systems,* Hilado, C. (Editor), page 75, Technomic, Westport, Connecticut, 1974.
21. Chiao, C.C., Moore, R.L., and Chiao, T.T., "Measurement of Shear Properties of Fiber Composites," *Journal of Composites,* Vol. 8, pages 161–169, March 1976.
22. Sumsion, H.T., and Rajapakse, Y.D., "Simple Torsion Test for Shear Moduli Determination of Orthotropic Composites," *Proceedings of 1978 International Conference on Composite Materials,* Toronto, April 1978, published by AIME, Warrendale, Pennsylvania.
23. ANSI/ASTM D3171–76, "Fiber Content of Reinforced Resin Composites," *Annual Book of ASTM Standards,* Part 36, page 746, 1978.
24. Jackson, W. (Hexcel Corp.), "Determining Fiber Fraction of Cured Epoxy/Graphite Composites," *SAMPE Twenty-third National Symposium,* page 160, Anaheim, California, May 1978.
25. Pike, R.A. and Novak, R.C. (United Technologies Research Center), "Pendulum Test Method for Assessing the Impact Response of Composite Materials," *Thirty-first Annual Technical Conference,* SPI, RP/C Institute, Section 12D, February 1976.
26. Rhodes, M. (NASA Langley), "Impact Fracture of Composite Structures," Paper AIAA–75–748, *ASME/AIAA/SAE Conference,* Denver, May 1975.

27. Greszczuk, L. and Chao, H. (McDonnell Douglas), CA–86–190971p (1977), Re ASTM–STP 617, 1977.
28. Herrick, J. (Fiber Materials, Inc.), "Multi-dimensional Advanced Composite for Improved Impact Resistance," *SAMPE Tenth National Technical Conference,* Vol. 10, New York, October 1978.
29. Labor, J. (Northrop), "Seminar on Defects in Graphite-epoxy Composites," held at Aerospace Corp., February 1979.
30. ANSI/ASTM D3529–76, "Resin Solids Content of Carbon Fiber-epoxy Prepreg," *Annual Book of ASTM Standards,* Part 36, page 881, 1978.
31. Hercules, Inc., "Test Methods for Determining Physical Properties of Carbon and Graphite Prepregs," HD–SG–Z–6006C, September 1974.
32. ANSI/ASTM D3530–76, "Volatiles Content of Carbon Fiber-epoxy Prepreg," *Annual Book of ASTM Standards,* Part 36, page 884, 1978.
33. ANSI/ASTM D3531–76, "Resin Flow of Carbon Fiber-epoxy Prepreg," *Annual Book of ASTM Standards,* Part 36, page 886, 1978.
34. Sanjana, Z.N., "Overage Indicators for Prepreg Products," *SAMPE Twentieth Symposium,* Vol. 24, page 330, San Francisco, California, May 1979.
35. Lobb, T.A., "Sensitivity of Mechanical Tests to Resin Variations in Graphite/Epoxy Prepreg Materials," *SAMPE Tenth National Technical Conference,* Vol. 10, page 475, New York, October 1978.
36. Hadad, D.W., "Chemical Quality Assurance of Epoxy Resins by Gel Permeation, Liquid, and Thin Layer Chromatography," *SAMPE Symposium,* page 301, 1977.
37. Chen, J.S., "Chemical Fingerprinting of Incoming Composite Materials," *Society of Manufacturing Engineers,* EM 78–403, 1978.
38. May, C., "Composite Matrix Quality Assurance," *SAMPE Twentieth Symposium,* Vol. 24, page 390, San Francisco, California, May 1979.
39. Delmonte, J., "Electrical Properties During Polymerization," *Journal of Applied Polymer Science,* Vol. 2, page 108, 1959.
40. Martin, B.G., "Some Techniques for Nondestructive Evaluation of Graphite Fiber-reinforced Composites," *Society of Manufacturing Engineers,* EM 78–402, 1978.
41. Vary, A. and Bowles, K.J. (NASA-Lewis Research Center), "Ultrasonic Strength Evaluation of Fiber-reinforced Composites," *NASA Technical Briefs,* Vol. 2, No. 4, page 484 (1977), NASA–TM–X–73646, E–9153, April 1977.
42. Daniel, I.M., Schramm, S.W., and Liber, T., "Ultrasonic Monitoring of Flaw Growth in Graphite/Epoxy Laminate Under Fatigue Loading," *Conference on Advanced Composites,* page 267, Technology Conference Associates, El Segundo, California, December 1979.
43. Mitchell, J.R., "Fundamentals of Acoustic Emission and Applications as an NDT Tool for FRP," *Thirty-fourth Annual Technical Conference,* SPI, RP/C Institute, Section 3F, February 1979.
44. Green, A.T. and Landy, R.J., "Acoustic Emission NDE for Advanced Composite Structures," *Conference on Advanced Composites,* page 228, Technology Conference Associates, El Segundo, California, December 1979.
45. Vary, A. and Lark, R.F., "Correlations of Fiber Composite Tensile Strength with Ultrasonic Stress Wave Factor," NASA–TM–78846, NASA-Lewis, Cleveland, Ohio, April 1978.
46. Ryder, J.T. and Wadin, J.R., "Acoustic Emission Monitoring of a Quasi-isotropic Graphite/Epoxy Laminate Under Fatigue Loading," *Conference on Advanced Com-*

posites, page 246, Technology Conference Associates, El Segundo, California, December 1979.

47. Wolff, E.G., Eselun, S.A., and Neubert, H.D., "Microcracking Effects on Dimensional Stability," *SAMPE National Symposium,* Vol. 24, page 1299, San Francisco, California, May 1979.
48. Hamstad, M.A. and Chiao, T.T. (Lawrence-Livermore Labs), "Acoustic Emission Produced During Burst Tests of Filament Wound Bottles," *Carbon Reinforced Epoxy Systems,* Hilado, C., page 280. Technomic, Westport, Connecticut, 1974.
49. Collins, R.M. (Grumman), "NDI Policy and Techniques for Advanced Composites," *SAMPE Conference,* Vol. 11, page 178, New York, November 1979.
50. Vary, A. (NASA-Lewis Research Center), "NDE of Fiber Reinforced Structural Composites," *SAMPE Conference,* Vol. 11, page 166, New York, November 1979.
51. McLaughlin, P.V., McAssey, E.V., and Dietrich, R.C., "Non-destructive Examination of Fiber Composite Structures by Thermal Field Techniques," *Thirty-fourth Annual Technical Conference,* SPI, RP/C Institute, Section 22D, February 1979.

11
Carbon/Graphite Fiber Composites in Aircraft, Space, and Automotive Applications

Throughout this book there have been numerous references to the aerospace and automotive industries to illustrate the adaptability of C/G fiber composites. In this chapter, the aerospace market, which is at present one of the largest potential applications for C/G products, will be examined in greater detail. The modulus to density ratio was highest available for structural materials and, hence, C/G fiber composites are found where maximum stiffness and lightweight are essential. There have been other benefits arising, such as the low order of thermal expansion coefficients under temperature extremes; enhanced fatigue life; and the relative inertness of the C/G products to corrosion. These advantages and some of the disadvantages will be reiterated in this chapter, which will emphasize applications, not only to the predominant aerospace industry but also to the automotive industry, which finds C/G composites very satisfactory. The price of C/G fibers have a long way to fall before very substantial uses will appear in automotive applications—some estimates place this well under $10/lb. Although the 1979 uses in aerospace industry will exceed 500,000 lb in 1979, and with added commitments from commercial aircraft, exceed 1,000,000 lb by 1982, efforts are underway to foster automotive uses to attain over 10,000,000 lb of C/G fibers production and achieve the lower manufacturing costs.

In the examination of forecasts of the use of advanced composites, the near future looks bright, and there is every indication that today's use of composites in military aircraft, for example, may grow from its present 5 to 10% of the weight of the aircraft to 30 to 55% by the middle 1980's on some military aircraft. This growth will be reflected not only in the C/G fibers, but also in the more abundantly used glass fibers and the versatile aramid fibers. The latter, for example, will reach a production capacity of 40,000,000 lb in 1981 (much of this will be used in tire manufacture). The aramids and glass fibers, it will be remembered, have demonstrated excellent hybridization qualities with the C/G fibers. During this expansion in the applications of advanced fibers to aircraft, there will be changes in the design and construction of aircraft manufacturing facilities to accommodate the specific needs for processing advanced fiber components. These facilities are significantly different from those plants concentrating on metal fabrication.

The C/G fiber composites are expected to assume an increasing role in these aircraft developments, as the experiences on major aircraft components for both military and commercial aircraft continue to demonstrate successful adaptation and utilization in service. There are other benefits for C/G composites on aircraft that are beginning to arise, such as reduced overhaul and maintenance costs compared to metals, which are more prone to cracks and corrosion in service.

There are disadvantages which currently exist for the C/G fiber composites—such as scatter in physical properties of developmental products—particularly on very high modulus materials; insufficient toughness in some instances, requiring the inclusion of glass or aramid fibers; limits on very high temperature oxidative stability of some of the fibers; flame resistance of some of the newer resin matrices developed for C/G fibers; and the price per pound. There are trade references purporting to show the large decrease in the price of C/G fibers with increasing volume. This is to be expected, though from the present scale of $20 to $30/lb to the $5 to $10/lb range, the inflation trends must be factored into consideration. Once an application has been accepted for production using certain raw materials and processing, the manufacturers of these materials and products cannot casually change procurement specifications even though such changes may accomplish significant improvements. Improvements, whether in properties or costs of the C/G fibers, will require a proving-in period to achieve commercial acceptance and establish design confidence of the consumer.

Herein are listed some of the major uses for carbon and graphite fibers as presented by United States manufacturers and by Japanese manufacturers. The United States has been the largest consumer of these fibers, importing some of the materials and technology from England and Japan. At the time of preparing this chapter, major suppliers of C/G fibers in this country, such as Hercules, Union Carbide, and Celanese have announced plans to establish or increase facilities for carbon and graphite fiber manufacture. At present most of these fibers are prepared from polyacrylonitrile (PAN) precursors, though pitch fibers from Union Carbide Corp. are gaining in acceptance. (See the earlier chapter on manufacture of C/G fibers.) PAN is produced in very large quantities for textile purposes by E.I. du Pont de Nemours, Courtaulds (England), and Toray Industries

(Japan). The pitch fibers, produced in the U.S. and Japan, could actively challenge the PAN-based fibers in the 1980's. Rayon, the first C/G fiber precursor, still maintains its role, though its increasing manufacturing costs have been some deterrent.

It is expected that in the early 1980's, the uses on commercial aircraft for C/G fibers will exceed military uses. DeVault[1] presented an analysis of end uses for carbon and graphite fibers in the United States for 1979 as follows.

AEROSPACE		INDUSTRIAL	
Aircraft	135,000 lb	Recreational uses and Sports	146,000 lb
Helicopters	27,000	Automotive and Miscellaneous	50,000
Missile	49,000		196,000
Aircraft brakes	40,000		
Engines	5,000		
	256,000		

Matsuda[2] indicated, at the same time as the above, that the consumption of high grade carbon and graphite fibers in Japan for 1978 was 220,000 lb and in Europe the corresponding consumption was 160,000 lb per year. Referring to the Japanese market he indicated that sports and leisure uses accounted for 85 to 90% of the production of high quality (high strength and high modulus) carbon and graphite fibers. The principal sports uses lay in golf clubs, fishing rods, and tennis racquets. The aerospace applications in Japan are small due to the absence of aircraft manufacturing facilities. Attention was also drawn to inexpensive pitch fibers by Kureha in Japan, whose use amounted to a little over 200,000 lb in 1978. In the next chapter on applications there will be a brief mention of the carbon fibers in concrete and related uses for construction.

From the broadest interpretation, advanced fiber composites may be considered as part of the reinforced plastics industry, and some of the major areas of the industry are shown in Table 11-1.[3] Reinforced plastics have made breakthroughs in a number of automobile components. Considerable interest is being evinced in the graphite-fiber reinforced all-plastic car, which could do for automobiles what glass fiber reinforced plastics did for boats in the 1950's and 1960's. Figures in the table are in millions of pounds of composite weight (a typical glass reinforced plastic—which accounts for the largest percentage—has about 35% resin by weight).

It is noteworthy that progress within the aerospace industry has not been dramatic with the glass fiber/resin composites. With the advantages of improved stiffness and improved fatigue resistance gained with graphite fibers, it is expected

Table 11-1. Markets for Glass Fiber Reinforced Plastics (in Millions of Pounds).[3]

APPLICATIONS	1979	1974	1969
Transportation, land	580	293	150
Marine	451	320	231
Construction	342	198	111
Anti-corrosion	233	174	76
Electrical	180	94	69
Consumer Goods	125	75	73
Appliances	130	82	27
Aerospace	23	30	33
Miscellaneous	77	75	89
Total	2141	1341	859

that aerospace volumes will increase both from C/G advanced fiber composites, as well as hybrid constructions of graphite fibers with glass fibers and aramid fibers. The types of applications and the extent of use in high performance aircraft of materials such as aluminum alloys, titanium, steel and advanced composites are reviewed by Paez.[4] New processing developments which improve the performance and cost effectiveness of metal structures were also noted. The paper also described fiber reinforcement of titanium. Such processes must be weighed in considering the role and advantages of advanced fiber composites.

AEROSPACE

As part of the program to develop confidence in graphite/epoxy advanced composites, the government and aircraft manufacturers have made available interim reports indicating the progress that has been made in fabricating and testing these structures. Because these reports are thorough, carefully organized, and demonstrate effective ways of using graphite/epoxy composites, some of the data are discussed below.

As an indicator of the dedication to advance composite technology, Northrop announced the establishment of a 325,000 ft^2 facility in early 1980 for the manufacture of advanced composites.[5] Integrated, flexible automation will be established for computerized control of temperature and cycles, as well as in the handling and use of advanced composites.

In a program sponsored by the National Aeronautics and Space Administration (NASA) at Lockheed Corporation, work has been continuing in the development of advanced C/G fiber/epoxy composite vertical fins (ACVF) for service use on L1011 transport aircraft. The ACVF to be developed will consist of the entire main torque box structure of the vertical stabilizer and does not include leading and trailing edges and rudder. The box structure extends from the fuselage production joint to the tip rib and includes front and rear spars; it is 25 ft tall with a base of 9 ft and represents an area of 150 ft^2. It consists of graphite fiber/epoxy resin composite system, with some aramid fiber interfacing between graphite and metal. The purpose is to gain a high level of confidence in the structural integrity and the durability of advanced composite primary structures. The torque box structure will have front and rear spars, skins with hat shaped stiffeners, and 10 horizontal ribs—all of graphite/epoxy.[6]

The indicated weight saving is 27.9%. Fabrication is developed about unidirectional graphite fiber tape material and tetraglycidylether of methylene dianiline.

In another development program of advanced composite materials of graphite/epoxy for inboard ailerons on the L1011 transport aircraft (under contract NASI-15069), the T300/5208 graphite fabric epoxy resins were used for fabrication and qualification testing.[7] Weight savings were clearly demonstrated (over 25%) over the metal counterparts. The C/G fabric is a bidirectional weave, 13 mils per ply nominal, and nominal 41% resin content in the prepreg.

Sakata and Ostrom[8] prepared a summary report on the utilization of advanced composites of aramid fibers and graphite fibers on commercial aircraft wing structures. The opinion was expressed that each aircraft manufacturer will require a composite wing development program to achieve technology readiness and an acceptable level of risk—since the data are not readily transferable from one company to another. NASA composites programs at three commercial aircraft manufacturers is illustrated in Fig. 11-1. Potential applications for the graphite fiber/epoxy composites to Lockheed's L1011 include the vertical tail, rudder, elevator, flaps, cowl doors, and ailerons—all at substantial savings in weight over the metal parts. The goal is a modified graphite fiber/epoxy material system with improved characteristics that will meet engineering and manufacturing requirements, and that at the same time will not invalidate the existing graphite/epoxy data base. There are aramid advanced fiber composites in the L1011 aircraft, such as on the overhead panels and side walls and overhead storage racks, with several years of excellent service and weight savings over earlier glass fiber laminates. Pre-impregnated, non-woven, graphite/epoxy tape is predominantly used for the current hardware development programs in the aircraft industry. The tapes are available from several commercial sources, for example, Fiberite, Hercules, and Narmco. They have usually been offered with 42 ± 3.0% resin content to permit adequate flow and penetration during fabrication.

The experiences with other commercial jet aircraft manufacturers continue to show promise in applications of the advanced fiber composites. Graphite fiber/epoxy composites on the vertical stabilizer for the DC-10 transport were studied at Douglas for the Langley Research Center.[9] In the initial study of a

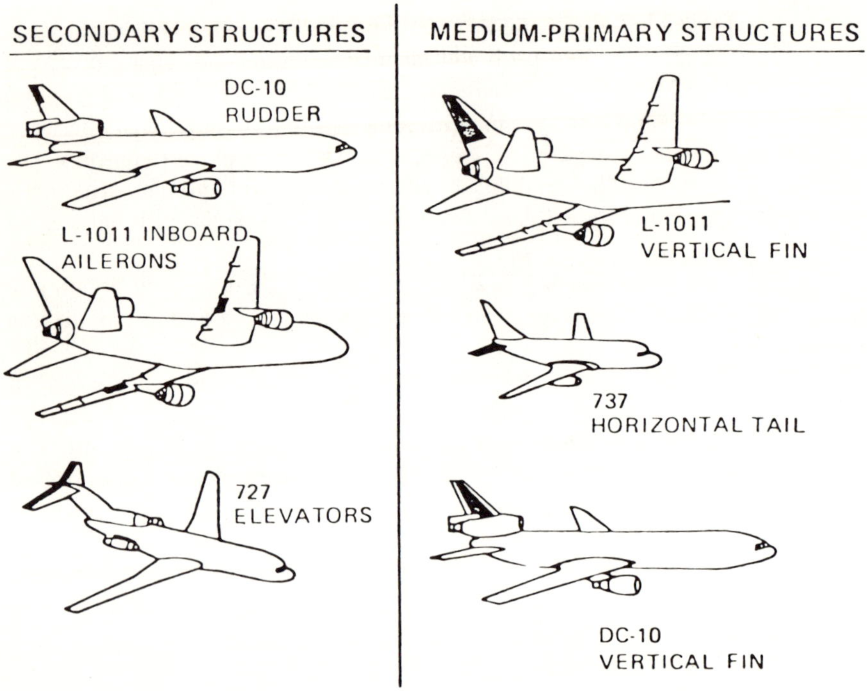

Fig. 11-1. NASA-ACEE composites programs at major commercial aircraft manufactures.[8]

"Thornel" 300/5208 composite vertical stabilizer, weight savings of 20% were anticipated (see Table 11-2).

Smillie[10] reviewed the earlier DC-10 graphite/epoxy composite rudder used on several commercial aircraft, which by 1978 had logged thousands of satisfactory flight hours. A study of potential composite material applications is shown in Fig. 11-2. The application of the advanced composites to the DC-10 airplane is illustrated in the accompanying photographs of the aft rudders and composite floor beams now in service on the DC-10 (see Figs. 11-2 a and b). Both applications have been certified by the FAA. The rudder is divided into upper and lower rudder assemblies, each consisting of forward and aft rudders. The upper-aft rudder has a span of 158 inches and a platform of approximately 34 ft^2. A rib stiffened design which lent itself to a one-step manufacturing operation was adopted. An increase in the spar-web thickness from 16 to 24 plies of C/G prepreg was necessary to eliminate any cracks in the front-spar laminates which might occur during cooldown of the component after curing. The conventional metal rudder weighed 91 lb, as compared with the graphite-epoxy rudder of 61 lb.

Table 11-2. Preliminary Weight Comparisons Composite Vertical Stabilizer—DC-10.[10]

	COMPOSITE STABILIZER ESTIMATE		METAL STABILIZER	
	kg	lb	kg	lb
Spar caps	150.0	330.6	158.4	349.2
Interspar skin panels	61.7	136.1	87.5	192.8
Spar webs	40.6	89.5	62.4	137.6
Interspar ribs	51.8	114.2	67.9	149.6
Access doors	16.6	36.6	18.5	40.7
Miscellaneous structure	13.4	29.5	28.7	63.3
Growth/Contingency	4.5	10.0	–	–
Box structure	338.6	746.5	423.3	933.2
Trailing edge skin and ribs	25.3	55.7	32.7	72.1
Total–Box and Trailing edge	363.9	802.2	456.0	1005.3
Weight reduction	92.1	203.1	–	–
Percent reduction	20.2	20.2	–	–

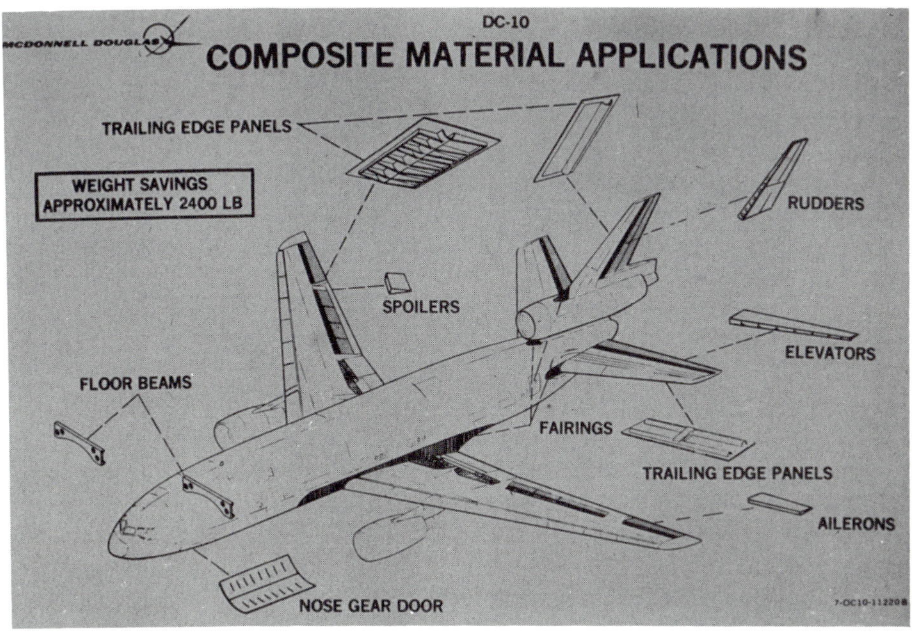

Fig. 11-2. Potential applications of graphite fiber composites on a DC-10. (*Courtesy Union Carbide Corp.*)

Fig. 11-2a. Production of upper aft rudders for DC-10 certified by FAA for service. (*Courtesy Douglas Aircraft Division, McDonnell-Douglas Corp.*)

Fig. 11-2b. Composite floor beam now in service on DC-10 has been certified for service by the FAA. (*Courtesy Douglas Aircraft Division, McDonnell-Douglas Corp.*)

Flight service experiences of 108 graphite/epoxy spoilers on Boeing 737 transport aircraft were obtained in a five-year study (four spoilers installed on 27 aircraft).[11] Some of the spoilers removed after one year of service were given visual, ultrasonic, and destructive testing.[12] No significant change of composite strength was reported. Motor nacelles for the Boeing 727 and on a DC-9–80 airplanes are now being made at Boeing and Rohr in advanced composites, principally with aramid and graphite/fibers. Forty percent weight savings over metal counterparts are anticipated. Similar planning is in progress for radar domes and engine nacelles for the Boeing 767.[13] The newer designs of commercial aircraft are also including the use of advanced fiber composite skins over honeycomb cores as flooring ("Nomex") on future commercial transport.

Increasing attention has been given to the hybrids of du Pont's "Kevlar" aramid fiber and graphite fibers. Unidirectional blends of the aramids with HM-S and A-S graphite fibers were achieved by alternating plies of the unidirectional tapes of the aramids with the unidirectional tapes of the graphite fibers. The aramids have a high strain to failure, which helps in achieving improved impact resistance for the hybrid composite.[14] Also examined were co-winding of dry fibers of the aramids with "Thornel" 300 graphite fibers in desired ratios side by side on an 18 inch drum covered with "Teflon" release film. These hybrid composites were developed principally for commercial aerospace applications. With HM-S graphite and aramid 49 fibers, hybrids of moduli of elasticity over 50 msi were attained.

Lear's new prop plane, called the Lear Fan (see Fig. 11-3) has an unusual design, with the engines built into the fuselage to reduce drag and one propeller —a "pusher" prop at the rear of the tail. The plane is being designed with graphite ("Thornel" 300)/Fiberite 934 composites, with aramid fibers where necessary for increased toughness. A weight reduction in excess of 700 lb through use of graphite composites is expected to give the airplane a range increase over 900 miles. The two 800 shaft horsepower engines will propel the aircraft at a speed of 350 miles/hour. Nine to ten passengers will be carried.[15]

Military aircraft are providing leadership in the adaptation of advanced fiber composites, as noted below:

F-14 200 lbs. of Boron fiber/epoxy (B/E)
F-15 200 lbs. of B/E and 43 lbs. of carbon fiber/epoxy (C/E)
F-16 250 lbs. of C/E composites
F-17 900 lbs. of C/E composites (see Fig. 11-3a)
F-18 900 lbs. of C/E composites

The STOL fighter aircraft (vertical/short take-off and landing) are expected to have in excess of 30% advanced composites in its structure by 1985 and the

Fig. 11-3. Over 75% (about 1,000/lbs.) of Airframe of the new Lear Fan is graphite epoxy. (*Courtesy Lear Fan.*)

general range as noted earlier will be about 30 to 55% in the mid-1980's for military aircraft. The AV-8B second generation of V/STOL aircraft is noteworthy by reason of its large usage of graphite fiber/epoxy composites in the structure. Wing spars are of sine wave web configuration made from graphite/epoxy woven fabric combined with unidirectional tape for stiffness. Wing weight savings amounted to 150 kg. Both NASA and the Air Force are funding research and development in the area of advanced fiber composites. From the many composite airframe and aerospace components designed and fabricated in the last ten years, weight savings of 25 to 40% have been predicted and demonstrated. Reduction of size and weight of airframe design will result in both system acquisition and lifecycle cost savings with smaller propulsion systems and lower fuel costs contributing to these cost reductions.[16]

Interest in applications to helicopters continues strong and some expect that the first all-composite aircraft flying may be a helicopter. There were early commitments to reinforced plastics for helicopter rotor blades by Boeing Vertol in Philadelphia, followed by Bell Aircraft and Sikorsky. It is now claimed that composite rotor blades are produced at lower overall costs than comparable size metal blades. Design freedom, resistance to corrosion and to crack propagation, improved fatigue life, and superior ballistics performance are some of the advantages claimed for the advanced composites over the metal. Boeing reports that, in tests, both metal and composite blades were fired at with 23 mm armor piercing shells. The metal blades subsequently fell apart in 3 minutes while the composite blades continued flying for an additional 60 hours. Designers claim

Fig. 11-3a. VF-17 Advanced composite components.

that the composite material provides alternate paths for stress which will reduce crack formation or propagation (which is catastrophic in the metal).[17] The blades described were produced from 6 inch wide epoxy impregnated glass tape for the H-46 Sea Knight helicopter.

Grumman Aerospace Company has an AF study contract for an aircraft that would be more than 50% composites in structure. Grumman produces the Navy's F-14 fighter aircraft, which has had a boron/epoxy horizontal stabilizer since 1970, one of the first composite primary structures on aircraft. More recently, graphite fiber/epoxy skins, accompanied by boron fibers formed hybrids in certain stress areas (compressive strengths of boron/epoxy composites are superior to compressive strengths of carbon/epoxy composites). Recent manufacturing techniques reflect automated procedures by tape controlled machines which guide the laser cutting of the profiles from the prepregs.

To demonstrate the advantages of thermoplastic matrix/graphite composites, full size elevator boxes for the YC-14 AMST aircraft were fabricated. These have a span of 19 ft and a maximum chord of 18 inches (see Fig. 11-4).[18] Union Carbide Corp. resin systems used in the composite were P-1700 and PKXA polysulfones, the latter a silane end-capped polysulfone. Cost savings were demonstrated over the graphite/epoxy composites during parts fabrication. For lay-up of larger components, resin impregnated graphite tapes and fabrics were selected over unidirectional tape because of the convenience in handling the fabric. Molded fittings were prepared from polysulfone/chopped graphite fiber.

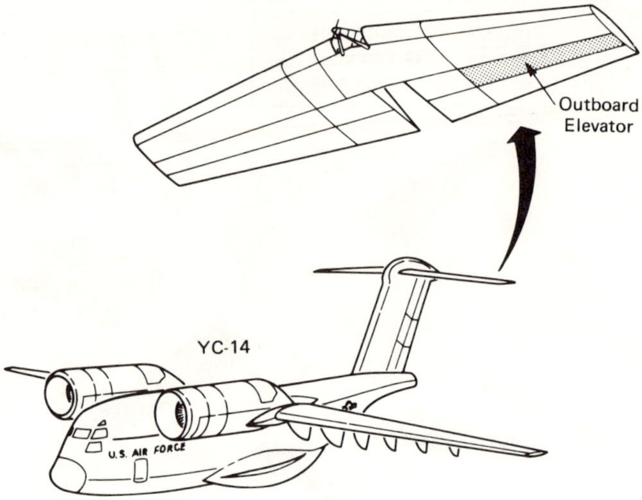

Fig. 11-4. YC-14 outboard elevator of graphite fiber composite.[18]

Advantages were also observed in the use of fusion bonding in some circumstances.

There are other examples of advanced composites being evaluated in military aircraft and commercial planes, such as horizontal stabilizers on the F-111; outer wing tips on the A-7, a graphite/epoxy composite with three years success; 50 rudders on the F-4 airplane; Sikorsky UTTAS military helicopter aramid applications on fairings, doors, access panels, and ballistic armor; De Haviland–DHC-7–STOL aircraft with numerous interior lightweight parts of advanced composites; over 100 lb of C/G fiber/epoxy composites on the Tornado, Mach 2 aircraft (England); Wing sections of C/G on the Jaguar aircraft; and numerous applications of Courtaulds' HM-S fiber composites on helicopters, engine pods, and naval aircraft; and applications of aramid fiber composites on the interior of business aircraft such as Canadair CL-600 Challenger and Cessna 441 and Citation 1. The U.S.S.R. has not disclosed their uses of advanced composites, though there are verbal accounts of synthetic organic fibers, such as aramids, and some technical papers on carbon fiber studies, the latter having been referred to in this text.

The Gossamer Albatross is in the record books as the first man-powered aircraft to cross the English Channel–a distance of 22 miles across the water–on June 12, 1979. It weighs about 70 lb, has a 96 ft wing span, and may be propelled at an average speed of 11 miles/hour, powered by a bicycle-like pedalling device and propeller requiring 0.25 horsepower energy (see Fig. 11-5).[19] An earlier model, the Condor, the first man-powered heavier-than-air craft, was the

Fig. 11-5. "Gossamer" man powered airplaine by MacCready in flight. (*Photo by Don Monroe.*)

first to fly a figure 8, 1.15 mile course on August 23, 1977 at Shafter Airport in Bakersfield, California. To go the greater distance it was necessary to achieve a significant weight reduction, and this was accomplished through the introduction of a graphite fiber/epoxy main spar for the wing span. "Mylar" polyester film covered the sail-plane-like wing and the pilot's compartment. The aerodynamic design and construction was under the supervision of Paul MacCready. An English man-powered model, the "Puffin 2" of Bristol University, also used carbon fiber reinforced structure, saving valuable weight, though the aircraft did not attain the success of the Gossamer Condor or Albatross.

High performance sail-planes have found that graphite/fiber composites offer stiffness and weight saving advantages over their equivalents in glass fiber composites. A 15 m sail-plane produced in Europe has main spars of C/G composites, which have proven satisfactory in service.

It is interesting to observe that some of the most popular construction items used on model airplanes for the hobbyist are spars of carbon fiber, with a low density core. These provide good stiffness and lightweight.

SPACE APPLICATIONS

Advanced composites with advantages of high stiffness to low density, good fatigue resistance, and excellent thermal dimensional stability have been receiving increasing attention on the part of designers for use in operational systems. In the chapter on electrical properties and applications, some of the C/G composites used in antennas systems of space vehicles have already been described. Their uses go considerably beyond this, as, for example, the graphite fiber epoxy composite for the large bay doors of the space shuttle orbiter (see Figs. 11-6 and 11-7). A change from aluminum to "Thornel" 300 graphite/epoxy prepreg (Fiberite Corp. and Hexcel Corp.) for door panel surfaces and the framework of interior ribs, and an aramid-fiber honeycomb core, have reduced the weight of the bay doors. Enclosure of the payload bay is accomplished by a series of 8 doors, each 15 ft long with a curved span of 13 ft. The 3200 lb doors weigh approximately 900 lb less than an equivalent aluminum alloy structure. This has substantially increased the payload capabilities on the spacecraft orbiter vehicle. The primary builder is Rockwell International Space Division, with the first units to be orbited in 1980.[20] There are further applications of advanced carbon fiber/carbon matrix composites for the high temperature exposure of the space shuttle's nose cap and wing leading edges. As described in the chapter on high temperature carbon/carbon composites, graphite/epoxy parts are carbonized and

Fig. 11-6. Space orbiter for the 1980's. Touchdown in an early trial. Graphite/epoxies are used to reduce weight. (*Courtesy Rockwell International Space Systems Group.*)

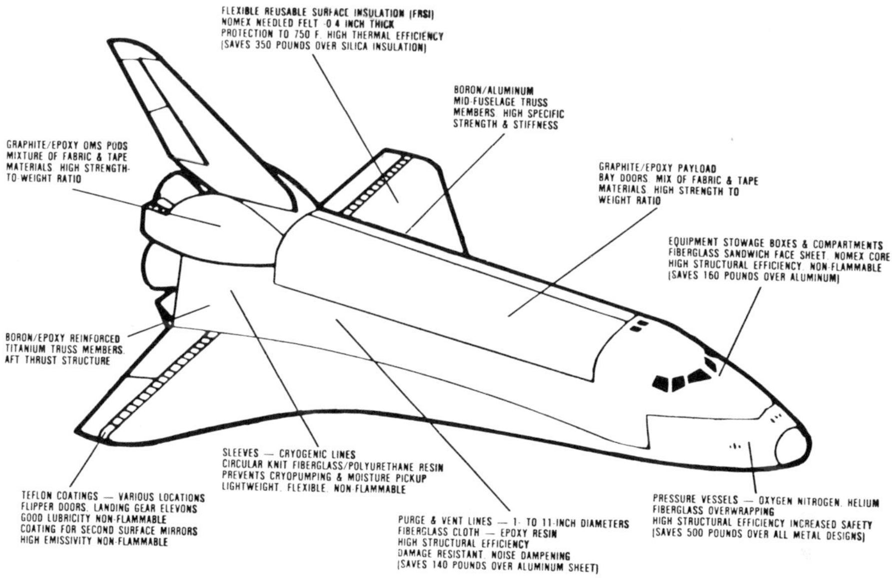

Fig. 11-7. Schematic illustration of advanced composites on the space shuttle.[20]

then reimpregnated with furfuryl alcohol resins several times to achive temperature capabilities up to 1000°F, and higher, for shorter time intervals. Carbon/carbon composites are also employed in aircraft brakes and offer potential for very high temperature applications beyond the capabilities of organic resin matrices.

Darms[21] reviewed the applications for advanced composites on the space shuttle. The United States space shuttle will be used as the launch vehicle for a majority of the space programs in the 1980's. It will provide transportation for low earth orbit insertion and recovery of satellites. It will also function as a base for the erection of large space structures for solar panels and antennas. In Fig. 11-8, an artist's rendition is shown of the orbiter vehicle positioned in space, with pultrusion equipment by Goldsworthy Engineering for forming beams in space from coated C/G fibers and other reinforcements—now undergoing a feasibility study. The 150,000 lb orbiter is capable of placing up to a 65,000 lb payload into orbit about the planet Earth. The shuttle's payload bay doors represent the largest application of composite materials on the shuttle, using approximately 2000 lb of graphite/epoxy cloth and unidirectional tape. Face sheets of the composite consists of a layer of 7 mil thick graphite fabric, oriented 45° to the length of the bay. The fabric is sandwiched between two longitudinal layers of 4 mil thick unidirectional tape to form the skin. An outer 2 mil thick lightning strike protection layer is co-cured with the outer skin.

Fig. 11-8. Artist's rendition of orbiter vehicle in space with pultrusion equipment for forming structural beams. (*Courtesy Goldsworthy Engineering Co.*)

	ALUMINUM	GRAPHITE FIBER/ALUMINUM	GRAPHITE FIBER/EPOXY
Coefficient of thermal expansion (10^{-6} in./in./°F)	12	2 to 3	0–1
Moisture (ground test) effect	None	None	Poor
Stiffness effectiveness, modulus/density	100	244	220–500
Thermal stability efficiency K_t/coefficient of thermal expansion	6	19.7	5
End fittings	Conventional	Potentially good	Special

Armstrong and Johnson[22] reviewed composite materials for spacecraft applications. They prepared a comprehensive list of typical components, reproduced in Table 11-3. Each component, its function, and environment is briefly

Table 11-3. Some LMSC Space Applications Using Graphite/Epoxy Composites.[22]

APPLICATION	FUNCTION	ENVIRONMENT	DESIGN CONSIDERATIONS	MATERIAL	CRITICAL DESIGN CONDITION	DESIGN APPROACH
Dimensionally stable structures	Provide stable platform	$70^\circ \pm 2^\circ$F active -100° to $+150^\circ$F Passive	Thermal dimensional stability	Hybrid GY-70/T-300 fabric	Thermal gradients	Precise Modulus and C.T.E. knowledge Tight manufacturing controls
Structural panels	Load carrying	-50° to $+150^\circ$F	Axial boost loads, Comp. shear, stiffness for high nat. frequency	HM-S/T-300 fabric	Axial compressive loads	Buckling analysis/panel tests
Deployment booms	Deployment of experiments	-50° to 150°F Cyclic	Ovalizing/load/frequency/stiffness Thermal cycling	GY-70/339 $[45/0/45/0_4)_s$	Microcracking and propagation	Analysis and testing
Aluminum fitting/graphite tube	Load transfer	-50° to 150°F Cyclic	Bending loads Thermal compatibility Interlaminar shear Reliability	Graphite tube. Fiberglass doubler. Hysol 9309 adhesive Aluminum fitting	Interlaminar shear in graphite plies adjacent to fitting	Finite element model for thermal conditions. Testing at temperature.
Spacecraft towers	Experiment support	-50° to 150°F	Combined compression and bending Frequency/stiffness Thermal dimensional stability Weight	HM-S $[+45/0_3)_s$ Chopped fiber and fabric fittings	Static load at fittings Tube stiffness	Analysis Static and dynamic test
Honeycomb panels	Solar array panels	200° to -50°F	Frequency/stiffness Dielectric Weight	"Kevlar" 49/GY-70 aluminum honeycomb core	Stiffness/weight ratio	Analysis Thermal and static tests
Box beam	Hardware support	50° to 150°F	Acoustic initiated load Thermal isolation Weight	T-300 fabric	Stiffness	Analysis Static test
Selectively reinforced magnesium beam	Load carrying	50° to 150°F	Bending Frequency/stiffness Weight	GY-70 ZK 60A-T5 magnesium	Stress concentrations in holes. Adhesion	Analysis
Graphite fittings/graphite tubes	Reduce thermal mismatch	-50° to 150°F	Axial load Thermal expansion compatibility	Chopped fiber T-300 fabric	Stress concentrations in fitting attachment holes	Thermal cycle test
Brackets	Tubular supports	-50° to 150°F	Coat Thermal compatibility Stiffness/frequency Static loads	T-300 fabric	Static load	Static load test Stiffness test
Lenticular ribs	Antenna ribs	-250°F to 100°F	Thermal dimensional stability Deployment stresses Stress relaxation and creep	HM-S	Deployment Stress relaxation	Analysis, development and functional test

described. This table addresses itself principally to C/G fiber/epoxy composites. In addition the following comparison was made of materials considered for dimensionally stable struts.

The graphite fiber reinforced aluminum alloy shows promise as constructional materials for the payload struts in space and will be reviewed in the chapter on high temperature matrices above 300°C. In the chapter on very high temperature matrices for graphite fibers, some of the experiences of graphite fibers with metals are examined.

Some of the carbon fiber applications in the space shuttle programs are handled in an inconspicuous manner, though they fulfill major roles. For example, Armco's Hitco organization has developed new high performance insulating materials of carbon fibers and a special heat resistant rubber to function as a liner for the large cannister-like rocket boosters for the United States' space shuttle. Each flight of the orbiting vehicle will require two rocket boosters, each measuring 125 ft long, to contain the solid fuel and protect the container as the fuel burns during flight. C/G fibers will also be found in the carbon/carbon composites used in the construction of the rocket nozzles which expel the booster's burning fuel as the rocket is driven skyward.

Various manufacturers are beginning to submit proposals for experiments that can be executed in an orbiting vehicle. The products range from drugs to fiber optics and a number of proposals have been reported.[23]

Hertz[24] has examined various effects of moisture on graphite/epoxy composites being considered for spacecraft structures. In particular GY-70 high modulus graphite fiber/X-30 resin matrix was examined at various temperatures and humidities. Among the dimensionally critical space structures were listed the following.

Optical supports
Antenna mountings
Gyro frames
Dipole antenna elements
Sensor support structures
Extendable booms and structures
Waveguide horns and filters
Telescope housings

For the spacecraft interiors, hybrid composites of graphite and aramid fibers were examined.[25] The polyimides and phenolic resin matrices were favored in this report. From a processing standpoint, the epoxy resins were more suitable for handling hybrid advanced fiber components.

Optical Support Structure for Space

A truss structure was designed which would provide a space stable support system for the secondary mirror of an optical telescope system.[26] Close attention was given to thermal stability. Graphite fibers included HM-S and AS-1, and "Thornel" 300 eight harness weave fabrics—all of them using Fiberite's 934 and Hercules 3501 epoxy resin. Materials which exhibit a thermal expansion coefficient close to zero were required for the developmental truss.

Similar conclusions were reached by Prunty in the design and fabrication of an optical bench and associated mirror support structural components. Stringent dimensional stability criteria dictated the use of graphite fiber/epoxy composites for the space telescope structure.[27] The precision support is required for a large optical assembly designed for X-ray observation of galactic and extra-galactic X-ray sources. The C/G fiber composite was 0.104 inch thick and comprised a 28-inch diameter cylinder and a 41.2 inch truncated cone, with an overall length 119 inches. The coefficient of thermal expansion was reported as ±0.10 micro-inches/inch/°F for the graphite fiber/epoxy selected.

MISCELLANEOUS AEROSPACE APPLICATIONS

There are many noteworthy programs on advanced components involving materials, as well as equipment and components, for aircraft and space application. Some of these are in a research phase and final conclusions have not been drawn and do not lend themselves to specific categories. Selected examples are presented below.

Composite Nozzle Flaps for Jet Engines

Using du Pont's NR-150 polyimide precursor solutions, graphite fiber composites have been flown in the demanding environment of a Pratt and Whitney F-100 engine for over 100 hours. The nozzle flaps have withstood more than 600°F in flight conditions on an F-15 airplane. These were being developed to replace titanium nozzle flaps.[28]

Graphite Fiber/Polyimide Honeycomb

Poesch reported on the development of and selection of materials for an aerospace honeycomb structure. Of the polyimide resins available at the time of this program, TRW's P-13N was selected for the development effort. The weight targeted for the honeycomb core was 2.0 lb/ft^3, requiring 0.001 to 0.002 inches per ply thickness. Sandwich panels were prepared with face sheets of graphite laminates applied on the honeycomb core. Face sheet-to-core bonding adhesive candidates were also examined on the basis of resin filleting tests, processing characteristics and high temperature bond strength. A panel was designed for operation at 500°F for two hours. Manufacturing procedures by corrugation and expansion of the honeycomb core were reviewed, and details were presented on techniques for preparation of honeycomb cores from graphite/polyimide composites.[29]

Polyimide Adhesives for Fan Blades

A Borsical blade shell is adhesively bonded to a solid titanium spar. This advanced composite fan blade for a turbofan-engine component used American Cyanamid's FM-34 polyimide adhesive and du Pont's NR-150 AG adhesive.[30] Processing details for satisfactory application to the fan blades were reviewed, and bond strengths reported up to 500°F both before and thermal aging. While this paper was not concerned with the bonding of advanced fiber composites, this is a good reference material on high temperature polyimide adhesives—which have a role to perform in some assemblies of advanced composites. Another example of graphite/polyimide engine fan blade is shown in Fig. 11-9.

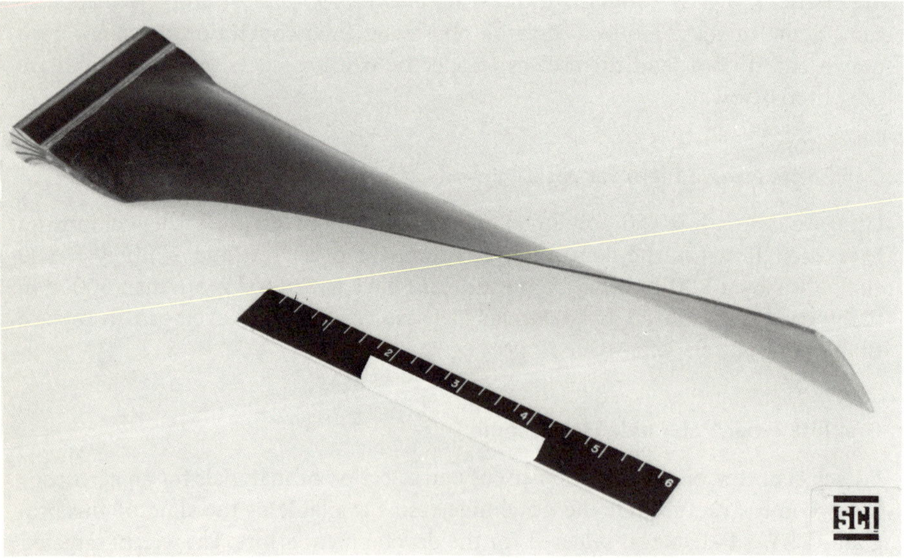

Fig. 11-9. Graphite/polyimide jet engine fan blade. (*Courtesy Structural Composites Industries, Inc.*)

Lightweight Aircraft Brakes

Armco's Hitco division has reported on the development of wheel and brake systems for the F-15 and F-16 fighter aircraft, F-18L prototype, and the B-1 bomber, as well as the Canadair Challenger and Gulfstream III executive jet aircraft.[31] Using C/G fibers and a carbonaceous matrix (C/G composites), it has been possible to develop aircraft disc brakes which have far outperformed their earlier metal counterparts. Not only is there substantial weight savings, but also increased durability in service—made possible by the special techniques of preparing carbon fiber/carbon composites for very high temperature service. Further uses of carbon fiber/carbon composites include solid propellant rocket motor components and re-entry heat shields—with a potential in the brake systems for automotive and truck service. Further details are given on aircraft brakes in chapter on high temperature resistant matrices.

Landing Gear Spring Blades

Schuetze[32] discusses the development of several landing gear spring blades for lightweight aircraft. A special sandwich structure provided high longitudinal stresses in the faces and high shear stresses in the sandwich core. The length of the spring blades made a high bending stiffness necessary, and hence C/G fibers were introduced into the composite. Furthermore, with the inclusion of aramid fibers, structural weight reduction could be maintained and dampening conditions improved in the hybrid composite.

Large Graphite/Epoxy Ring

A large graphite fiber/epoxy ring measuring 10 ft in diameter and 10 ft long was developed on a NASA contract to be cost competitive with an aluminum shell. Three 120° corrugated panels were prepared rather than one piece to reduce tooling costs. The skins applied 56% longitudinal orientation of the fibers and 44% oriented at 45° to the main axis. Doublers were used for stiffening and were prepared from "Thornel" 300 cloth and Narmco 5208 epoxy resin.[33] Assembly and tooling details were discussed at length in this report.

Composite Box Beam

The program entailed selective application of boron/aluminum and graphite fiber/epoxy as a unitary composite. A highly tailored hybrid laminate of boron/epoxy and graphite/epoxy was applied to each element of the structure to produce the most efficient all-composite wing box beam design. The optimum design entailed ribbed graphite fiber/epoxy substructures with mechanical fastened boron fiber/aluminum honeycomb sandwich upper and boron/graphite/epoxy hybrid plain panel lower cover. The optimized composite weighed 1536 lb, as compared to 2854 lb in aluminum.[34]

Engine Fan Frames From Composites

A contractor built two graphite/epoxy composite fan frames for an experimental engine as a weight and noise reducing measure.[35] The composite system employed A-S graphite fiber and Hercules' 3501 epoxy resin. A 25 to 35% weight reduction over a metal equivalent had been targeted. Ground tests with the engine proved satisfactory.

Aeroelastic Tailoring

The continuing development of advanced airframe technology is required in order to implement new structural concepts that are superior in cost/weight performance to more conventional structures. This concept is advanced by what has been called aeroelastic tailoring in promoting highly maneuverable aircraft technology (HiMat). As outlined by Lackman *et al.*,[36] both the wing and canard (smaller forward set of wings) surfaces are aeroelastically tailored to enhance the aerodynamic characteristics critical to the fighter aircraft mobility. This is achieved by orienting the advanced composite plies to control the change of the airfoil shape under load, as a function of the wing and canard bending, to attain the favorable airfoil shape in terms of life and drag for the specific flight conditions. The HiMat airfoils are a new and unique concept and in terms of aerodynamic efficiency (see Fig. 11-10) hold much promise. Because the concept includes so many high risk technical advances, it was decided to pilot the vehicle from ground controls during early flights. The increased lift from the combination of canards and wings increases maneuverability at both sonic and subsonic speeds. Advanced graphite fiber composites are prominent in this new design.

AUTOMOTIVE APPLICATIONS

The manufacturers and designers of advanced fiber composites expect that the large growth in high modulus and high strength carbon and graphite fibers will take place when automotive manufacturers have adapted these materials to automobile components on a production basis. National requirements for improved performance in automobiles on the basis of more miles per gallon (mpg) have been motivated by the decreasing supply of fuel from petroleum sources throughout the world. One of the more effective ways of improving the mpg performance of passenger automobiles is through a drastic weight reduction that

Speed	Supersonic
Maneuvering	8G Performance
Launch Weight	3400 lb.
Engine	General Electric J85-21

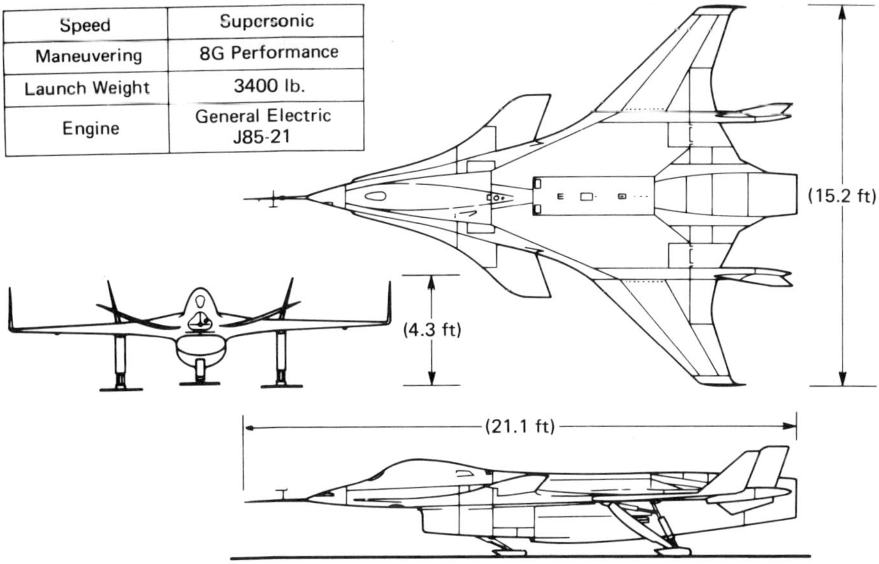

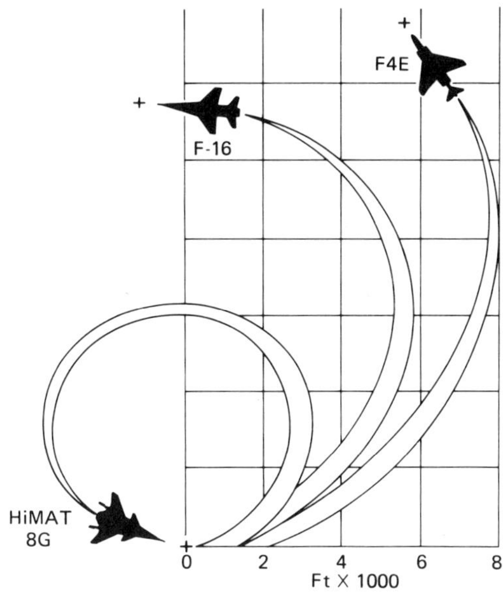

Fig. 11-10. HiMat airplane uses graphite composites.[36] *(Reprinted with Permission of Government by Data Publications–"U.S./R&D"–Brooklyn)*

will not sacrifice safety and other desirable features which we have come to expect in transportation. The experiences being gained by the aerospace industries in attaining substantial weight savings by replacing metal parts, where practicable, with advanced composites, is probably more relative to aerospace, where payload increase by weight reduction contributes to the economic well-being of the aircraft companies. In the automotive industry the reasons may not be as compelling, though our dependence on imported oil for fuel leaves us little choice at this time.

Fuel economy and lightweight vehicles have become vital concerns of the automotive industry, as a consequence of the Energy Policy and Conservation Act, which became law on December 22, 1975. Legislation mandates a production weighted Corporate Average fuel economy for passenger automobiles that will increase from 18 mpg in 1978 to 27.5 mpg in 1985. The ability to meet this requirement depends upon a substantial reduction of the average inertial weight of the car, as well as major gains in power plant efficiency. In a well-publicized program, Ford Motor Company exhibited, at the 1979 SAE Exposition in Detroit, an experimental automobile using graphite composites to a great extent. Relative weights of some comparable graphite fiber and steel components have been released by Ford[37] (see Table 11-4).

Among the problems that Kaiser directed attention to were the following.

1. Cost of raw materials. Advanced composites as developed for aerospace are prohibitively expensive. Hybrids involving lower cost glass will be necessary to offset the higher cost graphite fibers. Also, as pointed out earlier in discussions on thermosetting matrices, the faster curing and lower cost polyester binders are being actively considered.
2. Manufacturing techniques to adapt to the resin/fiber hybrids will have to be improved and speeded up.
3. Damage repair. Aerospace has been developing methods of repair and maintenance which are applicable to automotive use. In fact, advantages may lie in non-corroding composite materials over metals.

In Table 11-4, showing the weight savings in various metal components replaced by advanced fiber composites, data on the lightweight automotive vehicle appeared in a summary of Ford's graphite fiber/epoxy car, while the further data on heavier trucks appeared in a publication by Rockwell.[38]

There is increasing emphasis on the hybrids involving lower cost glass fibers to reduce the cost of an all C/G composite, which means trade-off in stiffness as graphite fiber volume is reduced in the composite.

In reiterating the automotive industry's desire to achieve fuel economy through vehicle weight reduction, glass fiber reinforced polyester and molding

Table 11-4. Lightweight Vehicle Program Graphite Component Weight Summary.[37,38]

	COMPONENT WEIGHT, (lbs)		
COMPONENT	STEEL	GRAPHITE COMPOSITE	REDUCTION lbs.
Hood	40.0	15.0	25.0
Door, right hand rear	30.25	12.65	17.60
Door guard beam	3.85	2.40	1.45
Suspension arm, front upper	3.85	1.68	2.17
Suspension arm, front lower	2.90	1.27	1.63
Transmission support	2.35	0.55	1.80
Driveshaft	17.40	12.00	5.40
Air conditioning, lateral brace	9.50	3.25	6.25
Air conditioning, compressor bracket	5.63	1.35	4.28
Further Examples of Weight Savings in Composites			
Heavy truck driveshaft	139	69	70
Lightweight truck drive shaft	35	18	15
Truck frame rail	259	127	132
Trailer axle housing	294	205	89

compounds were examined by Sanders.[39] Test results that were reported indicated that these materials would operate within the operating environment of the automobile. However, as already pointed out, the advanced composites with high modulus fibers which contribute to greatly improved fatigue, better dimensional stability, and greater stiffness represent decided improvements over the earlier glass reinforced plastics. To substantiate this position, carbon-glass hybrid materials for molding purposes, with a limit of 25% carbon fiber by weight, offered cost effectiveness. Sheet molding compounds of graphite fiber polyester surface skins and glass fiber/polyester cores have been offered.[40] Leaf spring applications were suggested for automotive use.

There have been several papers by automotive engineers addressing themselves to the problems of the design and fabrication of automotive components in graphite fiber reinforced plastics. These papers are reviewed here as representative of the state of the art, though much needs to be accomplished before the advanced fiber composites can claim significant inroads into the automotive industry. It is also interesting to note how the aerospace industries are cooperating in sharing their experiences with automobile manufacturers through joint authorship of recent papers. Referred to as a "Graphite Body-in-White Design," Fig. 11-11 illustrates the major body components fabricated in graphite fiber/350°F cure epoxy cloth and tapes for the body of the 1979 Ford LTD vehicle.[41] The advanced composites used were bidirectional woven cloth (10 X

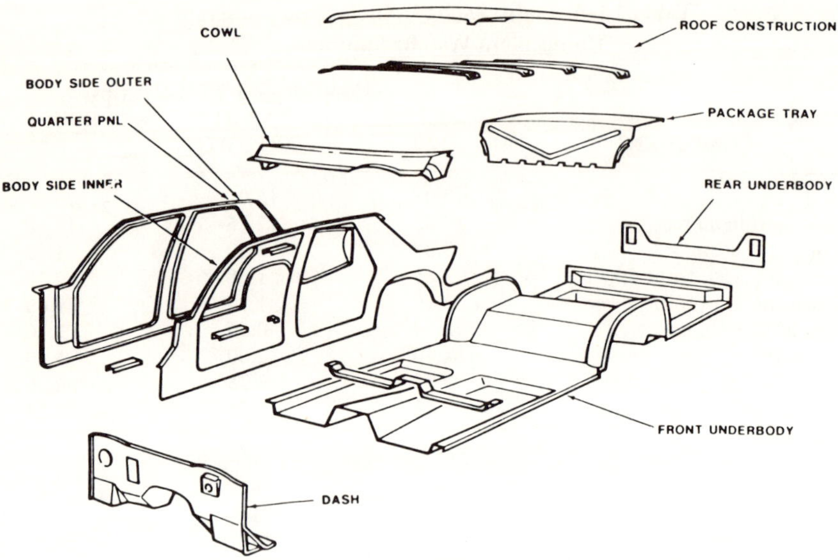

Fig. 11-11. Major "graphite Body-in-White" components.[41] *(Reprinted with Permission Soc. of Automotive Engineers)*

10-5 harness satin, tows of C/G 6000 filaments), used with high modulus graphite unidirectional tape, when required, for additional stiffness. The fiber types for woven cloth were of a high strength grade of C/G fiber ("Thornel" 300 or "Celion") and Hercules' high modulus HM fiber for the tapes. The baseline mechanical properties used in the design and analysis of the body-in-white were as shown in Table 11-5.

The body-in-white model refers, for example, to the basic automotive welded body shell, excluding add-ons, such as fenders and bumpers. The finite element stress analysis technique is used to compare the bending and torsional stiffness of the graphite fiber/epoxy body-in-white to that of the steel body-in-white. Bonnett *et al.*[41] used the "Argus" computer program to generate composite analysis and perform stress analysis of the models. Operated by Merlin Technologies, finite element models for the body-in-white components were generated. For the complete body assembly, 41 graphite fiber/epoxy parts weighing 160 lb, replaced 104 steel parts weighing 423 lb to afford an impressive weight reduction of 62%. Composite parts were fabricated by autoclave techniques used by aerospace industries. Prototype parts were incorporated into the 1979 Ford LTD experimental vehicle.

Sidwell *et al.*[42] offered additional data on the designing, fabricating, and testing of a graphite fiber/epoxy frame developed for the 1979 Ford LTD. "Thornel" 300 Union Carbide Corp. graphite fiber and "Celion" 3000 fiber was

Table 11-5. Baseline Mechanical Properties.[41]

	GRAPHITE CLOTH (90°)	HIGH MODULUS GRAPHITE/EPOXY TAPE	HIGH STRENGTH GRAPHITE/EPOXY TAPE
Ply thickness (in.)	0.011	0.005	0.005
Longitudinal modulus (E_{11} 10^6 psi)	10.3	30.0	21.0
Transverse modulus (E_{22} 10^6 psi)	10.3	1.45	1.2
Torsional modulus (G_{12} x 10^6 psi)	1.0	0.65	0.65
V_{12}	0.09	0.25	0.27
Density (lb/in.^{3})	0.056	0.056	0.056

woven and prepregged by Hexcel. The test program was confined to nondestructive loading. Torsion and bending tests were conducted to establish that targeted stiffness levels were attained. Among the conclusions reached were the following.

1. The experimental graphite/epoxy composite component exhibited higher stiffness while meeting the minimum strength requirements of steel.
2. Torsional stiffness and bending stiffness were unexpectedly high in the graphite/epoxy frame due to component integration and improved joint efficiency.
3. The weight reduction for the frame was 79 lb (27.7%) of the 1979 Ford steel frame.

The fabrication and tooling requirements for automotive components of graphite fiber/epoxy composites were emphasized in a paper by Cedar.[43] Included in the presentation were master tooling aid development, high temperature epoxy molds, component fabrication, component subassembly and body assembly. The three major subassemblies and the 20 graphite fiber reinforced plastic components that made up the body were bonded together in the final assembly fixture using Goodyear's urethane adhesive "Pliogrip" 6040/6045. Illustrations of the assembly of the prototype car were shown step by step in this paper. The production cure cycles, typical of aerospace practice, covered several hours of staged heating and cooling. A total of 98 molds (made of epoxy resin) were required for this program.

In the unidirectional C/G fiber/epoxy system provided by Narmco and Hexcel Corp.,[44] components were fabricated from either graphite cloth or cloth interleaved with unidirectional tape. Component verification was accomplished by processing test samples with the applicable component.

Segal[45] has described automotive and truck leaf springs produced from graphite/fiber composites. Motivations for their use in automotive and truck applications arise from the weight reduction and improved performance, as evidenced by lower noise and vibration. Drive shafts for automobiles are also cited as a good potential for graphite/epoxies, because the first natural frequency in bending is proportional to the square root of the specific stiffness (E/d). Development C/G composites are illustrated in Figs. 11-12 and 11-13, photographs of a drive shaft for automobiles and heavy duty truck springs manufactured from carbon fibers.

Yates *et al.*[46] discuss the design approaches possible with composites, which differ significantly from those often employed for metals. No longer is it possible to operate on the basis of a homogeneous isotropic material exhibiting well-defined elastic and plastic stress-strain behavior. In advanced composites the structure is decidedly anisotropic and the modulus of the material can be varied freely through the section and along the length of a component. The designer has more freedom than was previously available using isotropic materials. It is now possible to soften structures selectively and force loads away from highly

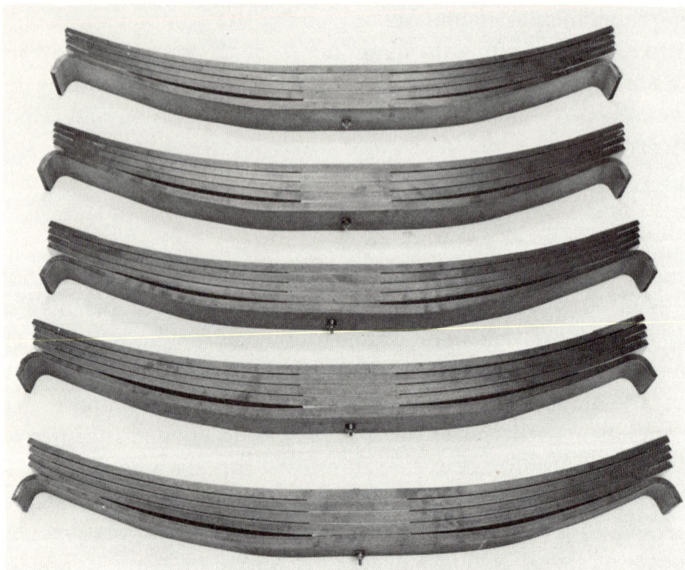

Fig. 11-12. C/G composite used in production of heavy duty truck springs. (*Courtesy Graftek Division of Exxon.*)

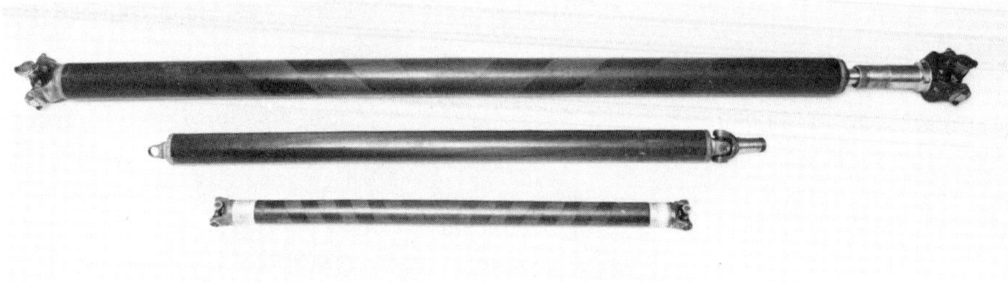

Fig. 11-13. C/G composites used in production of drive shafts. (*Courtesy Graftek Division of Exxon.*)

stressed areas, departing from traditional Mc/I bending stress distribution; to design one-piece structures without the previous concern with a myriad of parts and fasteners; to vary moduli and thickness easily; and to achieve cost effectiveness through hybridization. To implement their discussions, Yates *et al.*[46] used as an illustration a leaf spring for automotive purposes, carefully tailoring the design to optimize the choice of materials. The result was a tapered combination of glass and graphite fiber composites with a weight of 4.5 lb versus the baseline steel spring weight of 28.5 lb. Furthermore, the composite spring offers better fatigue characteristics than steel and lower noise transmission. A rotary pultrusion machine for producing curved leaf springs of composite materials has been recently announced.[47]

Not all of the automotive applications are prepared from graphite fiber advanced composites—many of the potential uses lie in molded parts, some of which use short lengths of high modulus fibers for stiffening purposes and for reducing mold shrinkage. An air conditioner mounting bracket was one of the first graphite fiber molded parts supplied to the automotive industry. Weighing less than 2 lb, the graphite fiber reinforced bracket replaced a maleable iron one weighing 7 lb. The actual molding compound contained 30% polyester resin, 20% continuous graphite, and 50% chopped glass. In another application (see Fig. 11-14), a molded part of carbon fiber/epoxy serves as a door hinge cover.

Automotive engineers see possibilities for the C/G fiber composites in four main areas of future vehicles: the engine, the drive train, chassis and suspension, and structural body components. New automobile designs with front wheel drives do not have the drive shaft considerations of rear drive vehicles. Estimates for the advanced composites of all types are up to 40,000 tons/year by 1985. This use will also depend on the price drop in the expensive graphite fiber—well below the current \$18 to \$20/lb level. Also to be considered are the energy needs of the raw materials and the finished parts, which have been estimated as shown in Table 11-6.[48]

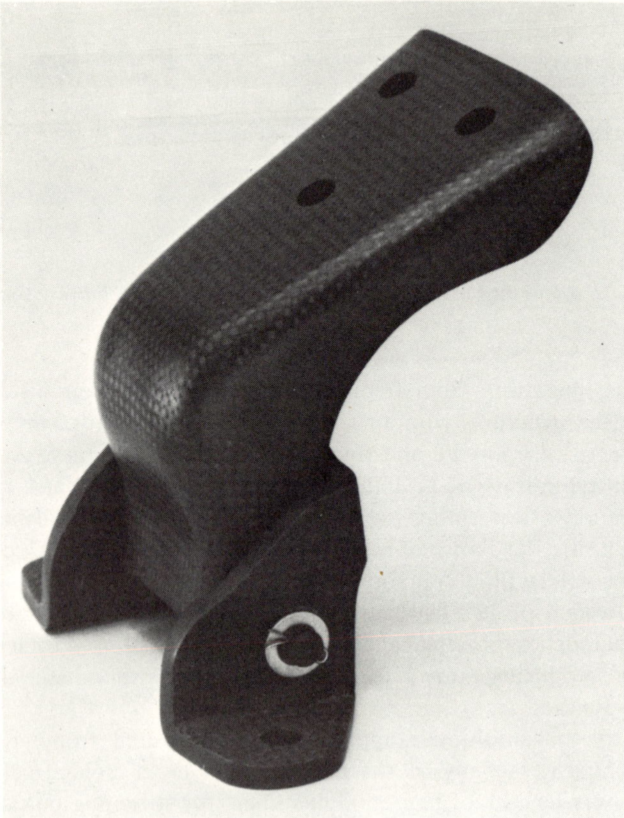

Fig. 11-14. Molded carbon fiber bracket for auto door hinge cover. (*Courtesy Union Carbice Corp.*)

Table 11-6. Energy Needs of C/G Epoxy Versus Metals.[48]

MATERIAL	RAW MATERIAL kwh/lb.	SCRAP FACTOR kwh/lb	FINAL PART kwh/lb
C/G epoxy	15	1.5	33
Steel	16	5.0	100
Aluminum	22	7.0	178
Titanium	88	7.0	700

Programs are continuing on automotive developments using composites. As Margolis points out, a 4000 pound passenger car of the 1970's operating at 15mpg, redesigned with lighter weight materials to 2500 pounds, achieves 27mpg.[49] Reviewed in this paper are new high strength metal alloys as well as composites, in leaf springs, wheels, frames, and body components of automobiles.

REFERENCES

1. DeVault, J. (Hercules), *SAMPE Conference on "Fibers Used in Composites,"* Newport Beach, California, October 18, 1979.
2. Matsuda, S. (Toray), *SAMPE Conference on "Fibers Used in Composites,"* Newport Beach, California, October 18, 1979.
3. SPI, RP/C Institute, *Chemical and Engineering News,* page 12, February 5, 1979.
4. Paez, C. (Grumman), "Present and Future Developments with Aerospace Materials and Structures," *SAMPE Tenth National Technical Conference,* Vol. 10, page 51, New York, October 1978.
5. Anon., *Aviation Week,* Vol. 111, No. 15, page 67, October 8, 1979.
6. Lockheed Corp., Report LR28843 (NASA–Langley Research Center, Contract NASI–14,000), January 22, 1979.
7. Lockheed Corp., Report LR28873 (NASA–Langley Research Center, Contract NASI–15069), January 22, 1979.
8. Sakata, I.F. and Ostrom, R.B. (Lockheed Corp.), "Utilization of Advanced Composites in Commercial Aircraft Wing Structure," NASA Reports 14,5381–1 and 14,381–2, August 1978.
9. Stephens, C.O. (Douglas Aircraft), "Advanced Composite Vertical Stabilizer for DC–10," Report ACEE–03–PR–8484 (Contract NASI–14869), June 1978.
10. Smillie, D.G., "Composite Structure Applications for Commercial Aircraft," *SAMPE Twenty-third National Symposium,* Vol. 23, Anaheim, California, May 1978.
11. Stoeklin, R.L. (Boeing), "Graphite Composite Flight Spoiler–Flight Service Evaluation," NASA–CR–132663, AR–1, July 1973, March 1975.
12. Stoeklin, R.L. (Boeing), NASA–CR–132682, October 1976.
13. Langston, P.R. (du Pont), *SAMPE Conference on "Fibers Used in Composites,"* Newport Beach, California, October 18, 1979.
14. Riewald, P.G. and Weben, C.Z., "Hybrid Composites for Commercial and Aerospace Applications," *Thirtieth Annual Technical Conference,* SPI, RP/C Institute, Section 14B, Washington, D.C., February 1975.
15. Tracey, R. (Lear Aviation), *SAMPE Meeting,* Palo Alto, California, January 26, 1979; and *Business Week,* page 148B, March 20, 1978.

15a. Warakomski, J., "AV–8B–Second Generation V/STOL," *Automotive Engineering,* Vol. 88, No. 1, page 61, January 1980.

16. Wynn, T.D. (General Dynamics), "Standard Approach to Design to Cost of Composite Airframe," *Society of Manufacturing Engineers,* EM78–401, 1978.
17. Anon., "A New Lift for Advanced Composites," *Modern Plastics,* Vol. 54, page 30, July 1977.
18. House, E.E., "YC–14 Thermoplastic Graphite Elevator," *SAMPE Twenty-fourth National Symposium,* Vol. 24, page 201, San Francisco, California, May 1979.

19. Anon., "The Gossamer Albatross," *Plastics Engineering,* Vol. 5, page 9, May 1979; and *Time,* Vol. 113, No. 26, page 45, June 25, 1979.
20. Anon., "It Boosts Payload of NASA Space Shuttle," *Modern Plastics,* Vol. 54, No. 6, page 58, June 1977.
21. Darms, F.J. (Rockwell), "Space Shuttle Advanced Composite Applications," *Advanced Composites Technology Conference,* Technology Conference Associates, page 181, El Segundo, California, March 1978.
22. Armstrong, H.H. and Johnson, R.R., "Organic and Metal Matrix Composites for Spacecraft Applications," *SAMPE Quarterly,* Vol. 9, page 13, January 1978.
23. Waldrop, M., "Industry in Space," *Chemical Engineering News,* page 24, February 20, 1978.
24. Hertz, J., "Moisture Effects In Spacecraft Structures," *SAMPE Twenty-fourth National Symposium,* Vol. 24, page 965, San Francisco, California, May 1979.
25. Hertz, J. (General Dynamics Corp.), "Lightweight Reinforced Plastic Laminates for Spacecraft Interior Applications," NASA-CR-147497, 63 pages, 1975 (C.A. 85-79164, 1976).
26. Smith, D.D. and Jones, R.E. (Boeing), "Space Stable Optical Support Structure," *SAMPE Journal,* page 4, September/October 1978.
27. Prunty, J. and Hall, R., "Application of Thermally Stable Graphite Composites to a Space Telescope Structure," *SAMPE Twenty-second National Symposium,* Vol. 22, page 718, San Diego, California, April 1977.
28. Anon., "News Item," *SAMPE Journal,* page 25, January/February 1979.
29. Poesch, J.G. (Hercules), "Development of Lightweight Graphite/Polyimide Honeycomb," Report H400-12-1-8, NASA Contract NAS-9-11368, July 1971.
30. Paul, R.D. and Winiarski, H.C. (United Technologies Corp.), "Polyimide Adhesive for Composite Fan Blades," *Adhesives Age,* Vol. 21, page 31, April 1978.
31. *Armco's Shareholder's Report for Second Quarter,* pages 6-7, 1979.
32. Schuetze, R. (West Germany), "Investigations on Fiber Reinforced Plastic Landing Gear Spring Blades for Light Aircraft," NTIS-N76-26291/4ST, January 1976.
33. Penton, A., Johnson, R., and Freeman, V.L. (McDonnell Douglas), *SAMPE Twenty-third National Symposium,* Vol. 23, page 137, Anaheim, California, May 1978.
34. Donohue, R.J. and Erbacher, H.J. (Grumman), "Composite Box-beam Optimization Program," *Thirtieth Annual Technical Conference,* SPI, RP/C Institute, Section 14A, Washington, D.C., February 1975.
35. Stotler, C. (G.E. Co.), "Advanced Composite Fan Frame for Quiet, Clean Short Haul Experimental Engines," *SAMPE Twenty-third National Symposium,* Vol. 23, page 1218, May 1978, Anaheim, California.
36. Lackman, L.M., Price, M.A., and Matoi, T.T. (Rockwell International), "Composite Applications to Aircraft Structures," SAE Paper 790146, Cobo Hall, Detroit, February 1979 (U.S., R&D, page 4673, August 1979).
37. Kaiser, R., "Technology Assessment of Automotive Uses of Advanced Composite Materials," *SAMPE Twenty-fourth National Symposium,* Vol. 24, page 904, San Francisco, California, May 1979.
38. Rockwell International, "Advanced Composites in Automotive Components," SP 7926, 1979.
39. Sanders, B.A. (General Motors), "Engineering Properties of Automotive Fiber Reinforced Plastics," Society of Manufacturing Engineers, EMR78-04 1978.
40. Kliger, H.S. (Celanese), "Designing with C-fiber Hybrid Composites," *SAMPE Twenty-third National Symposium,* page 10, Anaheim, California, May 1978.

41. Bonnett, R.E., Kulkarni, H., Lim, G., Beardmore, P., and Knight, R. (Ford Motor Co., and Vought Corp.), "Design and Fabrication of Automotive Components in Graphite Fiber-reinforced Composites," SAE Paper 790031, Cobo Hall, Detroit, February 1979.
42. Sidwell, D.R., Torczyner, R.D., Yates, D., Motwani, M.B., and Thomas, J.L., "Graphite-reinforced Plastic Passenger Car Frame," SAE Paper 790030, Cobo Hall, Detroit, February 1977.
43. Cedar, D.A., "Fabrication and Assembly of Graphite Fiber Reinforced Plastic Vehicle," SAE Paper 790029, Cobo Hall, Detroit, February 1979.
44. Kiefer, W.R., Mills, A.L., Bellinger, T.P., and White, G.H., "Fabrication of Automotive Body Components in Graphite Fiber-reinforced Plastics," SAE Paper 790028, Cobo Hall, Detroit, February 1979.
45. Segal, C. (Graftek), "Industrial and Commercial Applications of Advanced Fiber Composites," *Composites Technology Conference,* Technology Conference Associates, page 250, El Segundo, California, March 1978.
46. Yates, D.N., Moore, P.B., Tsou, F., Magee, C.L., and Harwood, J.J., "Designing with Plastics–The Need for Lighter Vehicles by 1985," *Plastics Engineering,* Vol. 33, page 35, July 1977.
47. Anon., *Plastics Technology,* Vol. 26, page 69, January 1980.
48. Wood, A.S., "Detroit's Energy Problem," *Modern Plastics,* Vol. 54, page 60, June 1977.
49. Margolis, J.M., *SAMPE Journal,* Vol. 16, no. 4, page 7, August 1980.

12
Industrial and Commercial Applications

The aerospace and automotive applications of C/G fibers received considerable attention in the previous chapter because they have been in the forefront of research and development to utilize their intrinsic properties and because of the prospects of large volume applications. Other industries, reviewed in this chapter, have found the qualities of C/G fiber unique to their applications, though they have not attained large volume utilization. The capacity to produce many millions of pounds of C/G fibers is not available yet, but the momentum appears to be gathering, and with the expected lowering of the costs of manufacture, prices should fall. There are many opportunities for C/G fibers in industrial and commercial applications, and these will be grouped in appropriate categories.

SPORTING AND RECREATIONAL USES

Of the 220,000 pounds of high performance C/G fibers used in Japan during 1978, 85 to 90% were used in sports and leisure applications. At the top of the list were graphite fiber composites employed in the manufacture of fishing rods and golf club shafts, which have benefitted from the light weight and stiffness, and at the same time the elastic qualities of the fiber reinforcement. Special pultrusion equipment for collimating the fibers and producing the tapered rods and tubes have been made for large productions, both in the United States and in Japan. In addition, the market has opened up on compression molded thermoplastics, C/G fiber reinforced housings for fishing reels, and other fiber reinforced accessories.

In this country and in Japan, the growing interest in high quality golf club shafts and frames for tennis racquets has been evident in the increasing demand for C/G in these products, replacing earlier metal or wood structures. Experienced players will recognize the advantages of graphite fiber golf clubs and racquets: improved stiffness and better strength to weight qualities. Golf clubs in particular are now penetrating the United States sports market, and special, modified pultrusion equipment is turning out the tapered C/G reinforced shafts.

Kuno[1] gives an up-to-date review of the adaptation of advanced materials technology to marine and sports equipment, citing many applications. For example, at the Nineteenth American Fishing Tackle Trade Show, 10 of the 21 manufacturers showed rods constructed of graphite and/or blends of glass fibers and graphite fibers. Kuno draws attention to marine and sporting goods markets in the United States (see Table 12-1).

Typical of the development effort being made is an improved tennis racquet frame using an expanded polyurethane plastic core, wrapped around with epoxy resin impregnated graphite fibers (modulus of elasticity at 30 to 40 msi).[2] The specification of this patent refers to the superior handling characteristics of the tennis racquets prepared in this manner. Additional manufacturing refinement is cited in the helical winding apparatus, such as the apparatus developed by Goldsworthy (U.S. Patent 3,579,402). Consumer reports on the available tennis racquets indicate that experienced and professional players prefer the graphite frame over the aluminum frame, though prices are approximately three times as much for the advanced composite frame (approximately $200–1979).

Among other sporting goods and accessories, trade references will be found to applications of graphite fibers as reinforcements on items such as ping-pong paddles, prepared from laminated wood and graphite fabrics; racquet ball and badminton racquets; cross country ski poles (manufactured in Norway), which are using poles manufactured of graphite fiber in place of glass fiber—by pultrusion process; cross country skis (modulus of fiber selected 30 msi); motorless sail-planes for flying; and all types of hockey and lacrosse sticks. Some of the sporting equipment, such as those experiencing high impact, will probably defer to hybrids (which offer better impact strength than the composites of carbon and graphite fibers). The above listing of sporting equipment suggests the stimulation and imagination which has captured the enthusiasm of sports enthusiasts for

Table 12-1. Marine and Sporting Equipment Markets (As of 1975 Sales, in Millions).[1]

Pleasure boats, motors, and accessories	$2320
Bicycles and supplies	958
Tennis equipment	666
Golf equipment	487
Fishing tackle	506
Snow skiing equipment	421
Archery equipment	130
Water skis	95

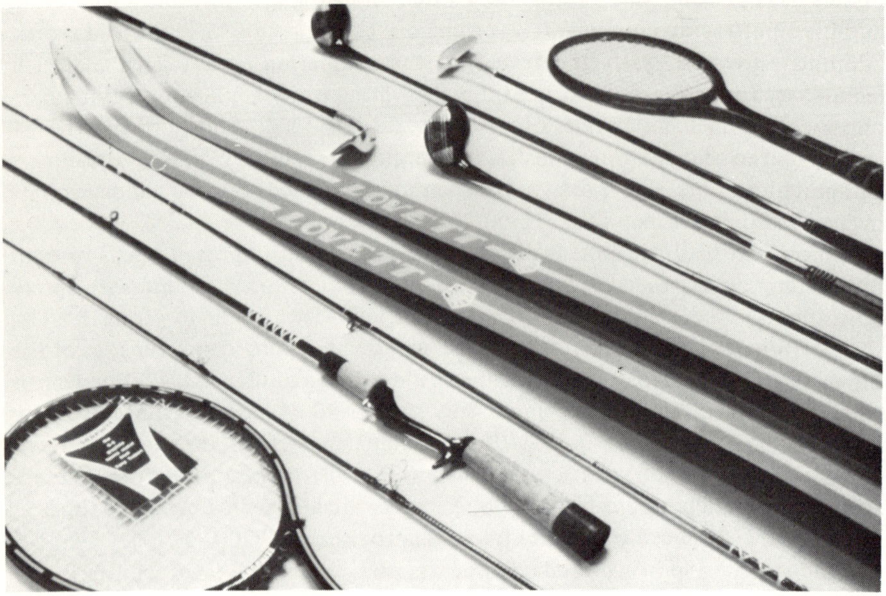

Fig. 12-1. Typical sporting goods using C/G fiber reinforcement. (*Courtesy Union Carbide Corp.*)

lightweight, high strength composites. A group of sporting goods currently using carbon fiber reinforcements is shown in Fig. 12-1.

In the past two decades, the design of alpine skis has made significant advance from wood/aluminum laminated construction to today's designs which utilize glass fibers, graphite, and aramid fibers for reinforcement. The influence that composite materials have had on ski design parameters and their impact upon construction methods used to manufacture modern alpine skis have been reviewed by Boehm.[3] Modern ski construction methods place high strength to weight ratio materials on the exterior of the ski. When the ski bends, the outer layers undergo the greatest strain and will then contribute most to the ski's elastic properties, provided they can carry the high stress levels. Lighter weight foams and honeycomb fill a large volume of the core section of the ski, where the stress levels are lower. Elasticity, torsional rigidity, vibration frequency, and vibration damping are performance related factors influencing the choice of materials.

MARINE APPLICATIONS

Marine applications, because of the strength and corrosion resistance of advanced composites are finding an increasing interest in advanced composites. The aramid fibers have been applied in small boats such as canoes and kayaks, though for the do-it-yourself individual, lower cost glass fibers and polyesters will continue to dominate the field of fabricating and repairing small boats. For the more difficult and special requirements, such as the masts for ocean racing sail boats, graphite fiber composites (because of stiffness and strength) have been used successfully, offering the desired stiffness and stability. Less satisfactory have been the graphite fiber rudders on small boats in a 1979 race between England and Ireland, where a number of failures occurred in the graphite composites. These have been attributed to inexperience in handling and lay-up of C/G composites. Protective top coats appear necessary, particularly when products are to be stressed frequently during salt water immersion.

A tri-maran construction has been built in England utilizing C/G Courtaulds fibers in the main cross-beams. The stiffness and strength requirements dictated the choice of these materials, about which the three main hulls are tied together.

The trimaran, "Great Britain IV," has an 11m graphite/epoxy composite as a cross-beam. Courtaulds' unidirectional graphite fibers were used, held cross-wise by a lightweight glass weft. Pultruded strips of graphite fiber and hybrid fabrics of carbon and glass were used on the bow section, outriggers, and deck panels.[4] Racing sculls featuring hybrid graphite composite skins and aluminum honeycomb core were stiffer, lighter weight, and lower cost than their wood counterparts.

As better familiarity is developed by marine specialists with the available reinforcements and resin matrices for marine components, no one group of materials will dominate the scene. Glass fibers and polyesters have had a large share of the multi-million pound market. The carbon and graphite fibers offer high modulus, high strength, low specific gravity, and are capable of enhancing the fatigue life of marine structures. In combinations with other materials, including wood products, the hybrid constructions will be tailored to meet the requirements of the individual marine design.

Hoover[5] reviewed marine applications of graphite composites, placing emphasis upon the masts and riggings. The speed of the sailboat is proportional to the overall weight and sail carrying ability—both of which are aided by advanced composites. Graphite cross tubes, battens, and masts (as long as 38 ft) are being used. Experimental hulls are also being studied of C/G fiber composites, though considerable background in these areas are evident from the glass fiber laminates and overlays. Part of the experimental efforts include hybrid laminates of mahogany veneer and plies of graphite on racing hulls.

Relevant to the present discussion are the details of one of the sailboats, a 33 ft keel sailboat now in production and constructed with layers of 100% carbon fiber fabric in its sandwich skin makeup.[6] It claims increased speed as a consequence of the considerable weight saving, accompanied by better feel and control at the helm. Principal characteristics of the Eagle yacht are as follows.

LOA–10.1 m
LWL–8.5 m
B. max.–2.42 m
Draft–1.83 m
Ballast–3500 lb
Displacement–5400 lb
Sail area–470 ft^2

PRESSURE VESSELS

High strength, high pressure, filament wound cylinders or bottles of advanced composites have enjoyed notable success on commercial aircraft. For example, there are 30 to 40 of them on a Boeing 747, fulfilling many functions of pneumatically operated devices, by providing a source of air or nitrogen pressure. They are manufactured from glass, aramids, carbon fiber, and hybrid fiber constructions–achieving weight savings of 10 to 15 lb per bottle or container over their metal counterparts. The techniques of producing the composite type of pressure containers were developed primarily upon glass fiber structures, and many thousands of gas pressure cylinders and bottles were manufactured. In more recent years, interest has been extended to newer fiber reinforcements which will permit higher pressures to be maintained, lower unit weights, and improved fatigue strength. Typical of the filament wound tanks are the gas storage tanks shown in Fig. 12-2. Also to be noted are metal tanks, which have been reinforced on the exterior with high strength fibers, circumferentially wrapped about the metal to give better performance.

Gerstle[7] described filament reinforced spherical pressure vessels, using a thin ductile bladder of aluminum, on which was wound high strength resin impregnated fibers of C/G and aramid. "Thornel" 400 and 75-S graphite fibers were used with an epoxy binder, and aramid fibers were understandably more compliant. Curves were reported to demonstrate the improvement in strength of the pressure vessels as a function of the volume fiber (up to 0.75 volume fiber were reported). The presence of the aluminum bladder (8 inches diameter), served not only as a mandrel for filament winding the vessel, but also it eliminated any leakage that could have occurred.

Graphite fiber/epoxy pressure vessels lined with 0.25 mm titanium metal have also been fabricated.[8] The vessels were 140 mm long and 102 mm in diameter. A special high strength graphite fiber was used, and the hoop wall was 2.76 mm thick. Aramid fibers were also used at the ends of the vessel to provide additional reinforcement. Burst pressures were in the range of 6500 to 6900 psi in most instances. Fatigue life of this composite pressure vessel was in excess

Fig. 12-2. Spherical gas storage tanks. (*Courtesy Structural Composites Industries, Inc.*)

of 2000 cycles, when pressurized to 50% of the expected burst performance. Composite performance was considerably in excess of titanium vessels, at lighter weight. Acoustic emission was employed during burst tests to study the effects of different winding patterns, artificial flaws, winding induced fiber fraying, and changes in resin matrices and fibers. Small effects produced by changes in these variables were greatly magnified as changes in acoustic emission. (See the test method chapter for techniques of acoustic emission.)

Hanson's[9] work on graphite fibers fabricated into NOL rings and filament wound biaxial cylinders provides useful information on filament wound pressure vessels, particularly at cryogenic temperatures. Materials used were "Thornel" 50 graphite fibers and ERL 2256 epoxy system. Longitudinal and hoop strains were measured on a 7.5 inch diameter filament wound cylinder. On the basis of specific tensile strength, graphite fiber filament wound cylinders were superior to glass fiber filament wound cylinders beyond 80 cycles service. See Fig. 12-3, comparing performance of glass fiber and graphite fiber pressure cylinders, and Fig. 12-4, a stress-strain diagram for a single cycle test of "Thornel" 50 graphite fiber/epoxy cylinder.

ENERGY STORING DEVICES

Advanced C/G composites have figured prominently in proposed designs for lightweight, high strength energy storage devices, such as flywheels—with poten-

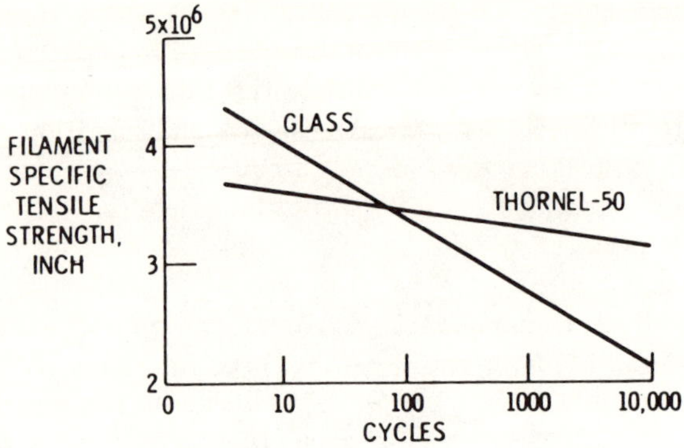

Fig. 12-3. Comparison of glass and graphite fiament cylinder fatigue life on the basis of special strength.[9]

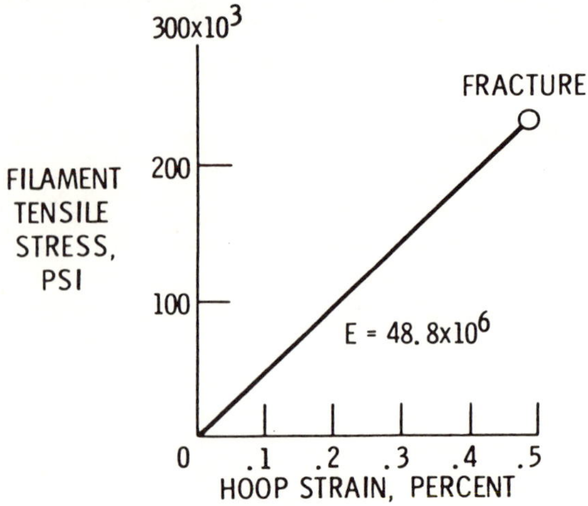

Fig. 12-4. Stress-strain diagram for single cycle test of Thornel-50 epoxy resin cylinder.[9]

tial applications in automotive industries. Composite material systems offer the potential of rotors with higher energy densities and greater safety than is possible with metals. The design and development efforts of Pratt and Whitney[10] initiated interest in flywheel rotors. Since then, several flywheel technology symposia have been sponsored by the United States government (DOE) and Lawrence Livermore Laboratories. A rotating body has kinetic energy proportional to the mass and the square of the spinning speed or angular velocity:

$$E_k = \frac{1}{2} I \omega^2$$

E_k is the kinetic energy; I is the moment of inertia; and ω is angular velocity. Lightweight composites rely on high rotational speed in preference to large mass for storing energy. Single rim and multiple rim thin wall cylindrical shell, rim spoke composite flywheels were stress analyzed and vibration tested.[11] They were designed to supplement the power of a 100 kw wind turbine generator during periods of low wind velocity. The single rim wheel was made from graphite fibers and epoxy resins at an 0.7 fiber volume loading. Multiple rim wheels were fabricated from glass, boron, and aramid fibers.

Reedy[12] described a circumferential wound rim method to fabricate prototype flywheels. Graphite fiber/epoxy rims with an inner to outer radius of 0.775 ratio were prepared. An analysis of constant thickness rims identified this ratio as having the best storage capacity for a C/G fiber epoxy composite. A "wagon-wheel" aluminum hub design was used with bands as spokes, overwrapped with aramid 49 epoxy composite. The bands were in tension along their entire length. At 32,000 rpm, the flywheel delivered 0.5 kwh of energy with a 3 to 1 speed reduction. The rim inside radius was 7.625 inches; the maximum hoop stress was 206 ksi at 40,000 rpm and the weight 16.4 lb. Properties of circumferentially wound composites are shown in Table 12-2.

Table 12-2. Properties of Circumferentially Wound Composites.[12]

	GRAPHITE/ EPOXY	ARAMID 49/EPOXY	S-GLASS/ EPOXY	E-GLASS/ EPOXY
Hoop tensile modulus (10^6 psi)	18	11	7.8	6.0
Radial tensile modulus (10^6 psi)	1.5	0.7	3	2
Poisson's ratio	0.27	0.34	0.25	0.25
Hoop tensile strength (ksi)	220	200	226	150
Radial tensile strength (ksi)	4.8	2.3	5.8	4.0
Density (lb/in.^{3})	0.054	0.05	0.072	0.071
Specific hoop stiffness (10^6 in.)	333	220	108	80

Applications of flywheels as energy accumulators for road vehicles have been studied in West Germany, with the graphite fiber/epoxy composites demonstrating the most promise.[13] Studies are underway at Daimler-Benz in which the automatic drive with internal combustion engines is supplemented by a flywheel accumulator. In their use on road vehicles, results have been encouraging. The authors of this paper rate the available materials for the energy accumulators as follows.

MATERIAL	THEORETICAL ENERGY/ DENSITY (Wh/kg)	ENERGY/DENSITY ATTAINED WITH TODAY'S MATERIALS IN SHORT-TIME TESTS	ENERGY/DENSITY AFTER 10^7 LOAD CYCLES
Glass fiber	222	132	37
Carbon fiber	247	134	107
Aramid fiber	306	130	50
High tensile strength strength steel	90	90	36

Spin tests on composite flywheels reached speeds in excess of 2500 ft/second on pin-wrapped flywheels. Favorable rotor performance was estimated in the neighborhood of 43 Wh/kg. The rotors failed by ripping bands from the rim with the rim remaining intact as it decelerated. It was pointed out that the flywheel would provide energy for a regenerative braking system. This stored energy could be used for vehicular acceleration, allowing the engine to operate at its most economical speed.[14] A hoop wound, high strength graphite/epoxy rim was used in the tests.

Carbon fiber reinforced high speed rotors have also been examined at the University of Birmingham.[15] Tip speeds as high as 2 km/second have been reported, with some graphite fiber/epoxy matrix composites sustaining tip speeds of 1.5 km/second for many hours. This performance has not been attained in metals, according to the authors, and may be improved upon by tapering rods and discs. Limiting speed for isotropic materials, to use a guideline, is the square root of two multiplied by tensile strength/density.

Penn and Chiao[16] point out that, for any flywheel, the maximum energy content per unit weight (specific energy) is limited by the material strength and the shape of the flywheel. This is given by the relation

$$E_{sp} = K_s \frac{\sigma}{\rho} .$$

σ/ρ is the material strength to density ratio and K_s is a shape factor–highest with uniform stress distribution. (See Fig. 12-5 for typical flywheel cross-sections and shape factors.) Authors point out that other key elements of the flywheel research are the rotor hub, vacuum chamber seals, bearings supporting the flywheel axle, and the coupling system. Any spinning body with high angular momentum has gyroscopic forces which are felt by the vehicle carrying the flywheel.

Flywheel Configuration		Typical Relative Shape Factor (K_S)*
1. Flat Disc (with hole)	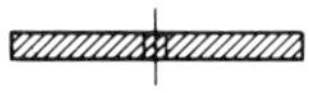	0.30
2. Modified Constant Stress Disc (no hole)	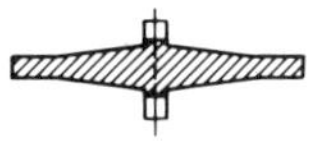	0.90
3. Constant Stress Disc (no holes)		1.0 (OD → ∞)

*An isotropic material may have a shape factor as high as 1.0 while that of an orthotropic material cannot exceed 0.5.

Fig. 12-5. Flywheel cross sections and shape factors.[16]

Toland[17] summarized the current status of flywheel developments, drawing attention to the many programs that have been sponsored in United States government funded laboratories. He felt that the optimum designs have not been attained and demonstrated in tests. As pointed out in aerospace applications of graphite composites, materials/geometry tailoring will permit optimization. Because of higher rotor speeds involved in advanced composite applications to flywheels, there will be additional considerations of bearings, seals, and vacuum technologies. To date, he cited only one moderately successful demonstration of a full scale (1kWh) rotor—a multiple thin ring designed at Garrett Airesearch in Los Angeles.

LOW FRICTION

In the chapter on thermoplastic matrices for C/G fibers, attention was directed to the lubricity of some of the highly filled compounds as bearing surfaces, and significant wear tests conducted by manufacturers of these molding compounds. For lower coefficients of friction, tetrafluoroethylene powders have also been included, though advantages of carbon fiber filled compounds will include stiffer, higher modulus materials, improved fatigue life, and low order of shrinkage during molding of components, as well as the low friction for bearing surfaces. Other literature references are cited below to indicate the growing interest in the use of carbon fiber reinforced composites as bearing surfaces.

Carbon and graphite yarns and rovings are also used in many industrial and aerospace applications because of strength, wear resistance, chemical inertness, and thermal and electrical conductivity. Prepared from rayon precursors, the carbonized fibers are combined into large units to provide plied yarns which are used in pump and motor packings. Finely divided graphite and/or tetrafluoroethylene are applied to these yarns to lower surface friction.[18] Graphite felts, varying in density from 3 to 60 lb/ft^3 after carbonization, are available from 1/8 inch to 6 inches in thickness.

Further data[19] were presented on wear and frictional properties of bearing materials made from C/G fiber filled thermoplastics, with the advantages in the carbon fiber filler over glass fibers as shown in Table 12-3. There are large reductions in wear rate and substantial increases in the limiting-pressure-velocity values, especially under high loads. Theberge *et al.*[19] felt that intrinsic abrasiveness decreases with an increasing amount of graphitic order in the fibers. In Table 12-3, data on wear testing were developed on thrust washers (face I.D.–0.900 to 0.905 inch, face O.D.–1.120 to 1.125 inch) operating at 50 ft/

Table 12-3. Wear and Frictional Properties of Carbon Fiber Reinforced Thermoplastics.[19]

BASE RESIN	FIBER CONTENT (% by WEIGHT)	FIBER TYPE	WEAR FACTOR AT EQUILIBRIUM $10^{-10}\ \frac{in.^3\text{-min.}}{ft\text{-}lb\ hr}$	COEFFICIENT OF FRICTION STATIC/DYNAMIC (40 psi/40 psi, 50 fpm)
Polysulfone	30	Glass	90	0.22/0.18
Polysulfone	30	Carbon	75	0.17/0.14
Polyphenylene sulfide	0	–	540	0.30/0.24
Polyphenylene sulfide	30	Carbon	160	0.23/0.20
Nylon 6/6	0	–	200	0.20/0.28
Nylon 6/6	20	Glass	78	0.23/0.28
Nylon 6/6	20	Carbon	40	0.20/0.20
Nylon 6/6	30	Glass	75	0.25/0.31
Nylon 6/6	30	Carbon	20	0.16/0.20
Nylon 6/6	40	Glass	70	0.26/0.18
Nylon 6/6	40	Carbon	12	0.13/0.18
Ethylene tetrafluoroethylene	0	–	6000	0.50/0.40
Ethylene tetrafluoroethylene	30	Glass	20	0.40/0.26
Ethylene tetrafluoroethylene	30	Carbon	17	0.24/0.22

minute under 40 psi load and mating against 12 RMS SAE 1040 steel. The wear factor appears as a dimensional constant in the table.

The high temperature performance of molded bearing surfaces will depend upon the temperature limits of the matrix binder in the composite.

Molded graphite fiber polyimide bearing studies indicate a good potential for bearing surfaces which will operate up to 600°F with low coefficients of friction. There is an advantage over the polytetrafluoroethylene (PTFE) as a bearing surface, in that the resistance to creep, and the fatigue resistance contributed by graphite fibers, will enhance performance. Wear resistance of steel on the graphite composite is quite low. The dry lubrication characteristics of graphite fiber reinforced composites have been demonstrated to provide low friction wear characteristics due to the fibers' lubricity. Graphite reinforced nylon 6/6 thermoplastic compounds have exhibited friction characteristics approximately equal to PTFE coated materials.

In studies upon the unlubricated bearing surface of C/G composites, Brown and Blackstone[20] drew some significant conclusions, some of which are summarized herewith.

1. Usable bearings capable of sustaining heavy loads may be fabricated from graphite fiber reinforced epoxies and polyimide resin matrices. The best overall performance came from the polyimides.
2. At 250°F and below, TFE lined bearing surfaces had lower friction and wear than the graphite fiber reinforced plastics.
3. Graphite fibers significantly increased thermal conductivity from the bearing surfaces.
4. At temperatures of 350°F, the performance of graphite/polyimide approached that of the TFE lined materials.

Self lubricating composites utilizing carbonized phenolic resin matrix plus 59% volume of graphite fibers offer high temperature capabilities[21] –possibly even better than the polyimides noted above. Advantages of the higher volume of fiber loading over the lower amounts were evident in wear tests under high loads. The low coefficient of friction of graphite fiber/epoxy composites has been used to advantage in structures for supporting machine tools.[22] Furthermore, the low deflection under loads, the low coefficient of thermal expansion, and improved vibration dampening qualities were deemed of advantage over metal predecessors.

Graphite fiber reinforced polyphenylene sulfide (PPS) has been selected for a production ball valve, not so much for its lubricity as for its high mechanical properties, which allowed installation into mechanical piping for the chemical process industry.[23] Test results on a 2 inch valve were presented and shown to be as effective as the more expensive, corrosion resistant metals such as titanium and "Hastelloy." The PPS binder permitted exposure of a variety of corrosive considerations, such as arise in treatment of waste fluids, water desalinization, SO_2 gas treatments, and treatment of various acids and alkalies. It was approved for use at temperatures up to 300°F and internal pressures of 200 psi (see Fig. 12-6, a photograph of valve from Babcock and Wilcox,[23]). This product was featured in a recent (December 1979) chemical engineering trade show in the Coliseum in New York City.

CARBON FIBER REINFORCED CONCRETE AND CERAMICS

The utilization of low cost and low strength carbon fibers for the reinforcement of portland cement concrete has demonstrated considerable promise over the use of other fibrous reinforcements. Kureha of Japan has successfully introduced the carbon fibers into this important activity, pointing out the superiority of these fibers over glass fibers, for example, under prolonged aging in alkaline conditions[24] (see Fig. 12-7). In these applications, emphasis was placed upon low cost carbon fibers derived from pitch. These fibers are being used in pack-

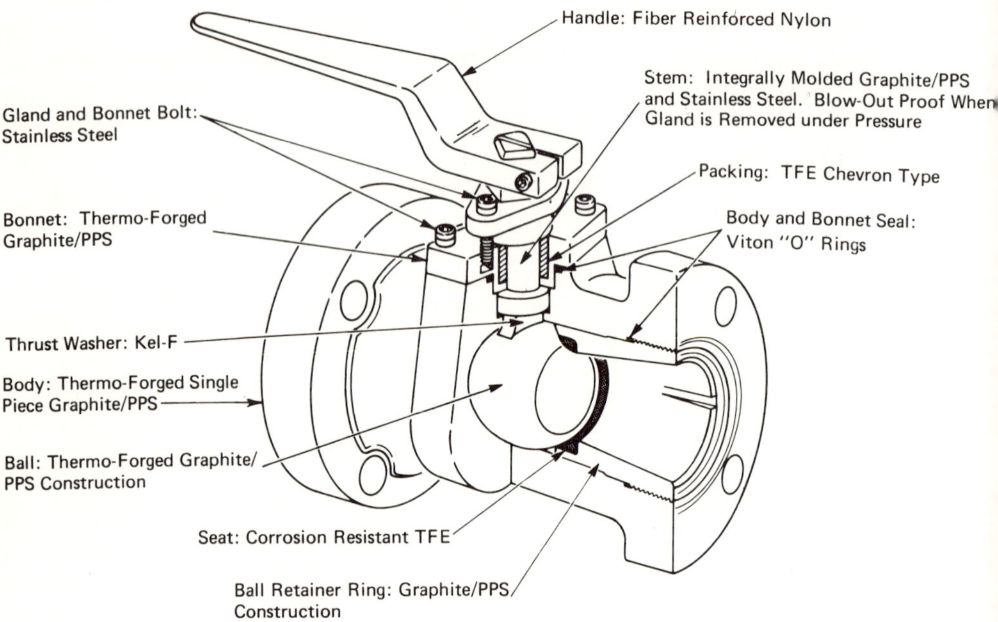

Fig. 12-6. Graphite fiber reinforced PPS ball valve. *(Courtesy-Babcock & Wilson.)*

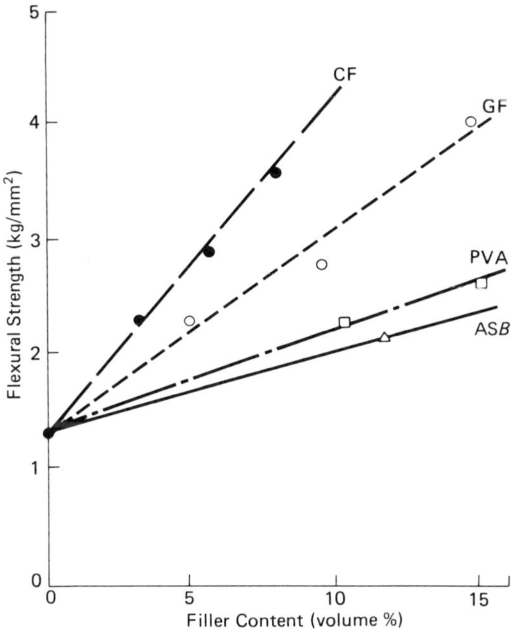

Fig. 12-7. Reinforcement effect of Kureha carbon fiber on Portland cement (water ratio: 0.35, steam aging, 1 week data). CF = Kureha carbon fiber; GF = glass fiber; AS = asbestos; PVA = polyvinyl alcohol.[24]

ings for pumps, special electrodes, heat resisting insulating materials, and reinforcements for engineering plastics. High performance, higher cost carbon fibers produced from other raw materials and processes are being used in high performance composites, according to Okuda.[24]

With respect to carbon fiber additions to concrete for reinforcement, there is considerable interest in this market because of the prospect of reducing cracks in concrete upon aging. Glass fibers, metal wires, asbestos fibers, sisal fibers, treated wood fibers, and many others, available at low cost, are being considered. Presumably, in the last analysis under prolonged aging, the materials which will be selected are those that will maintain their qualities under the stressed, alkaline environment, and the carbon fibers should rate high in this regard. This subject has been reviewed by Hoff,[25] and a selected bibliography has been prepared on fiber reinforced cement and concrete. In this report there are several references to carbon fibers, two of which are noted in the references at the end of this chapter.[26,27]

Fiber reinforced cement products, such as exemplified by the widespread application of asbestos/cement (A/C) water and sewer pipe, have been used in the United States for 50 years. These products are being sold throughout the Near East due to their resistance to high sulfate soils. Over 50 million pounds of assorted A/C products were exported from the United States in 1977 (value about \$16 million).[28] Increasing environmental concern with asbestos fibers will prompt interest in other chemical resistant fibers (such as C/G fibers).

MISCELLANEOUS APPLICATIONS

There are a number of noteworthy developmental applications of C/G fiber reinforcements or woven cloth of carbon fiber which have been noted in the technical press. To indicate the range of interests, a few examples of recent origin are described briefly in the following paragraphs.

An application of a specially treated carbon fiber cloth into a 3.5 m long electrode appeared in an electrostatic precipitator. A conductive composition prepared of 0.05 to 0.30 mm long graphite fiber and finely powdered graphite particles 0.09 to 0.3 mm were dispersed in an aqueous emulsion of a phenol/formaldehyde resin, applied to a carbon fiber cloth and cured into an electrode. The electrode had low density, high strength, and corrosion resistance to sulfuric acid, hydrofluoric acid, and hydrochloric acid. Its performance was satisfactory in tests on removing sulfuric acid mist from an industrial gas (velocity 0.8 m/second, potential 52 to 65 volts, at 100°C).[29]

Carbon fiber reinforced materials have been used for prosthesis devices and for internal bone plates in Germany.[30] The possibility of tailoring the required elastic and strength properties of internal fixation plates made of C/G fiber composites offers density and property advantages over the metal plates. C/G reinforced composites have been known to be biocompatible, and adaptable to the mechanical requirements of bone. Mechanical data of human bones are in the range of 80 to 150 MN/m^2 for flexural strength, 7 to 20 GN/m^2 for Young's modulus, and 8 to 15 MN/m^2 for torsional shear strength—values which are readily matched by C/G fiber composites. Carbon fiber reinforced prepreg of AS-3501 from Hercules and resin matrix 914-C from Ciba-Geigy appeared prominently in the experiments. There are also reports of porous graphite devices for use as a bone replacement.[31] Graphite was produced, according to this patent, with ordered circular pores of 100 to 1000 microns diameter over the surface. The pores facilitate the penetration of further bone growth tissue.

Continuous lengths of C/G fiber may be found in filament wound tubings (some manufacturers plan production rates of 360 ft/minute) and assorted profiles of pultruded shapes. The photographs in Figs. 12-8 and 12-9 demonstrate the results of some of these undertakings. These miscellaneous products could encourage new industrial uses of advanced composites.

Fig. 12-8. Miscellaneous pultrusions of advanced composites. (*Courtesy Goldworthy Engineering Co.*)

Interest in developing substitutes for asbestos garments for protection against heat and fire led to the examination of carbon fiber fabrics from polyacrylonitrile precursors. Protective garments of this material including some fabrics coated with metal, are worn by firemen and by iron and steel workers to protect them from the splash of molten metals. This use was illustrated in a recent national publication,[32] which directed attention to the developing market for asbestos fiber substitute materials.

A major drawback of these industrial applications is the price of C/G fibers. The desired properties are inherent to the materials, but their present high prices prompt consideration of other materials. Through the use of hybrid constructions, it has been possible to develop acceptable products using, for example, glass fibers to replace the more costly carbon fibers. Notwithstanding, their basic heat resistance, their high ratio of modulus to density, their good chemical resistance, and their good fatigue life are stimulating attributes of C/G fiber composites. While they may not in themselves dominate the industrial picture, more applications are being considered as a result of their contributions.

Fig. 12-9. Assorted graphite fiber tubes. (*Courtesy Graftek Division of Exxon.*)

REFERENCES

1. Kuno, J., "Marine and Sports Equipment," *Plastics,* page 15, November 1979.
2. Cecka, A.M. and Dano, P. (Fansteel, Inc.), U.S. Patent 4,061,520, "Method of Making Composite High Strength to Weight Structure," December 6, 1977.
3. Boehm, H.C., "Influence of Composite Materials on Alpine Ski Design," *SAMPE Journal,* Vol. 15, page 14, September/October 1979.
4. Short, D. and Summerscales, J., "Hybrids–A Review," *Composites,* Vol. 10, page 215, October 1979.
5. Hoover, J.J., "Marine Applications for Graphite Composites," *SAMPE Twenty-third National Symposium,* Vol. 23, page 1230, Anaheim, California, May 1978.
6. *Bulletin on Carbon Fiber Composite Yachts,* Eagle Yachts, Bristol, Rhode Island, 1979.
7. Gerstle, F.P., "Analyses of Filament Reinforced Spherical Pressure Vessels," ASTM, STP 546, pages 604–631, 1974.
8. Hamstad, M.A. and Chiao, T.T., "Graphite/Epoxy Pressure Vessels Lined with Ultra-thin Titanium," *SAMPE Twenty-second National Symposium,* Vol. 22, page 312, San Diego, California, April 1977.
9. Hanson, M.P., "Tensile and Cyclic Fatigue Properties of Graphite Filament Wound Pressure Vessels," *SAMPE Symposium,* Vol. 15, page 249, April 1969.
10. Morganthaler, C.F. and Bond, S.P., "Composite Flywheel Stress Analysis and Materials Study," *SAMPE Twelfth National Symposium,* Anaheim, California, October 1967.
11. Chamis, C. and Kiraly, L.J., *Proceedings of the Flywheel Technology Symposium,* Washington, D.C., 1975 (CA-87-69320t to CA 87-69325y, 1977).
12. Reedy, E.D., "A Composite Rim Flywheel Design," *SAMPE Quarterly,* Vol. 9, No. 3, page 1, April 1978.
13. Hanselmann, G. and Han, E., "Flywheel Energy Accumulators for Road Vehicles," *SAMPE Twenty-third National Symposium,* Vol. 23, page 740, Anaheim, California, May 1978.

14. Reedy, E.D. and Street, H.K., "Composite Rim Flywheels, Spin Tests," *SAMPE Quarterly,* Vol. 10, page 36, April 1979.
15. Barker, A.J., Moon, P.B., and Rolls, M.P., "Carbon Fiber Reinforced Plastics as Material for High Speed Rotors," *Composites,* Vol. 10, No. 3, page 149, July 1979.
16. Penn, L. and Chiao, T.T., "Fiber Composites for Energy Storage Flywheels," *Annual Technical Conference,* SPI, RP/C Institute, Section 12F, 1976.
17. Toland, R.H., "Current Status of Composite Flywheel Development," *SAMPE Twenty-third National Symposium,* Vol. 23, page 856, Anaheim, California, May 1978.
18. Bulletin, "Graphite and Carbon Yarns and Rovings," Polycarbon, Inc., Los Angeles, 1976.
19. Theberge, J., Robinson, R., Badner, J., and Arkles, B., "Carbon Fiber Reinforced Thermoplastics," *SAMPE Twentieth National Symposium,* Vol. 20, page 72, April 1975.
20. Brown, R.D. and Blackstone, W.R., "Evaluation of Graphite Fiber Composites for Use in Unlubricated Sliding Bearings," ASTM, STP 546, pages 457–476, 1974.
21. Lavik, M. and Hopkins, V., "Development of Self-lubricating Composites Utilizing Carbonized Phenolic Matrix," Midwest Research Institute, Project AF-7343, MRI-4014E, December 1975.
22. Knight, C.E. *et al.,* U.S. Patent 4,072,084, "Graphite Fiber Reinforced Structure for Supporting Machine Tools," February 7, 1978.
23. Leopold, P. (Babcock and Wilcox), "Advanced Composite Graphite/PPS Valve," *SAMPE Twenty-third National Symposium,* Vol. 23, page 651, Anaheim, California, May 1978; and *Chemistry and Engineering News,* page 21, December 19, 1977.
24. Okuda, K. (Kureha Chemical Industry Co.), "Pitch-based Carbon Fiber Reinforcements," *Conference on Advanced Composite Technology,* page 26, by Technology Conferences Associates, El Segundo, California, March 1978.
25. Hoff, G.C., "Selected Bibliography on Fiber Reinforced Cement and Concrete," AD/A-032082, U.S. Army Engineers, Vicksburg, Mississippi, June 1976.
26. Briggs, A., Bowen, D.H., and Kollek, J.J., "Mechanical Properties and Durability of Fiber Reinforced Cement Composites," *International Carbon Fiber Conference,* The Plastics Institute, Paper No. 17, London, February 1974.
27. "Carbon Fiber Cement Composites," *Civil Engineering and Public Works Review,* Vol. 67, No. 789, pages 357–361, April 1972.
28. News Item, "ISO/TC 77 on Products in Fiber Reinforced Concrete Hold Plenary Meeting," *ASTM Standard News,* Vol. 7, page 39, April 1979.
29. Antonov, A.N. *et al.* (U.S.S.R.), "New Carbon-polymer Materials for Electrodes of Electrostatic Precipitators," *Chemical Abstracts,* CA91–109325g, 1979.
30. Fitzer, E., Huttner, W., Class, L., and Kinzl, L. (University of Karlsruhe), "Torsional Strength of Carbon Fiber Reinforced Composites—As Internal Bone Plates," *Fourteenth Biennial Conference on Carbon,* American Carbon Committee, page 58, Penn State University, June 1979.
31. Estherly, W.P., Robbins, J.M., and Rosson, D.E., U.S. Patent 4,070,514, "Method of Fabricating Graphite for Use as a Skeletal Prosthesis," January 24, 1978.
32. Anon., *Business Week,* page 98, December 3, 1979.

13
High Temperature Resistant Matrices for Service Above 300^0C (Metal Matrices and Carbon/Carbon Composites)

As service temperatures exceed 300°C, organic polymers such as the polyimides will not function adequately as binders or matrices for C/G fibers. There are several options available for preparing composites which will function above 300°C: metal matrices, such as aluminum or magnesium; carbonaceous binders formed from pyrolysis of pitch or synthetic resins, followed by reimpregnation and repyrolysis; pyrolytic carbon formed from chemical vapor deposition; and, lastly, glass and/or ceramic binders for C/G fibers. In particular, the carbon/carbon composites formed from carbonaceous binders and carbon fibers are fulfilling unique roles in applications above 300°C.

METAL MATRIX COMPOSITES

Metal matrices have been examined closely in recent years in conjunction with heat resistant fibers, such as C/G filaments, glass fibers, and boron fibers. While metal matrices have not received the attention accorded to pitch and resin binders in advanced composites, metals derive benefits from C/G fibers, including high strength to density and stiffness to density ratios, improved fatigue resistance, and lower coefficients of thermal expansion. Metal matrix composites have the potential to offer many of these same advantages at temperatures in excess of 300°C. Studies and developments are still proceeding, and the full capabilities of metal/graphite fiber composites still have to be demonstrated. The metal matrix composite, once it has been formed, will require fabricating techniques more similar to those of metal processes than those of the advanced fiber/resin composites which utilize the early processing thermoplasticity of the resins to fabricate large and complex structures. Techniques also must tailor material distribution to achieve specific design criteria. The most likely early applications for metal matrix composites appear to lie in space programs, where some of the physical properties of graphite fiber with aluminum and magnesium appear to surpass the resin/fiber composites, which in the absence of atmospheric corrosion may be an overriding consideration. The boron fiber/aluminum composite has been accepted for important components on the forthcoming space shuttle, to cite an early recognition of these products.

The stability of graphite fibers in the presence of a metal may be examined from the standpoint of chemical reactivity or the absence of carbide formation. Aluminum carbide has a lower density (about 12%) than the aluminum metal. At temperatures near their metal's melting point, aluminum, molybdenum, tantalum, and tungsten have been shown to form carbides with ease.[1] The free energies of formation are large and negative. Using the same criterion, Morse[1] felt that iron, nickel, magnesium, and cobalt would also be candidates. Hence, this would indicate that special non-reactive coatings would have to be devised for incorporating graphite fibers into some metal matrices. Binary and ternary alloy systems were also considered by Morse. By nature of its crystalline structure, graphite may form compounds such as K_8C with alkali metals which are electron donors. This fact has added significance in the light of a recent patent,[2] which prepares graphite fibers for inclusion in metal matrices by immersing the fibers into an alloy of sodium and potassium, liquid at room temperature. In an inert atmosphere this treatment enhanced the wettability and infiltration of the graphite fibers by liquid metals. After alkali metal treatment, the fibers are immersed into molten baths of aluminum, magnesium, copper, tin, lead, or zinc. Removed from the molten bath, and on cooling, the graphite fiber reinforced metal is fabricated into desired structures.

The wettability of the graphite fibers for impregnation by liquid metal has also been enhanced by a molten tin-titanium alloy, or a molten copper-tin-titanium alloy of molten sodium infiltrates.[3] After removal from the molten infiltrate, the yarn of graphite fibers is immerse into a metal impregnate of aluminum or magnesium (melting points, 600° and 650°C) or their alloys. The metal infiltrate is leached out or displaced from the graphite yarn, and the interstices are filled with the impregnated metal.

Pfeifer[4] reviewed the technology of liquid metal infiltration (LMI), which has emerged as the most viable candidate for preparing graphite fiber/aluminum composites. A preference is shown for 201 casting aluminum alloy containing copper, 4.7%; manganese, 0.3%; titanium, 0.3%; and silver, 0.5% (which yields higher mechanical properties with the C/G fibers). Typical problems that arose in the metal infiltration process were related to non-uniformity and matrix-rich zones or fiber-rich zones and irregularities in the graphite fiber bundles being treated. Further processing controls of temperatures and wettability of the graphite fiber surface improved the quality of the composites. A schematic flow chart Fig. 13-1, summarizes some of the details of the process. A continuous flow is illustrated, whereby the C/G fiber is precoated with silicon carbide, titanium boride, or titanium boride mixed with titanium carbide—to enhance the wettability of the metal impregnant into the graphite fiber yarn. The titanium-boride barrier coating provided the most consistent results upon graphite fibers from a rayon precursor. The titanium boride coating was formed from a chemical vapor deposition of $TiCl_4$ and BCl_3 and a small amount of zinc as a reducing agent. Much of the recent work on LMI of graphite fibers has used titanium boride as a barrier coating to liquid aluminum alloy.

As described in a recent patent,[5] a graphite yarn containing 11,000 individual fibers of 50×10^6 psi modulus was exposed to a vapor reaction of 0.38% $TiCl_4$, 0.21% BCl_3, and 0.80% zinc, the balance of the weight being an inert argon gas. The gas mixture was maintained for 30 minutes at 650°C to provide a coating 200 Å thick. The fibers coated with TiB_2 were transferred under argon to a molten bath of the aluminum alloy for two minutes. Materials were hot pressed in a graphite die under vacuum at 600°C for five minutes to form a composite plate.

Withers and Alexander[6] placed emphasis on boron fibers in the development of advanced composites with metal matrices. The price of boron fibers has not fallen significantly due in part to the scarcity of the fine tungsten wire cores,

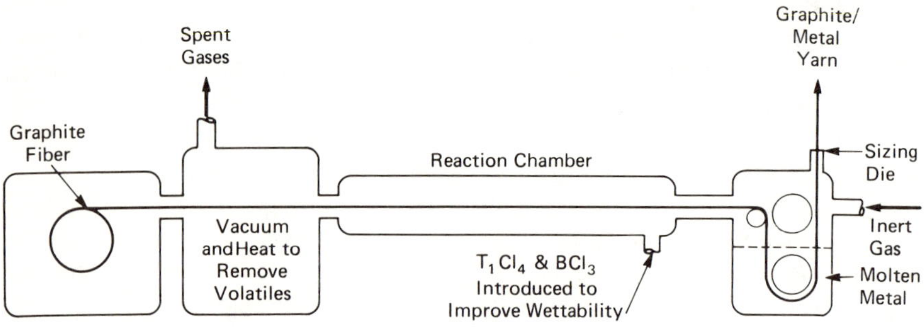

Fig. 13-1. Schematic process of graphite fiber tow impregnated with metal.[4] *(Reprinted With Permission of American Institute of Aeronautics and Astronautics)*

and attention has been diverted to the more available C/G fibers. Boron filaments have shown stability in aluminum and in magnesium at temperatures below 600°C. This report suggested the replacement of the tungsten core in boron fibers with a low cost carbon or graphite substrate. Also examined were filaments of silicon carbide, which is more thermally stable than boron, and experimental filaments of B_4C, B_4Si, and TiB_2. Elevated temperature properties of silicon carbide filaments were 300 ksi tensile strength at room temperature and 200 ksi tensile strength in air at 1200°C.

Silicon carbide coatings formed on C/G fibers by vapor deposition processes from alkyl silanes as the source of silicon have been described by Hoffman.[7] Silicon carbide fibers formed by this same process showed tensile strengths up to 500 ksi and moduli of elasticity of 62 msi, with a specific gravity of 3.2. The coatings from this product on C/G fibers are formed by thermal decomposition of vapor on a heated substrate. This will improve the stability of C/G fibers in molten aluminum and magnesium, markedly reducing the rate of reaction that takes place at high temperatures between the carbon and the metal. Basically, the filament must not lose strength as a result of exposure to the elevated temperature cycle required for fabrication, said temperatures arising from the molten metal or the high temperature sintering and pressing operations. The latter fabrication technique must be designed to minimize breakage of fiber reinforcement.

Advantages were claimed for coatings of a mixture of silicon oxide and silicon carbide on C/G fibers. These coatings provided protection for the fibers when they were dispersed in a solid matrix of aluminum, magnesium, titanium, and nickel or their alloys.[8]

Silicon carbide whiskers deposited upon C/G fibers have also had an important role in enhancing the physical properties of resin/fiber composites.[9] Interlaminar shear strength of the carbonaceous fibers in epoxy composites was increased markedly, though at some loss in tensile strength.

Titanium carbide coatings were formed on graphite fibers to approximately 0.1 micrometer in thickness. Deposition took place in the appropriate chemical vapor with temperatures varying from 1365° to 1750°K for up to 150 seconds. "Thornel" 50 graphite fiber (rayon precursor) that was so treated was infiltrated with aluminum alloy and hot pressed in vacuum to produce small bars for flexure testing. Strengths as high as 90% of the rule-of-mixtures were obtained. Titanium carbide coated fibers appeared to suffer much less reaction than the uncoated fibers.[10]

It has been shown that some types of graphite fibers are degraded more by the liquid metal infiltration process than are others. Degradation was assumed when the aluminum carbide (Al_4C_3) content in the composite exceeded 2000 ppm. Sullivan and Raymond[11] presented evidence that the oxidation treated graphite fibers experienced the greatest amount of attack by molten aluminum and formed the highest amount of aluminum carbide on the surface of fibers. This led to a much greater loss in strength of graphite/aluminum composites than those fibers not given the high temperature oxidation of the graphite fiber. There was less attack on graphite precursors from rayon or pitch than experienced on polyacrylonitrile (PAN) precursors.

High strength graphite/aluminum composite bars of round and rectangular cross-section were produced by a pultrusion process. "Thornel" 300 (PAN-based) fibers were used and infiltrated with 6061 and 356 aluminum alloy melts. Graphite aluminum composite wire preforms were shaped in this manner.[12] Pfeifer[4] also produced wire preforms which were placed in regular arrays for diffusion bonding at temperatures of 550° to 560°C, 3500 psi, for 10 minutes. His baseline consisted of 30 (v/o) eight strand T50 graphite fibers in a 201 aluminum alloy matrix. More complex shapes were produced by utilizing the metallurgical techniques of hot isostatic pressing at high pressure upon the graphite/aluminum preforms in a pressure vessel. A schematic illustration of pultrusion of continuous graphite/aluminum composites is shown in Fig. 13-2.

Properties of graphite fiber reinforced metal composites reflect the experimental stage of the activity and only incomplete data are available. There are several good candidate metals and a wide choice of C/G fibers, accompanied by different fabricating techniques which will influence final physical properties. Hence, data are being accumulated upon the more promising systems—some of which have appeared in a few scattered reports. Table 13-1 presents data assembled by Pfeifer on the average tensile strengths of different graphite/aluminum composites in precursor wire form.

Table 13-1. Average Tensile Strength of Graphite/Aluminum Composites in Precursor Wire Form.[4]

ALUMINUM ALLOY	PRECURSOR FIBER TYPE	COMMERICAL DESIGNATION	VOLUME FRACTION (%)	AVERAGE TENSILE STRENGTH (ksi)
201	Rayon	Thornel 50 (8 end)	30	125
202	Rayon	Thornel 50 (8 end)	28.7	114
201	Rayon	Thornel 50 (single strand)	41	150
202	Rayon	Thornel 75	27	116
202	PAN	Thornel 300	29	35
201	PAN	Thornel 300	45	50
201	PAN	Stackpole Panex	23	18
201	PAN	Hercules Type A	37	34
201	PAN	Hercules HM-S	24	113

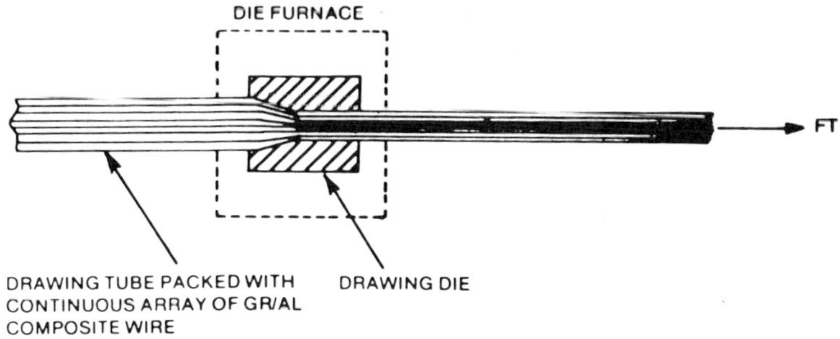

Fig. 13-2. Pultrusion of continuous metal/graphite composites.[4] *(Reprinted with permission AIAA)*

It is interesting to note in Table 13-1 that the rayon precursor graphite fibers and the example of high modulus graphite included (HM-S) showed the greatest strengths in the "precursor" graphite/aluminum wires. Other properties were shown for the "Thornel" 50 fiber/201 aluminum alloy—typical of which are the following for 30% fiber volume composite panels:

Tensile modulus—19 to 20 msi
Longitudinal tensile strength—86.5 ksi
Transverse tensile strength—18.8 ksi
Ultimate compressive strength—68 ksi.

Data were also developed by Davis and Sullivan.[13] Of the various graphite fibers examined and for volume percentages up to 45%, approximately 22×10^6 psi was the highest modulus reported for the graphite/aluminum composites. A typical room temperature tensile stress-strain plot for the "Thornel" 50/201 baseline graphite/aluminum and "Thornel" 300/356 composites are shown in Fig. 13-3. The reduction in longitudinal tensile strength for the composite materials, graphite and aluminum, is approximately 10% below room temperature values when tested at the temperature range of 250° to 350°C.

Amateau[14] reported upon the fatigue behavior of aluminum alloy (6061 and 201) and graphite fiber ("Thornel" 50) composites prepared by liquid metal infiltration. S/N fatigue curves were run up to 10^8 cycles. Specimens were circular (diameter approximately 4 mm). They underwent sinusoidal loading at room temperature with a stress ratio of 0.1 and cyclic frequencies of 10 to 30 Hz. The fatigue strength of the liquid metal infiltration process composites was approximately 50% greater than those prepared by hot-pressed graphite/aluminum composites.

Reinforcement of aluminum I-beams with graphite/aluminum composite wire has been reported.[15] "Thornel" 50 graphite/6061 aluminum alloy wire reinforce-

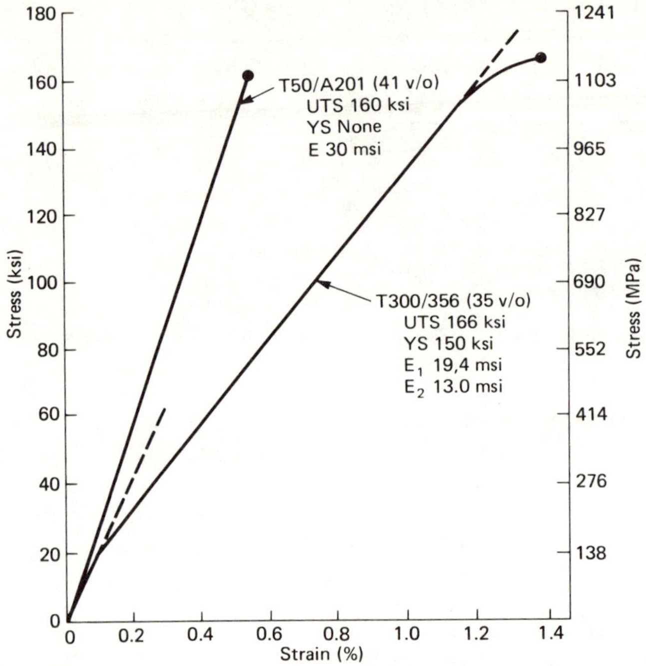

Fig. 13-3. Stress/strain behavior of rayon and polyacrylonitrile (PAN) Graphite Aluminum composite wire.[4] *(Reprinted with Permission American Institute of Aeronautics and Astronautics)*

ments were hot isostatically pressed on the flanges of the I-beams. Metallographic examination showed that good consolidation was achieved. Naval applications for graphite fiber reinforced aluminum alloys and the plans for utilizing these materials on hardware and missiles were reviewed by Pfeifer.[16] One of the programs was the development of an 0.10 inch diameter graphite/aluminum wire and its consolidation into a 1/4 inch wide ribbon. Further data bases were being developed, including the influence of environment. The prospects of obtaining good environmental protection are made possible by the consolidation of aluminum protection foils on exterior surfaces at the time of fabricating of the graphite/aluminum composites.

In describing aluminum/carbon fiber composites, Hancox and Wells[17] used C/G modified epoxy composite to reinforce an aluminum alloy hat section rather than aluminum infiltration into the carbon fibers. A more flexible matrix resin was obtained by adding a polyurethane to the epoxy matrix (Ciba's MY720/DADS system; see epoxy matrix chapter).

Graphite/aluminum and graphite/magnesium composites involving high modulus graphite filaments possess moduli of elasticity values approaching 50 msi. On the basis of specific property values, the elastic modulus of the graphite/magnesium composite is the highest of any known metallic material.[18] The high strength and moduli are retained at high temperatures (above 300°C); the thermal conductivity is superior to the resin matrix advanced composites; joining techniques to other metal components are better; and the fatigue strengths are improved over the base metal. The incorporation of Union Carbide Corp.'s VS0054 (100 msi modulus) graphite fiber into a magnesium matrix produced a specific modulus of 850×10^6 inches. The properties of graphite/aluminum composites with high modulus graphite fibers are as follows (GY-70 graphite fiber has a PAN precursor and VS0054 graphite fiber has a pitch precursor).

PROPERTY	GY70SE/201 ALUMINUM	VS0054/201 ALUMINUM
Fiber loading (v/o)	37-38	48-52
Tensile strength (ksi)	115-120	150-155
Tensile modulus (msi)	30	50
Wire diameter (in.)	0.028 (8 strand)	0.025 (2 strand)

Rubin[19] reviewed the applications of the advanced composites made principally from fibers of graphite, borsic, boron, and silicon carbide with metal matrices. Borsic is a modification wherein a surface coating of silicon carbide is deposited over boron to provide higher temperature capability. When borsic fibers are used in a titanium matrix, the uniaxial composite retains 68% of its transverse tensile strength up to 530°C. Graphite/aluminum composites have been used on tactical missile skins, stiffeners, rocket launch tubes, armor components, and lightweight field erectable assault bridges. Lead acid storage batteries are mentioned as a promising application, where graphite fibers dispersed in a lead matrix have increased strength and creep resistance. The use of graphite fibers in conjunction with copper alloy bearings have enhanced performance. It has been demonstrated that, as the fiber ends become exposed, wear and erosion resistance of the bearing surfaces are increased. The potentials for graphite fibers with metal matrices are numerous and will begin finding their roles in the industrial fields during the 1980's.

The metal matrix composites for advanced structure show considerable promise, with carbon/graphite fibers contributing significantly to low thermal expansion and high specific stiffness. Mechanical properties are summarized by Rubin in the table shown below. Properties are dependent upon the volume of fiber loading. Both carbon/graphite fibers and the higher density fiber reinforcements are included in the table for comparison.[19a]

Table 13-1A. Mechanical Properties of Metal-Matrix Composites.[19a]

FIBER	MATRIX	REINFORCEMENT, VOL%	DENSITY, lb/in.^{3}	UTS(L), psi	E(L), $\times 10^6$ psi	UTS(T), psi	E(T), $\times 10^6$ psi	DATA SOURCE
GT 50	201 Al	30	0.086	90	24	7	5	Aerospace
G GY 70	201 Al	34	0.086	95	30	4.5	5	Lockheed[b]
G GY 70	201 Al	30	0.088	80	23	10	6	Lockheed[b]
G HM Pitch	6061 Al	41	0.088	90	47	(15)	(7)	Aerospace
G HM Pitch	AZ31 Mg	38	0.066	74	43	–	–	Lockheed[b]
B on W, 5.6-mil Fiber	6061 Al	50	0.090	200	34	20	23	GD Convair
Borsic	Ti	45	0.133	184	32.5	67	27	TRW
G T 75	Pb	41	0.270	104	29	–	–	Aerospace
G T 75	Cu	39	0.220	142	35	–	–	Aerospace
FP	201 Al	50	0.130	170	31	(20)	20	DuPont
SiC	6061 Al	50	0.106	215	33	(20)	20	DWA Composites
SiC	Ti	35	0.142	175	38	75	30	DWA Composites
SiC Whisker	Al	20	0.101	50	15	50	15	DWA Composites
B_4C on B	Ti	38	0.135	215	33	>50	>20	DWA Composites

[a] Rule-of-mixture predicted values given are in parentheses.
[b] AFML Contract No. F33615-Tl-C-5190

CARBON FIBER/CARBON MATRIX COMPOSITES (CARBON/CARBON COMPOSITES)

There is a large background of information on the carbon/carbon composites used on high temperature applications for military and commercial purposes. They are probably best known for applications to cases for rockets, high speed aircraft, satellite structures, rocket nozzles, missiles, and aircraft brakes—where very high temperatures are encountered because of service conditions or re-entry temperatures in returning to the earth's atmosphere. The processing of carbon/carbon composites involves pyrolyzing of advanced fiber/resin composites at temperatures decomposing the resin, followed by further impregnation and pyrolysis to yield a matrix of high char strength carbonaceous resins, bonding high strength and high modulus C/G fibers. This technology has opened a new group of applications requiring good performance at temperatures in excess of 300°C.

Development efforts on carbon/carbon composites go back to the early 1960's, as carbon fibers began to appear commercially, and steps were taken to use these to reinforce large graphite components such as rocket nozzles. There were significant physical improvements in higher strength and resistance to the erosion of high velocity, high temperature gases from the rocket nozzle. Dense, low void components were sought after, through repeated impregnations with heated and liquefied natural pitch products and coal tar products which have served as precursors for the manufacture of bulk graphite. It is not necessarily desirable to select the strongest and stiffest C/G fibers for carbon/carbon composites, because these do not lend themselves to weaving and knitting processes used to prepare some of the fibrous bases used in carbon/carbon composites. In addition, there are efforts to replace the carbonaceous pitches from petroleum or coal tar origin with resin matrices having low pressure impregnation capabilities, capable of yielding high char strength residues after multiple pyrolysis and impregnations. These materials may also be augmented by chemical vapor deposition techniques to form high heat resistant pyrolytic graphite or carbide structures within the composite. In short, measures are being taken to utilize more effectively the unique qualities of carbon and graphite, because of their good physical and chemical properties at both low and very high temperatures.

The preparation of carbon/carbon composites has several alternatives which may be followed. In the selection of C/G fiber reinforcement, there are many types to choose from, including those from cellulosic or rayon origin, the PAN precursors, and, more recently, high strength and high modulus pitch fibers. The properties of these fibers have been brought out in discussions of their mechanical properties in advanced composites employing resin matrices. Germane to this discussion is the fact that today, C/G fibers from rayon (such as "Thornel" 50) continue to maintain a prominent position in the carbon/carbon composite development, though for improved stiffness, higher modulus fibers are commercially available and for improved toughness, lower modulus fibers–including those prepared through carbonization–only have the advantages of better processability in fabricating the fiber structures used in carbon/carbon composites. The methods of developing the carbonaceous matrix in the carbon/carbon composites are numerous and include the following.

1. Molten pitch from coal tar sources or petroleum sources is impregnated into the C/G fiber structure under heat and pressure. This may be followed by pyrolysis and subsequent re-impregnations.
2. Resin matrices which have been known to have high char strength after pyrolysis, such as those from certain grades of phenolic resins and furfuryl alcohol resins will penetrate effectively into thick fibrous structures. Pyrolysis followed by re-impregnation cycles are necessary.
3. Chemical vapor deposition (CVD) from a gaseous phase (usually of methane and nitrogen, sometimes with a small amount of hydrogen), forms upon heated substrates such as the C/G fibers, high strength pyrolytic graphite. This process may be intercalated into either of the first two processes to enhance physical properties of the composites of carbon/carbon.
4. Impregnation of a still porous carbon/carbon composite prepared by the steps above with liquid monomers capable of forming heat resistant structures represents an additional refinement. The choice of such monomers is very limited, though penetrating liquids of tetraethyl silicate coupled with strong mineral acid catalysts will yield a silicon-oxygen network with good heat resistance. Silicon resins may also serve this purpose.

Attention will be focused upon the first three techniques outlined above for forming carbon/carbon composites and references will be made to recent technical publications which describe the processing details, particularly as they influence final properties. Because many of the research efforts on carbon/carbon composites have been funded by United States government sources, there will be references to nose cones for space re-entry vehicles and to rocket nozzles, where the most demanding conditions prevail for high strength, heat resistant products. From a longer range viewpoint, one must recognize the prospects of lightweight carbon/carbon composites in the foundry for handling molten metals; in aircraft and truck brakes, where high energy dissipation must take place rapidly; in high temperature bearings in steel mills and chemical processing plants; in components in nuclear power plants; and in rocket and missile developments.

PITCH MATRIX BLENDS

Patent literature (see the patent abstracts in Chapter 1) has cited various combinations and processing of pitch from coal tar and from petroleum to impregnated C/G fibrous bases. More recently, the singular advantages of mesophase pitch base, as in the production of C/G fibers, have been recognized. Among the matrices mentioned is the 15V Allied Chemical's coal tar pitch and Ashland Oil's petroleum pitch A-240, which appear in various reports. The high pressure impregnating carbonization (HPIC) cycle was designed to densify carbon/carbon composites by impregnating with pitch and carbonizing at around 100 MPa (14.5 ksi) in a hot gas autoclave. Comparison of densification at 100 MPa with low pressure impregnation (3 MPa) indicates a doubling of the coking yield, according to Weber *et al.*[20] A typical HPIC cycle is illustrated in Fig. 13-4. A full pressure of 100 MPa was applied for about 16 hours on a three dimensional C/G woven billet. Though lower impregnation pressures could be used, two to three times as many cycles would be required to achieve the same amount of densification. The standard graphitization cycle for nose tip cones follows the following time and temperature cycle in an argon atmosphere.

1. Room temperature to 600°C at 300°C/hour.
2. 600° to 1000°C at 20°C/hour.
3. 1000° to 2500°C at 70°C/hour.

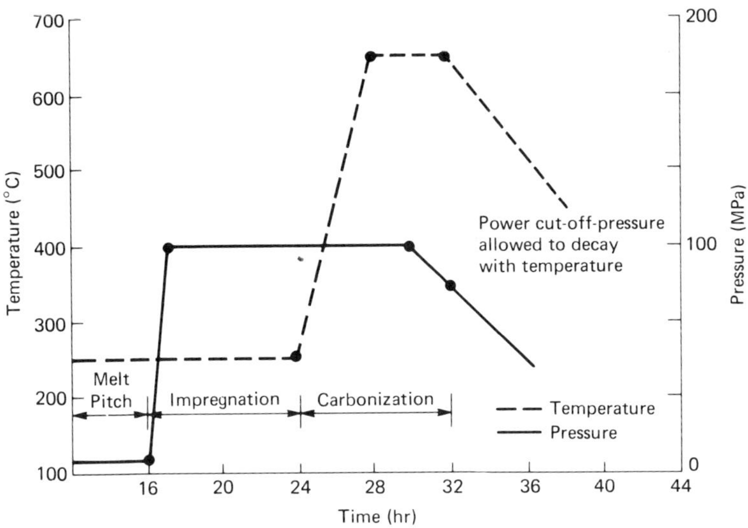

Fig. 13-4. Standard HPIC cycle at approximately 100 MPA for impregnation and carbonization.[20]

4. 2500° to 2700°C plus 0° to 25°C at 100°C/hour.
5. 30 minute soak at 2700°C (plus 0° to 25°C).
6. Cool furnace and unload.

The very high graphitization temperatures may be detrimental to the tensile strength of some of the carbon pitch fibers, according to Johnson *et al.*[21] In their paper, they identified processing parameters and their effect on woven three dimensional billets, 2 inches by 2 inches by 5 inches, which had been impregnated with V-15 pitch. The densification of a three dimensional carbon/carbon billet requires the infiltration of the pitch into relatively large void pockets located in between fiber bundles. A lower 2400°C graphitization temperature has significantly increased the tensile strength of a three dimensional pitch fiber carbon/carbon billet, using 1.5 ksi pressure impregnated carbonization cycles.

A short explanation on the differences between a two dimensional and a three dimensional billet follows. A two dimensional square block is a series of alternating layers of graphite cloth (composed of warp and fill graphite fiber yarn bundles) and an impregnating matrix. The graphite fiber cloth is processed by graphitizing rayon, PAN, or pitch cloth—as explained earlier. Subsequent carbonization at 650°C and graphitization at 2600°C are practiced. Tape wrapping or rosette lay-up permits the adaptation of unidirectional C/G fibers. Fabricating the three-dimensional fiber block involved the placing of axial (Z) graphite rods in a holding jig, with the X (warp) and Y (fill) graphite fiber bundles manually inserted and packed down for tight fiber layer spacing, while automation is accomplished in winding three dimensional cone mandrels.[22] For both two dimensional and three dimensional blocks, the number of impregnating, curing, carbonization, and graphitization cycles will determine billet density and porosity. The range of billet density is 1.45 to 2.05 g/cc with an objective of 10% porosity or less at 1.85 to 1.95 g/cm^3 density. Generally, the two dimensional carbon/carbon billets are used in nozzles as short inlet and throat rings and as long exit cones, while three dimensional carbon/carbon composites are used in the nose inlet throat area of nozzles, demonstrating more resistance to erosion. Delaminations and cracks are less prevalent in the more costly three dimensional structures. See also Figure 13-11b later in this chapter.

The development of the three dimensional reinforcement concept originally consisted of non-interlaced yarns in three mutually perpendicular (orthogonal X, Y, Z) directions. Later developments replaced the X and Y yarns by woven fabrics.[23] Results show that the three dimensional composites are superior to the best bulk graphite (ATJS) because of higher strength and lower coefficient of expansion (C.T.E.) (see Fig. 13-5). Ablative performance is at least equivalent to ATJS.

The latest three dimensional billets prepared from woven fabrics, pierced by

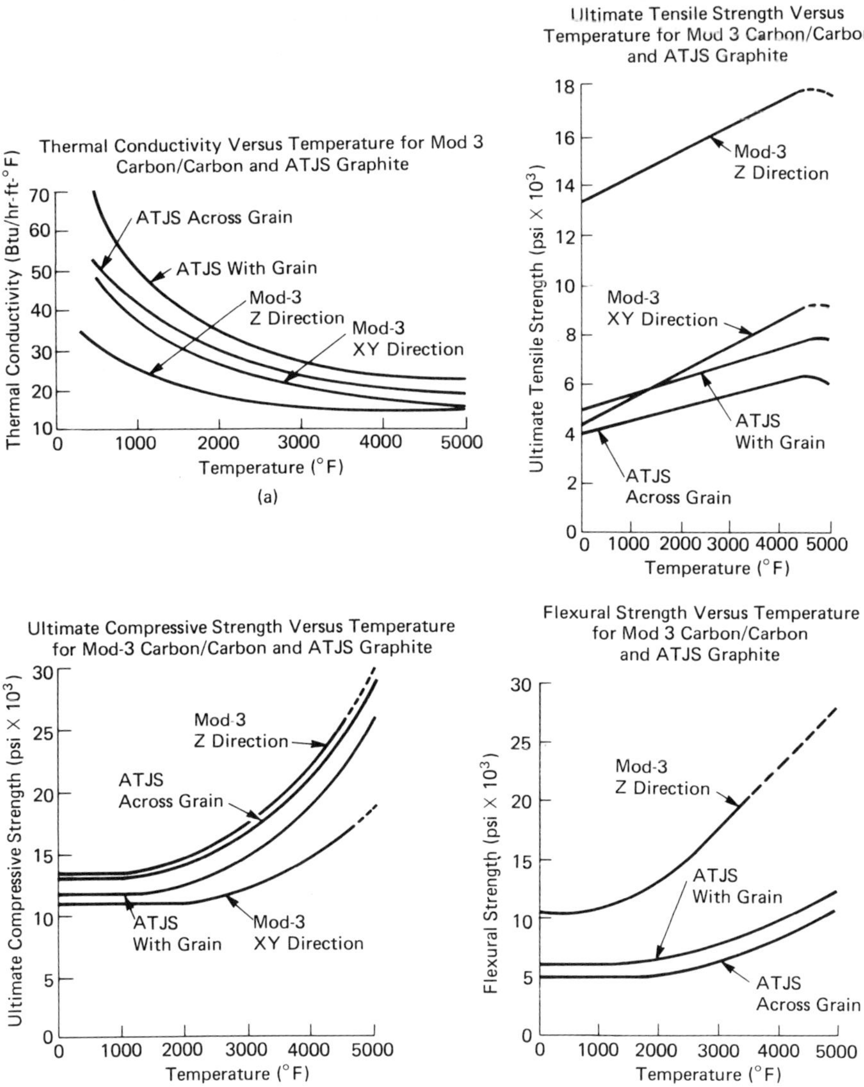

Fig. 13–5. Mod-3 carbon/carbon and ATDS graphite.[23]

yarns perpendicular to the planes of the fabrics, show good resistance to matrix micro-cracking and propagation. Impregnation by phenolic resins and/or coal tar pitch, which have high carbon yields, followed by carbonization and graphitization (at 2704°C) for several cycles, yields a bulk density of 1.70 g/cc^3. Properties reported by McAllister and Taverna[24] are given in Table 13–2.

Table 13-2. Properties of Mod 3 Three Dimensional Billets of Carbon/Carbon Composites.[24]

	AT 25°C	At 2700°C
Tensile strength, ksi (MN/m^3)	15 (103)	10 (69)
Tensile modulus, psi × 10^6 (GN/m^2)	6 (414)	1.5 (10.3)
Flexural strength, ksi (MN/m^2)	14 (97)	15 (103)
Flexural modulus, psi × 10^6 (GN/m^2)	4 (276)	2.5 (172)
Coefficient of thermal expansion at 1090°C (10^{-6} cm/cm/°C)		0.5
At 2000°F (10^{-6} in./in./°F)		0.3
Thermal conductivity, BTU/hr/ft/°F (Kcal/hr/M/°C)	At 260°–32 (48) At 2500°C–14 (21)	

The demanding performance from the heat shields and nose tips of re-entry vehicles has found added structural integrity in the carbon/carbon composites reinforced with high strength graphite fibers. Another example is the development of carbon/carbon composites for aircraft brakes. As pointed out by Kirkhart,[25] 10 to 20 million ft-lb of energy are absorbed by a braking system weighing 1/10 of the engine weight. The carbon/carbon brakes have three to five times the capability of steel brakes. Reinforcement of C/G fibers from rayon and PAN precursors was examined. Fabricated carbon composite discs for the brakes of military and jet aircraft are now in service on military and commercial aircraft. Each pound of carbon absorbs nearly one million foot pounds of energy for a maximum energy stop. The c-c composite brakes weigh about 1/3 of the steel brakes.

Fitzer and Heym[26] studied the mechanical properties of the most promising unidirectional fiber/matrix composites at temperatures up to 500°C in air. Commercial unidirectional tapes of C/G fiber/epoxy ("Fiberdux" 911) and boron fiber in an aluminum matrix were procured and cured under pressure. For comparison purposes, polyimide ("Keramid" 601) impregnated C/G fibers were used; a carbon/carbon composite was prepared by liquid infiltration of a molten coal tar pitch, baked up to 550°C under nitrogen and then carbonized to 1400°C and reimpregnated and recarbonized four times for maximum strength; and the carbon fiber reinforced aluminum was prepared by precoating the fiber with nickel or silver wetting agent, followed by liquid metal (aluminum) infiltration at approximately 700°C. The physical properties are reported in Table 13-3. All composites were prepared and tested in same laboratory. The performance of flexural strength of the fiber reinforced composites as a function of temperature are shown in Fig. 13-6.

In a later report, Fitzer *et al.*[27] showed the increase of carbon/carbon composites flexural strengths as a function of the pitch impregnation recarbonization

Table 13-3. Mechanical R.T. Properties of Unidirectional Composites.[26]

MATRIX	EPOXY	POLYIMIDE	G/Al	B/Al	C/C
Fiber volume fraction (%)	55	56	45	49	50
Flexural strength (MN/m^2)	1580	1270	1150	1640	950
Flexural modulus (GN/m^2)	115	126	240	168	154
Percent Ultimate strain	1.5	1.1	0.45	1.0	0.6
Density (10^3 kg/m^3)	1.45	1.45	2.16	2.5	1.65

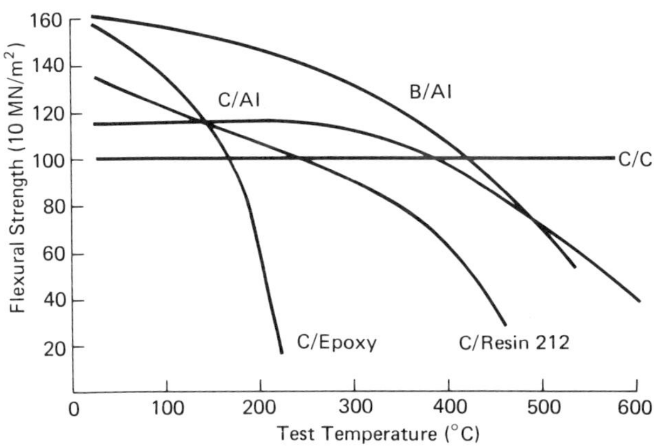

Fig. 13-6. Flexural strength of unidirectionally fiber reinforced composites as function of test temperature.[28]

cycles. The data on repeated graphitization after each cycle are also shown in Table 13-4.

The processing of cokes at high temperatures shows dramatic reduction of coefficients of thermal expansion at heat treatments above 700°C for A-240 petroleum pitch and above 600°C for V-15 coal tar pitch over a wide range of processing pressures.[28]

Table 13-4. Flexural Strength of Carbon/Carbon Composites.[27]

	AFTER CARBONIZATION AT 1000°C	AFTER GRAPHITIZATION AT 2600°C
Start	200 MN/m^2	180 MN/m^2
1 cycle	430	240
2 cycles	650	400
3 cycles	820	530
4 cycles	1000	620

RESIN MATRICES FOR CARBON/CARBON COMPOSITES

Interest in synthetic resin matrices for the preparation of carbon/carbon composites continues to be strong, for the following reasons.

1. Low viscosities at low processing temperatures and at low pressures favor the synthetic resin systems over the petroleum or coal tar pitches.
2. The synthetics offer better purity and identifiable chemical structures than the products of natural origin. Composition of the pitches varies with location of the source and methods used to extract the pitch.
3. High carbon content resin systems are available and are convertible to high temperature resistant carbonaceous products.

Information in the synthetic resin systems and their properties during and after high temperature pyrolysis is limited. These systems are not usually prepared for commercial use with that objective in mind. However, early tests during the 1960's revealed that certain thermosetting polymers from furfuryl alcohol and from phenol-formaldehyde resin possessed high "char strength"—that is to say, after pyrolysis, the loss in properties was not as large as occurred for other resins. As an indication of the other extreme, thermoplastics such as polystyrene would disappear at very high pyrolysis temperatures, with negligible char residue. As in the case of the pitches previously described, the synthetic thermosetting resin matrices are decomposed at pyrolysis temperatures, leaving behind a carbonaceous residue, with significant strength. Subsequent reimpregnations and pyrolysis of C/G fiber reinforced resins will leave an increasing char deposit, and ultimately the strength will approach that of the impregnated but non-pyrolyzed structure. At that time there exists a carbon/carbon composite of graphite fibers surrounded by a carbonaceous substance, usually from highly cross-linked thermosetting polymers.

Shrinkage during cure and pyrolysis will tend to be large for two important reasons: 1) the cure of high polymers involves polymerization shrinkage, which tends to be greater the more rapid the cure temperatures are applied, and 2) the furfuryl alcohol and phenolic resin polymers which show promise are condensation polymerization type polymers, and, during cure, water is formed as a by-product, contributing to more porosity and more shrinkage. In either event, the presence of a large volume concentration of C/G fibers in the composite will markedly reduce these shrinkage considerations.

Siebold[29] reviewed the carbonization of phenolic resins with the goal of processing C/G fiber reinforced composites into carbon/carbon composites. Castings of Monsanto's SC-1008 phenolic resins were pyrolyzed using five different cycles, the maximum temperature up to 2300°F. Thermal conductivity increased continuously to 2300°F and appeared to correlate directly with bulk density. Slow pyrolysis to 1350° and 1830°F was recommended for best property balance of the pyrolyzed phenolics. Weight loss, bulk density, and linear shrinkage are shown in Fig. 13-7 for the phenolic resin after different temperatures of pyrolysis.

The impregnation of porous carbon structures with furfuryl alcohol (FA) resin systems, followed by carbonization of the FA, has been described by Schmitt and Schreyer.[30] Using a maleic anhydride catalyst for the furfuryl alcohol, a porous reticulated carbon foam is impregnated with FA, cured at 100°C, and then carbonized at 900°C–1100°C in an inert atmosphere for a number of successive cycles. Low density thermal insulators were prepared in this manner.

On a speculative note, it is quite likely that some of the polybenzimidazoquinazoline and acetylene terminated polyimides with a heterogeneous aromatic chain will have the heat resistance and high carbon char yield important to graphite fiber carbon composites. To be noted is that some liquid epoxy resins are compatible with large quantities of coal tar pitches (for example, the low temperature hardening compositions of epoxy/coal tar used for highway surfacing) and would have better processability over the 100% pitch materials. Many thousands of pounds of these combinations have been used on highways.

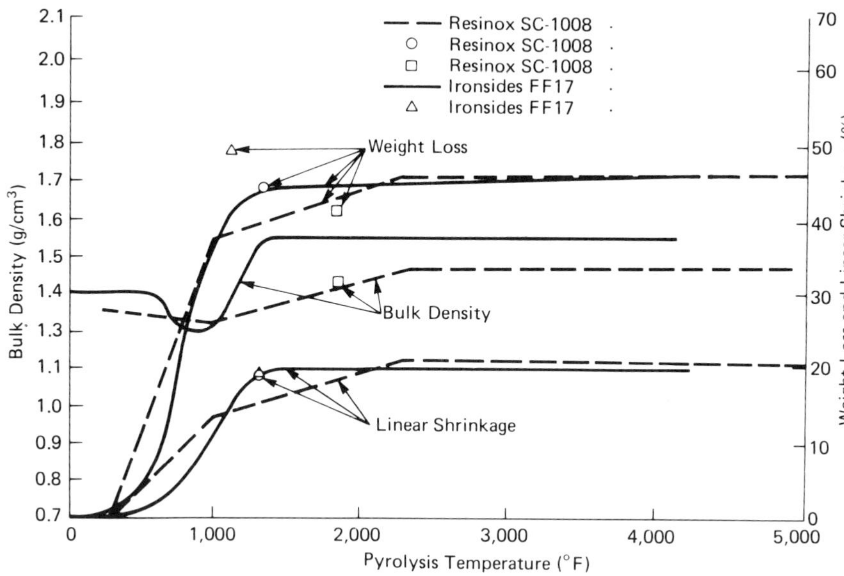

Fig. 13-7. Bulk properties of Resinox SC-1008 and Ironsides FF17 phenolic resins as a function of pyrolysis temperature.[29]

There are noteworthy examples of carbon/carbon composites appearing in government reports, as in the fabrication of a 15 inch long, 9 inch diameter cylinder.[31] Cylinders were prepared with phenolic resin impregnated graphite cloth, accompanied by four furfuryl alcohol (with oxalic acid catalyst) impregnations and repyrolyzation. Cured specimens were placed in a pressure vacuum chamber, first under vacuum to impregnate with a furfuryl alcohol resin system and then under positive pressure of 80 psi. Post-cure and pyrolization cycles involved a slow rise of temperature to 800°C over a 60 hour period. Graphitization took place under argon gas for two hours at 2500° to 2600°C. Typical processing characteristics were as follows.

Initial laminate flexural strength–22,000 psi.
After first impregnation–weight gain 22.3%; after pyrolysis–net gain 10.7%.
After second impregnation–weight gain 13.6%; after pyrolysis–net gain 5.9%.
After third impregnation–weight gain 11.6%; after pyrolysis–net gain 7.4%.
After fourth impregnation–weight gain 7.5%; after pyrolysis–net gain 3.2%.
Weight loss from graphitization–0.8%.
Flexural strength of carbon/carbon composite–12,000 psi.
Interlaminar shear strength–1400 psi.
Apparent porosity–12.0%.
Water absorption–8.5%.
Apparent specific gravity–1.62.

The physical properties of this early carbon/carbon composite were not very exciting–except for one overriding consideration: here was a material that would operate at temperatures in excess of 500°C and maintain its structural integrity.

Carbon/carbon composites were employed to fabricate 38 mm internal diameter tubes for use in a nuclear reactor environment.[32] The tubes were wound from epoxy impregnated C/G tape at a winding pitch of ±87° and ±57°, the latter providing the equiaxial strength required in the final tube. The tubes were carbonized at 900°C in argon over a four day cycle, which reduced the epoxy mass by 50%. Reimpregnation with liquid furfuryl alcohol resins and carbonization was repeated four times, with final graphitization in argon at 2750°C for two hours. Considerable voids were found in final products and were the subject of microscopic study.

Carbon/carbon composites formed from a 60% carbon char resin and a square weave of 99% carbon fabric showed a weight loss of 16 to 20% when carbonized at 840°C.[33] After three impregnations with furfuryl alcohol resin/oxalic acid catalyst and three carbonization cycles, laminates were obtained with more than 60% of non-carbonized control mechanical properties up to 480°C.

Eitman *et al.*[34] examined fiber interactions between C/G fibers of rayon,

PAN, and pitch origin, and the carbonaceous binder. In addition to the fiber variables, three carbonaceous matrices were used—two from coal tar pitch and the other a pyrolyzed phenol/formaldehyde resin. It was concluded that as a consequence of carbon/carbon composite processing conditions, extensive changes took place on the in-situ properties of the C/G fibers. In the program, unidirectional fiber samples were prepared in the form of one inch diameter rods. Each composite was densified by successive impregnation, pyrolysis (at 1000°C), and graphitizations (at 2800°C). The table of data showed the increase in density and the decrease in porosity during the successive processing treatments. Considerable densification increase was observed for the C/G fibers from rayon and PAN origin. This was accompanied by a substantial increase in the modulus of elasticity of the fibers. The results on the composites were inconclusive as to the influence of matrix carbon binder on the different fibers.

In considering the effects of pyrolyzed phenolic resin binders in carbon/carbon composites, the practical applications of carbonized phenolic resin in ablator heat shields for thermal protection should be remembered. For many years, these structures have performed effectively as an expendable element during the high temperature re-entry of an orbiting vehicle into the earth's atmosphere. In this work, it was found that graphite fiber orientation has a significant influence upon the performance of the ablator.[35]

Ablative plastic composite materials were investigated and developed for long-time heating environments by General Electric Co.[36] The desirable material performance goals were low surface ablation, insulative ability, and low weight without asymmetrical ablation, char instability, or spallation. Quartz and carbon fibers were used, including hybrids of both, together with phenolic and polyimide binders to form the structure. Best results were observed on carbon/quartz bifilament tapes. In interpreting the ablative materials in the light of carbon/carbon composites, it should be recognized that the ablative materials undergo carbonization during their performance in service and are considered expendable. The carbon/carbon composites have been pyrolyzed before service, developing stronger structures for rocket nozzles and nose cones which are structurally stable and superior in high strength and resistance to thermal stress failure, and with a crack arrest mechanism by the built-in graphite yarns.

CHEMICAL VAPOR DEPOSITION (CVD)

The use of chemical vapor deposition (CVD) techniques in the infiltration of carbonaceous structures makes possible substantial improvements in the properties of the carbon/carbon composite structures. CVD may be used in lieu of, or in addition to, impregnation of C/G fibers with pitch or synthetic resin matrices. The versatility of the CVD technique is also evident in the diversity of the products which may be formed. In addition to pyrolytic graphite, there are carbides of titanium, silicon, and boron, for example, and borides of silicon and titanium—all of which lend themselves to the CVD process, and the potential for attaining substantial improvements in physical properties of carbon/carbon composites.

"Pyrolytic carbon" (PC) and "CVD carbon" are general terms relating to the carbon material that is deposited at approximately 1100°C on a substrate by the thermal pyrolysis of a carbon-bearing vapor. "Pyrolytic graphite" (PG) on the other hand, refers to carbon deposited from a hydrocarbon gas over a temperature range 1750-2250°C. The electrical, thermal and mechanical properties of PG are quite anisotropic, depending upon the direction in which they are measured. As noted earlier, in Chapter 1, Maxim's patent disclosed CVD in 1880 as a technique for improving the quality of filamentary materials used in electrical light lamps. The most economical gaseous source for the deposition of pyrolytic graphite is natural gas, which normally contains 85 to 95% methane, 2 to 5% acetylene, and 2 to 5% other hydrocarbons. Gas, such as hydrogen, is added to the mixture to facilitate processing.

Infiltration by pyrolytic carbon or pyrolytic graphite of porous substrates such as carbon felts, woven carbon/graphite products and three-dimensional fabrics prepared by tufting or needling of woven carbon fiber products, takes place in furnaces such as that shown in Fig. 13-8. Some of the problems associated with resin and pitch impregnation, such as volatile elimination and organic material shrinkage, are obviated by the internal surface coating process of porous areas. Carbon growth in pores ceases when PG or PC is deposited on the surface of the solid at a much faster rate than within the porous structure, thereby closing the surface pores. It then becomes necessary to remove, by grinding or machining the surface layers of PC or PG to reopen the pores for further infiltration cycles of CVD. The additional CVD cycles improve physicals of the resulting higher density carbon-carbon composites.

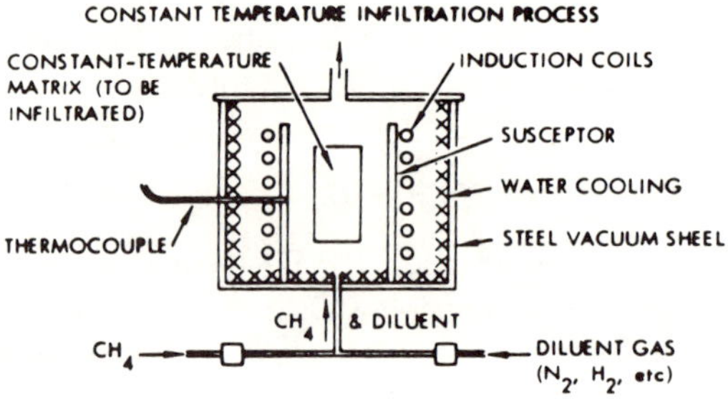

Fig. 13-8. Operation at Super-Temp Company.[37]

Kotlensky discusses isothermal infiltration, which is preferred to thermal gradient processes.[37] For deposition of pyrolytic carbon, processing takes place at 950 to 1150°C at pressures of 1 to 150 torr, and a typical cycle may require 60 to 120 hours. Pyrolytic graphite is generally deposited at 2150°C and deposition rates may vary at 3 to 10 mils per hour; whereas pyrolytic carbon may deposit at a lower rate of 0.01 to 0.1 mil/hour. There are relationships between density and structure with temperature and pressure, as shown in the accompanying Fig. 13-9. The selection of temperature, pressure, gas flow, and furnace geometry affect deposition rate of pyrolytic carbon and pyrolytic graphite. Processing measures avoid sooty isotropic carbon, which is not readily graphitizable.

Carbon felts prepared from rayon precursors (fiber diameter about 10 microns and spacing between fibers 10 to 50 microns), offer a good base with which it

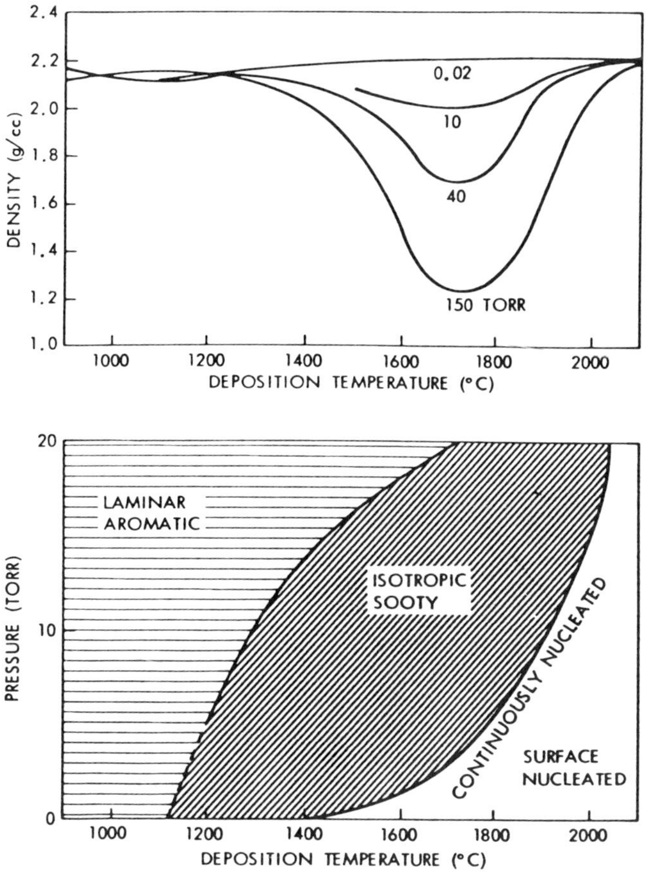

Fig. 13-9. Relationship of density and structure with temperature and pressure.[37]

is possible to prepare larger structures. The original density of 0.1 to 0.2 g/cm^3 may theoretically go to 2.0 g/cm^3 by CVD, though practically this is not achieved. For higher mechanical properties, C/G fiber fabrics offer good prospects. Infiltration by CVD requires considerable skill and experience–and only a few organizations have these capabilities. As the layers of C/G fabrics are stacked together, one would hope for a complete welding or consolidation by the deposited PC or PG. Because of the build-up of pyrolytic carbon or graphite on outer exposed faces, the inside areas may be sealed from the CVD deposition. Hence, the necessity for re-exposing inside porosity by machining (for additional infiltration cycling) and the avoidance of delamination.

Composites composed of a carbon felt substrate and chemically vapor deposited carbon matrices have shown a high degree of thermal stress resistance when exposed to some degree of heat-treatment. The felt substrates are formed by the pyrolysis of viscose rayon felt, whereas the higher strength C/G fibers are not as easily felted by reason of their strength and stiffness. Improved performance is realized with PAN fiber, which is felted and carbonized. Densities fell within a broad range of 0.10 to 0.20 g/cm^3. The felt was infused with equal gas flows of methane and hydrogen at 1325° to 1350°C and then heat treated up to graphitization temperatures. Physical properties were as follows:[38] (See Table 13-5 comparing properties of carbon fiber felts.)

Table 13-5. Properties of Carbon Fiber Felts.[38]

	FLEXURAL STRENGTH	FLEXURAL MODULUS
PAN precursor fibers	88.5 ± 13.0 (12,600 ± 1850)	16.3 ± 3.2 (2.33 ± 0.45)
Rayon precursor fibers	63.0 ± 4.0 (8,900 ± 560) MN/m^2 (psi)	9.9 ± .17 (1.41 ± 0.17) GN/m^2 (10^6 psi)

Thermal expansion and thermal conductivity measurements for the two carbon/carbon composite felts are shown in Figs. 13-10 and 13--11, which compare rayon with PAN precursor fabrics.

Relationships between structure and strength for CVD carbon infiltration substrates have been examined.[39] Studied were the effects of fiber volume, type of fiber, tufting, sewing (detrimental to carbon fabric) and three-dimensional woven fabrics. Numerous photomicrographs provided much useful information. In the following table Table 13-6 are shown the mechanical properties of CVD carbon infiltrated graphite and carbon cloth.

Table 13-6. Mechanical Properties of CVD Carbon Infiltrated C/G Cloth Substrates.[39]

MATERIAL	DENSITY PRIOR TO INFILTRATION (g/cm^3)	FINAL DENSITY g/cm^3	FIBER VOLUME %	EDGE FLEXURAL STRENGTH (psi)	FLATWISE ORIENTATION FLEXURAL STRENGTH (psi)	INTERLAMINAR SHEAR STRENGTH (psi)
Graphite cloth (G-1550-8 H/S)	0.504	1.63	33.6	24,000	18,700	4000
Carbon cloth (GSCC-8, 8 H/S)	0.505	1.55	33.7	14,800	8,700	1370
Carbon Cloth Square Weave	0.497	1.65	31.1	18,200	19,400	6000
Hand Tufted Graphite Cloth (G1550, -8 H/S)	0.456	1.66	30.4	12,000	9,000	2800

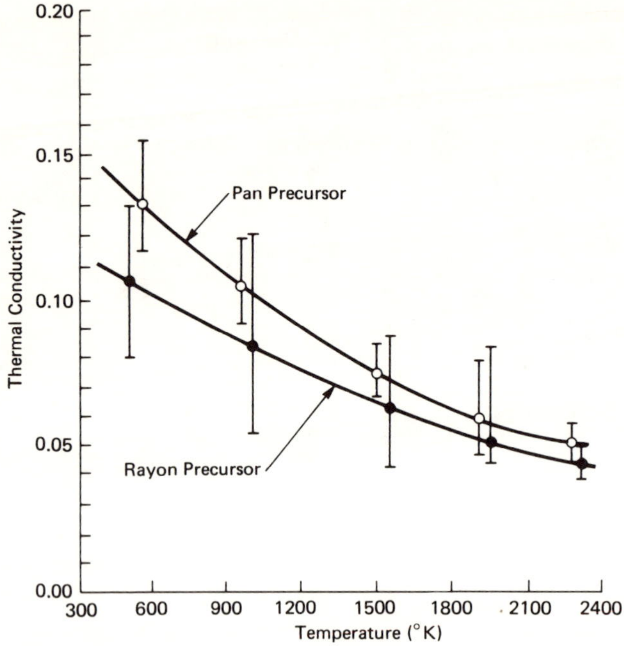

Fig. 13-10. Thermal conductivity of heat treated PAN- and rayon-based CVD/felt composites with a rough laminar matrix micro-structure. (Average and range of six samples each; measured out-of-plane.)[38]

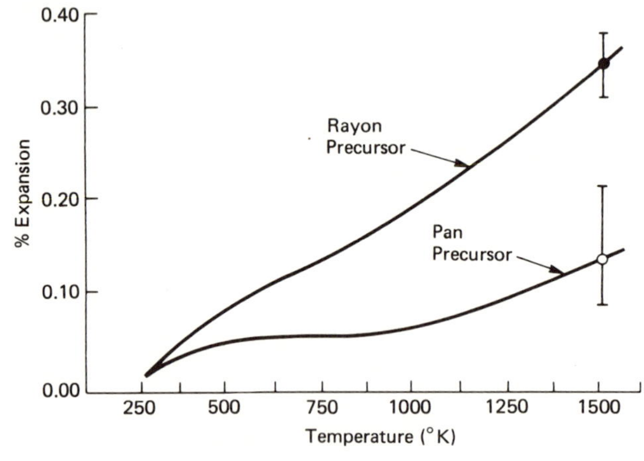

Fig. 13-11. Thermal expansion of heat treated PAN- and rayon-based CVD/felt composites.[38]

Pyrolytic graphite produced by pyrolysis of methane or acetylene at reduced pressure was deposited on graphite fibers at 1800° to 2200°C. Properties are markedly anisotropic (see Table 13-7).

Table 13-7. Properties of Graphite Fibers with Pyrolytic Graphite Deposits.[40]

PROPERTY	TEST TEMPERATURE °C.	DIRECTION $a-b$	DIRECTION c
Density (mg/m^3)		2.05–2.20	
Tensile strength (MN/m^2)	R.T.	70.0	3.4
Compressive strength (MN/m^2)	R.T.	70.0	310.0
	2500	70.0	414.0
Young's modulus (GN/m^2)	R.T.	28.0	–
	2500	10.0	–
Coefficient of Thermal Expansion (m/m/°C.)	0–2200	0.5×10^{-6}	23×10^{-6}

The pyrolytic graphite, if removed from a substrate (such as a flat graphite panel), is actually self-sustaining and has been used to reproduce high temperature, oxidative resistant structures. The pyrolytic graphite is not a fibrous material and actually replicates the base upon which it has been deposited. For more complete information on pyrolytic graphite, there is an excellent article by Smith and Leeds.[41]

The solid rocket motor nozzle throat for the Missile X first stage is the largest carbon/carbon composite ever made (by Thiokol Corp.: 500 lb, 92 inches diameter). They have been made from woven or wound cylindrical preforms. Resistance to thermal stresses is outstanding. The final structure is obtained by impregnation of the preform with molten pitch or resin or by infiltration by the CVD process (or combinations of both)—both requiring multiple cycles and further carbonization. Costs are high—about $85 per ounce for the final unit.[42]

Also of current application interest are the aircraft disc brake pads of carbon/carbon advanced composite, which have outperformed steel pads in their ability to bring to a halt large jet aircraft and the orbiting space shuttle. The high temperature heat dissipating qualities of these composites, and their light weight, are outstanding. CVD processing techniques have contributed much to their success. Reports indicate that CVD infiltration of carbon fiber charred carbon resin composites enhance performance, and that multiple processing measures should be considered. A development illustrating the use of C-C composite brakes appears in the Fig. 13-11a.

Warren and Coulbert[43] discussed the infiltration of pyrolyzed resin carbon composites with pyrolytic carbon obtained through the use of CVD. Over 50%

Fig. 13–11a. Typical carbon-carbon composite disk brake assembly as used on Douglas DC–10 aircraft. (*Courtesy Super Temp Division of B F Goodrich Co.*)

strength increase in the carbon/carbon composite has been achieved with as little as a 10% weight addition of pyrolytic carbon in the charred resin system. They discuss the shrinkage problems associated with the pyrolyzed resin systems that do not contain CVD carbon; stresses which arise from pressure curing, including carbon/graphite fiber breakage and disorientation; and generally low reliability and low ultimate strengths. They indicate that CVD, particularly in combination with the other processes, provides additional property flexibility and the prospects of marked improvement in mechanical properties.

Pyrolytic graphite formed by CVD, reinforced with C/G filaments into carbon/carbon composites, has demonstrated attractive erosion resistant characteristics for rocket nozzles under firing conditions. One disadvantage has been the resistance of excessive thicknesses to delamination. Tests at firing temperatures and pressures indicate that carbon/carbon composites erode at a rate of 1 to 3 mils/second. Also studied were various textile configurations and controlled filament placement and concentration for increased strength and environmental resistance.[44]

Multidirectional reinforced carbon/carbon composites have offered improvements in strength and modulus. Figure 13-11b shows a schematic view of a 3D, orthogonal weave preform. Fiber Materials Inc., have developed special preforms with fiber orientations in multiple directions. Physical properties of the composite may be enhanced by the appropriate design of weaving parameters such as fiber orientation, volume fraction of the fiber in required directions, fiber spacing, woven density, yarn packing efficiency, and fiber selection.[45] Significant advantages arise in thermal stress resistance and improved fracture strength.

Both pressure impregnation of woven 3D preforms with pitch and CVD infiltration treatments are used to improve density, thermal conductivity, and physical properties of the final composite. Cylindrical preforms have also been fabricated. These are used in the manufacture of carbon/carbon rocket nozzles (see Fig. 13-11c).

Interest in carbon densification processing has increased in recent years because of aerospace applications. A report has been prepared on the different micro-structures of a family of carbon felt and carbon matrix.[46] Isothermal processing at 1000° to 1300°C was used with a methane gas diluted with nitro-

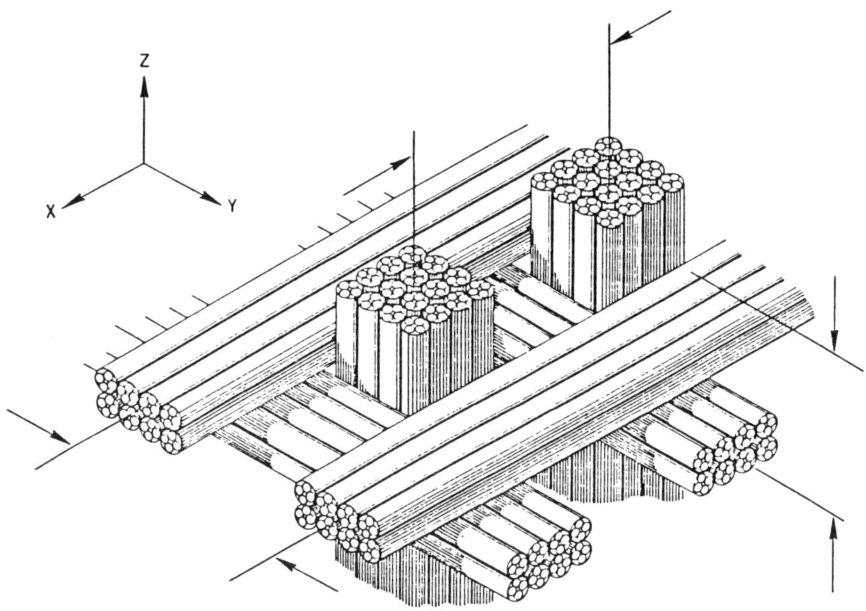

Fig. 13-11b. Schematic of 3-D orthogonal weave. (*Courtesy Fiber Materials Inc.*)

Fig. 13-11c. Rocket nozzle of carbon-carbon composite. (*Courtesy Fiber Materials, Inc.*)

gen, sometimes with hydrogen addition, for deposition on a carbon felt. Existence diagrams were developed to show how to obtain the desired micro-structures under defined conditions. Electrical and thermal properties were reported on some of the carbon/carbon composites.

Rowe conducted a detailed study on the effect of weave spacing on properties of carbon/graphite fiber reinforced carbon composites.[47] Thermal, mechanical, ablation, and micro-structural properties were determined. Binders from pyrolyzed phenolic resins and CVD pyrolytic graphite were compared. The phenolic resin pyrolysis was accompanied by furfuryl alcohol resin impregnation and repyrolysis, as described earlier. Carbon matrices were enhanced by Super-Temp Company (Santa Fe Springs, California) by isothermal vapor deposition, involving repeated deposition of carbon from methane at 1100°C under reduced pressure for 150 hours. It was concluded from results of density and porosity measurements that the pyrolyzed phenolic resin composites are more readily densified than CVD composites. Also, the tensile and compressive properties of both materials are affected by fiber distribution, but not weave spacing, that CVD parts are stronger (36% in tension and 85% in compression).

As noted earlier, the combination of a pitch or a carbonaceous resin binder with C/G fibers forms the carbon/carbon composite after due pyrolysis. The fortuitous introduction of CVD pyrolytic graphite has made possible substantial improvements in physical properties, particularly when used in addition to the above binders. There is room for growth and improvement in these remarkable carbon/carbon composites, which are superior to many metals in physical and thermal properties at very high temperatures.

GLASS MATRICES

Using 7740 glass (Corning Glass Works borosilicate glass) as a matrix, composites were prepared with a variety of graphite fibers containing from 50 to 70% volume of fiber. A collimated monolayer tape is prepared by passing the graphite fibers through a slurry of finely divided glass powder with a small amount of binder, followed by winding on a drum. The dried composite is stacked and placed in a graphite die. Hot pressing is performed at temperature exceeding 1473°K at pressures of 3.5 to 14 MPa in an inert atmosphere to form the glass matrix composite. Exceptional dimensional stability was observed in both axial and transverse directions. The negative coefficients appearing in Table 13-8 reflect a shrinkage in the specimens indicated up to 575K, when expansion ensued for the high modulus graphite reinforced glass matrix. Good flexural strength retention was observed up to test temperatures of 600°C for HM-S graphite/7740 composites. Transverse coefficients of thermal expansion were at a much lower order of magnitude for the glass matrix base (7740).

Prewo[48] emphasizes the need for materials that can provide exceptional dimensional stability in the presence of thermal gradients and adverse environments. He cited optical mirrors, precision support structures, and stable space platforms using graphite fiber and glass matrix composites.[49] Properties of thermally stable composite (TSC) are based on a borosilicate glass designated as 7740 by Corning Glass Works and Hercules HM-S graphite fibers. Thermal expansion characteristics of 0°/90° HM-S/7740 composite are shown for -100°

Table 13-8. Coefficients of Thermal Expansion at 295° to 423°K.[48]

VOLUME PERCENT GRAPHITE/MATRIX	LONGITUDIANL (10^{-6} K^{-1})	TRANSVERSE (10^{-6} K^{-1})
60% GY-70/7740	-0.29	7.6
60% HM-S/7740	-1.0	3.6
60% T-300/7740	0.38	4.3
62% GY-70/epoxy	-0.99	28.8
62% HM-S/epoxy	-0.54	27.0
62% T-300/epoxy	0.02	25.2
50% GY-70/Mg	1.8	21.6
50% SiC/Ti	5.8	7.2
50% B/Al	5.8	19.1

to 300°C in Fig. 13-12. The data for TSC is comparable to or lower than fused silica in thermal expansion. Advantages are also claimed for the improved thermal conductivity of the 7740 glass matrix. Flexural strengths of the 0/90 cross-ply HM-S reinforced 7740 glass were reported at 45 ksi at both 22° and -123°C. Authors suggest potential development products of graphite fibers and glass matrices in a number of diverse composites.

Using monofilaments and filamentary yarns of silicon-carbide fiber, Prewo has also reported good flexural strength and toughness to 700°C for borosilicate

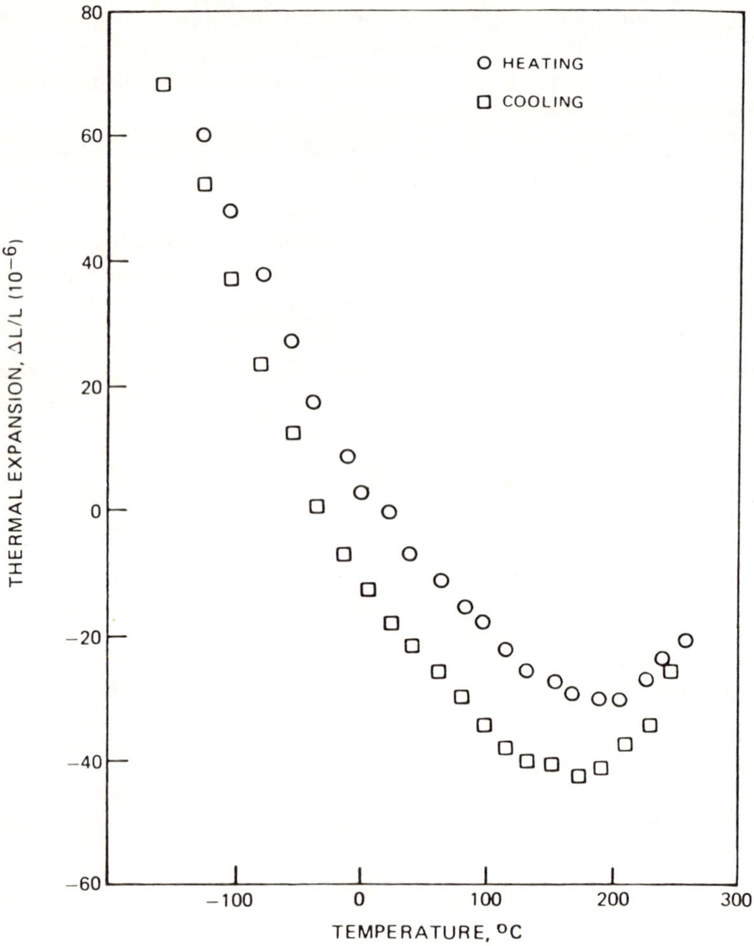

Fig. 13-12. Thermal expansion of 0°/90° HMS reinforced 7740.[49]

glass matrices.[50] Composite thermal expansion of the SiC/glass matrix showed independence of fiber orientation. Tensile strength of 3400 MPa and moduli of elasticity of 415 GPa were obtained with the monofilaments of silicon carbide fiber.

As the picture begins to develop, it is apparent that the matrices available for the utilization of carbon/graphite fibers are many, and processes are varied for achieving high temperature performance. Through examination of their strengths and weaknesses, and consideration of salient cost factors, it becomes possible to make good engineering decisions.

REFERENCES

1. Morse, B.K., "Graphite Fibers as Reinforcements in a Metal Matrix," Technical Report AFML-TR-66-346, December 1966.
2. Plevitt, A. and Ban, D.H., "Method of Making Metal Impregnated Graphite Fibers," U.S. Patent 4,157,409, June 5, 1979.
3. Pepper, R.T. and Kendall, E.G., "Metal Impregnated Graphite Fibers," D0944 Al Report, patented November 6, 1973.
4. Pfeifer, W.H., "Graphite/Aluminum Technology Development," in *Hybrid and Select Metal Matrix Composites,* Renton, W.J. (Editor), Chapter 6, American Institute of Aeronautics and Astronautics, 1977.
5. Lackman, W.L., Penty, R.A., and Jahn, A.F. (Fiber Materials, Inc.), U.S. Patent 3,894,863, July 15, 1978.
6. Withers, J.C. and Alexander, J.A., "High Modulus Filaments for Metal Matrix Reinforcements," ASM/ASTME Paper Westec-7-75, Los Angeles, California, March 1967.
7. Hoffman, P. (Avco), "Latest Developments in Boron, Silicon Carbide, and Graphite Fibers," *Fourth Annual SAMPE Seminar,* Newport, California, October 1979.
8. Pepper, R.T. and Zack, T.A. (Fiber Materials, Inc.), "Graphite Fiber Metal Composites," U.S. Patent 4,072,516, February 7, 1978.
9. Simon, R.A. and Prosen, S.P., "Carbonaceous Fiber Composites," NOL-TR-68-132, October 1968.
10. Imprescia *et al.,* "Carbide Coated Fibers in Graphite Aluminum Composites," NASA-CR-2711, LX-6181-PR, July 1976.
11. Sullivan, P.G. and Raymond, L., "Graphite Fiber Morphology: Effect on Behavior in Graphite-aluminum Composites," *Tenth SAMPE Symposium,* Vol. 10, page 466, New York, October 1978.
12. Gigerenzer, H. and Strempek, G.C. (Fiber Materials, Inc.), "Fabrication of Discontinuous Graphite-aluminum Composites by Pultrusion," NTIS Report A-A039-959/25T, February 1977.
13. Davis, L.W. and Sullivan, P.G., "Exploratory Development of Graphite-aluminum Composites," Final NETCO Report under Contract N00024-75-C-5002 to Naval Sea Systems Command, February 1976.
14. Amateau, M.F. and Hanna, W.D. (Aerospace Corp.), "Fatigue Behavior of Aluminum Alloy–30% Graphite Fiber Composites," Report ATR-75 (9450)-1, March 1975.
15. Gigerenzer, H. and Strempek, G.C., "Selective Reinforcement of Aluminum I-beam with Graphite-aluminum Composite Wire," NTIS-AD-A031 437/7ST, April 1976.

16. Pfeifer, W.H. and Kinna, M., "Naval Applications for Graphite Fiber Reinforced Aluminum Alloys," *SAMPE Conference,* Vol. 8, page 428, New York, October 1976.
17. Hancox, N.L. and Wells, H., "Aluminum (Carbon Fiber Hybrid Composites," *Thirty-third Technical Conference,* SPI, RP/C Institute, Section 12C, February 1978.
18. Paprocki, S., Kizer, D., and Meyerer, W., "Ultra High Modulus Fibers—For Graphite Aluminum and Graphite Magnesium Composites," *SAMPE Twenty-fourth National Symposium,* Vol. 24, page 1451, San Francisco, California, May 1979.
19. Rubin, L., "Applications of Metal-matrix Composites—The Emergency Structural Materials," *SAMPE Twenty-fourth National Symposium,* Vol. 24, page 1236, San Francisco, California, May 1979.
19a. Rubin, L., "Applications of Metal-Matrix Composites," *SAMPE Journal,* Vol. 15, no. 4, page 4, August 1979.
20. Weber, G.W., Young, K.R., and Taylor, A.J., "Densification of Carbon-carbon Composites at 30 ksi," *Thirteenth Biennial Conference on Carbon,* American Carbon Committee, page 68, Irvine, California, July 1977.
21. Johnson, A.C., Pinoli, P.C., and Keller, R.L. (Lockheed Missile), "Optimization of Carbon/Carbon Processing," *Fourteenth Biennial Conference on Carbon,* American Carbon Committee, page 238, Penn State University, June 1979.
22. Laramee, R., Lamere, G., Prescott, B., Mitchell, R., Sottosanti, P., and Dahle, D. (Thiokol), "Carbon/Carbon Material Application to Solid Rocket Nozzles," *Thirteenth Biennial Conference on Carbon,* American Carbon Committee, page 74, Irvine, California, July 1977.
23. Rolincik, P.G. (Avco), "Mod-3 Pierced Fabric Reinforced Carbon/Carbon Composites," *Eleventh Biennial on Carbon,* American Carbon Committee, page 343, Oak Ridge, Tennessee, June 1973.
24. McAllister, L.E. and Taverna, A.R. (Avco), "Development and Evaluation of 3 Dimensional Graphite Carbon/Carbon Composites," *SAMPE Seventeenth National Symposium,* Vol. 17, Los Angeles, California, April 1972.
25. Kirkhart, F., "Carbon Aircraft Brakes," *Twelfth Biennial Conference on Carbon,* American Carbon Committee, page 279, Pittsburgh, July 1975.
26. Fitzer, E. and Heym, M. (West Germany), "Carbon Fiber Reinforced Composites for Applications at Elevated Temperatures," *Thirteenth Biennial Conference on Carbon,* American Carbon Committee, page 128, Irvine, California, July 1977.
27. Fitzer, E., Huttner, W., and Manocha, L., "Influence of Process Parameters on Mechanical Properties of Carbon-carbon Composites With Pitch as Matrix Precursor," *Fourteenth Biennial Conference on Carbon,* American Carbon Committee, page 240, Penn State University, June 1979.
28. Evangelides, J.S. and Rellick, G.S. (Aerospace Corp.), "Properties of Coke from Binder Pitches as Functions of Processing Temperature and Pressure," *Fourteenth Biennial Conference on Carbon,* American Carbon Committee, page 405, Penn State University, June 1979.
29. Siebold, R.W., "Carbonization of Phenolic Resins," *SAMPE National Symposium,* Vol. 20, page 327, San Diego, California, April 1975.
30. Schmitt, C.R. and Schreyer, J.M., "Carbon Composite Structures," U.S. Patent 3,632,385, January 4, 1972.
31. Forcht, B.A. (LTV Aerospace Corp.), "AHP Cylinder Fabrication—LTV Carbon-carbon Composite," Technical Report AFML-TR-70-255, March 1971.
32. Mills, R.W. and Peggs, J.D., "Voids in Filament Wound Graphite/Graphite Composites," *Eleventh Biennial Conference on Carbon,* American Carbon Committee, page 333, Oak Ridge, Tennessee, June 1973.

33. CA 71-92321X, Vol. 71 (1969), Mackay, H.A., "Characterization of Carbon Fabrics and Filaments," U.S. A.E.C., SC-RR-68-651, 30 pages, 1969.
34. Eitman, D.A., Greszczuk, L.B., and Jortner, J., "Fiber Matrix Interactions in Carbon-carbon Composites," *SAMPE Symposium,* Vol. 19, page 346, May 1974.
35. Sheldahl, R.E. and Wright, G.F. (Sandia), "Effect of Fabric Orientation on the Ablation Performance of Carbon Phenolic," Report Conference 760513-4, 1975.
36. Juneau, P.W., Metzger, J.W., Curtis, F.P., and Penton, R.M. (General Electric Co.), "Ablative for High Heat Loads," Report 71SD2066, Contract F-33615-69-C-1503, September 1971.
37. Kotlensky, W.V. (Super-Temp Co.), "CVD Carbon Infiltration of Porous Substances," *SAMPE National Symposium,* Vol. 16, Anaheim, California, April 1971.
38. Pierson, H.O., Northrop, D.A., and Smatana, J.F. (Sandia Corp.), "Development and Properties of PAN-based Carbon Felt-pyrolytic Carbon Composites," *Eleventh Biennial Conference on Carbon,* American Carbon Committee, page 275, Oak Ridge, Tennessee, 1973.
39. Bauer, D.W. and Kotlensky, W.V. (Super Temp Co.), "Relationship Between Structure and Strength for CVD Carbon Infiltrated Substrates," *Twenty-third Pacific Coast Meeting,* American Ceramic Society, San Francisco, California, October 1970.
40. Parmee, A.C., "Pyrolytic Graphite Coated Rocket Nozzles," *SAMPE Fourth National Conference,* page 483, Palo Alto, California, October 1972.
41. Smith. W.H. and Leeds, D.H., "Pyrolytic Graphite," *Modern Materials,* Vol. 7, page 139. Academic Press, New York and London, 1970.
42. Stedfeld, R., "Carbon-carbon Lightweight Composite for Hot Spots," *Materials Engineering,* Vol. 91, page 13, January 1980.
43. Warren, J.W. and Coulbert, C.D. (Super-Temp Co.), "Introduction to Carbon Composite Materials," Bulletin issued by Super-Temp Co., Santa Fe Springs, California, 1971.
44. Macri, B.J. (Atlantic Research Corp.), "Development of High Temperature Resistant Materials for Use in Naval Ordnance," Contract N00017-73-C-4306, October 1973.
45. Cook, J.L., and Crawford, J.A., "Multidirectional Substrates for Advanced Composites," 76th Annual Meeting of the American Ceramic Society, March 1975.
46. Pacault, A., Loll, P., and Delhaes, P. "CVD Carbon Composites," *Thirteenth Biennial Conference on Carbon,* American Carbon Committee, page 167, Irvine, California, July 1977.
47. Rowe, C.R. (Naval Ordnance Laboratories), "Effect of Weave Spacing on the Properties of 3-D Orthogonal Carbon-carbon Composites," *SAMPE National Symposium,* Vol. 19, page 359, Buena Park, California, April 1974.
48. Prewo, K.M., Dicus, D.L., and Bacon, J.F., "Graphite Fiber Reinforced Glass Matrix Composites for Aerospace Applications," *SAMPE Twenty-fourth National Symposium,* Vol. 24, page 61, San Francisco, California, May 1979.
49. Prewo, K.M. (United Technologies), "New Dimensionally and Thermally Stable Composite," *Conference on Advanced Composites,* page 1, Technology Conference Associates, El Segundo, California, December 1979.
50. Prewo, K.M., and Brennan, J.J., "High Strength Silicon-Carbide Fiber Reinforced Glass Matrix Composites." *Journal of Material Science,* Vol. 15, pages 463-468, 1980.

14
Manufacturing and Processing Techniques for Carbon/Graphite Fiber Composites

Manufacturing and processing techniques for C/G fiber advanced composites have reflected the developmental nature of some of the products. As industrial interest grows, more attention is being given to automated processes and more permanent tooling for larger productions. Fortunately, glass fibrous reinforcements have established a large experience processing base, which will guide the utilization of carbon and graphite fibers. In this chapter, emphasis will be placed on those methods conventionally used in aerospace industries in the United States. Throughout this book, there have been frequent references to quality control and inspection methods, and the reader is advised to examine the chapter on testing methods for more specific details.

The ramifications of manufacturing and processing of advanced composites are diverse, due to the numerous resin matrices that are used as binders, the types of C/G fibers, and the hybrids that are used for reinforcements. In fact, this subject could be expanded well beyond the framework of this chapter. In attaining an overall perspective of C/G fiber technology, current manufacturing and processing techniques are reviewed herein.

MATERIAL SELECTION

Carbon and graphite fibers are made up of many thousands of fine, wispy filaments which are easily broken and put into disarray if mishandled. They are usually sized by the manufacturer to improve handling characteristics (which may involve compatible resins for enhancing impregnation capability). The filaments are grouped together into tows or yarns, though the physical distinctions are not standardized.

The yarns (usually 10,000 filaments or less) may be twisted together to enhance bundle integrity. Larger bundles, sometimes called tows, generally consist of 10,000 or more filaments (up to 160,000) and are usually not twisted. The continuous lengths of fiber are supplied on spools to manufacturers engaged in further processing.

Discontinuous Fibers

Chopped 1/4 to 1/2 inch carbon fibers are the least expensive fibers and most generally are used as reinforcements to molding compounds. While they can be used with all types of resins, thermoplastic resin matrices are usually used as matrices (see the chapter on thermoplastic resin matrices). The C/G fibers impart reduced shrinkage, greater stiffness, improved fatigue, and higher electrical conductivity to the molding compound.

Chopped carbon fiber reinforcements may be used to replace glass fibers in spray lay-ups, where the resin matrix and fibers are sprayed separately to fabricate parts, co-mingling in the streams impinging upon the mold. More finely divided milled fibers are available (average about 300 microns in length), as well as longer fibers up to any specified length. The economics of component parts for automotive use are dictating hybrids of lower cost from chopped glass fibers and chopped carbon fibers.

Short chopped fibers of carbon and graphite are used to manufacture fibrous webs such as mats, papers, and felts—useful for reinforcing components with complex curvature or which require isotropic properties. Care should be taken to confine fibers within the immediate area of processing.

Continuous Fibers

Collimated, unidirectional fibers are used to prepare C/G fiber tapes—usually 3 to 30 inches wide, with a nominal thickness of 5 to 6 mils. A lesser number of filaments in the tows will be used in preparing thinner tapes. Tapes are generally processed in the range of 10 to 30 ft/minute by prepreggers. The resin matrix is usually applied by a hot melt process, as a partially cured thermosetting epoxy resin is softened and pressed into the collimated fibers. Ultimately, when parts are assembled from prepregs, the molding and curing cycle will advance the resin to an insoluble state. In Fig. 14-1, a 20 inch wide carbon fiber tape is being processed.

More recent polyimides (for example, PMR-15 and PMR-11) are applied as liquid monomers and require cure advancement after impregnation of fibers.

Fig. 14-1. Preimpregnated unidirectional tape being processed. (*Courtesy Fothergill and Harvey, England.*)

Some thermosetting resin matrices are applied from solution, necessitating solvent removal.

Woven fabrics permit significant cost savings in the molding operations by reducing the labor of positioning reinforcing fibers. Fabrics are generally described according to the type of weave and the number of yarns per inch—first in the warp (W) direction (parellel to fiber length) and then perpendicularly in the fill (F) direction. A plain weave is a construction in which one warp end is repetitively woven over one fill yarn, and under the next. It generally exhibits the most stable construction and has uniform properties in both directions.

Satin weave fabrics, such as the eight harness satin (one warp end passes over seven fill yarns and then under one fill yarn), are more pliable than plain weave fabrics, and conform more readily to compound curves. Strengths are higher because of the greater linearity of the fiber reinforcements. PAN-based carbon fabrics are usually made by weaving carbonized yarns, while pitch-based fibers can be woven at intermediate processing, then converted to a high modulus product while in the fabric form.[1] There are many combinations possible with other fiber reinforcements, such as glass fibers, to lower cost and improve impact strength. Emphasis, for example, may be placed on the continuous warp fibers of graphite or carbon, while the fill may comprise a minimum of glass fibers. In some instances, a small percentage of fine chopped fibers may be added to the woven product to improve overall surface appearance during the molding operation, as in sheet molding compound (SMC) production. There are no limits on distributions of C/G fibers, and, in hybrid combinations with other fibers, effective reinforcement for the resin matrices is achieved.

Special knitted and woven constructions have been developed, some with resin impregnation for molding purposes. The three dimensional fabrics developed for carbon/carbon composites have already been described in applications to high temperature rocket components. Knitted and woven carbon fiber tapes, some prepared with lubricants from polytetrafluoroethylene, are also used as low friction packings to replace asbestos.

Resin Matrices

Resin matrices for the manufacture of advanced composites of carbon and graphite fibers have been discussed in considerable detail in earlier chapters. On the basis of testing and field evaluation over several years, certain epoxy and polyimide products appear prominently in aerospace and aircraft specifications. These are not intended to inhibit new developments, but they do stress the necessity of thorough testing before commitments are made to aircraft structures. For automotive and other industrial applications, molding compounds using polyester resins, in combinations with graphite fibers and other fiber reinforcements, are gaining in popularity.

Much can be written about the molding and processing of the synthetic resins, though because this is not the main thrust of this book, only the salient features of resin matrices which influence manufacturing and processing of advanced composites are noted below. These are features with which the processor must be familiar in order to use the materials successfully.

Thermosetting resin systems comprise a basic resin polymer and its curing agent. When they are combined, the ensuing chemical reactions will influence:

1. The shelf storage life and the working life of the system at various temperatures;
2. The temperature range for completing cure of the resin system and its fibrous reinforcement;
3. The pressure to be used for forming and shaping parts; and
4. Post-curing, as is frequently required for advanced composites.

Curing temperatures and molding pressures of the thermosetting resin systems influence the production methods and the tooling selected. In addition, the manufacturing and storage history of the resin matrix system will affect the processing of composites. Overage materials will not produce good parts, and checks are taken to ascertain that the resin matrix is still within its useful storage life and temperature, before fabrication.

MATERIAL HANDLING

Prepregs of C/G fibers and resin are usually shipped in a refrigerated truck and must be stored under refrigeration in the fabrication plant until ready for use. Prepregs of C/G fiber/resin systems are usually packaged in sealed polyethylene film, and when removed from the refrigerator must be allowed to reach room temperature before removal from their sealed film containers. Whether collimated unidirectional tapes, woven fabrics, or felt are used depends upon the design and functional requirements of the composites.

The choice of resin matrix and the character of the fiber reinforcement are generally specified and controlled by the procuring agency. A typical batch acceptance test by an aircraft manufacturer is shown in Table 14-1, which illustrates the careful detail with which a typical graphite fiber tape is evaluated ("Thornel" 300 and Narmco's 5208 epoxy). The average thickness per ply of this unidirectional composite tape is 0.0054 inches.

For comparison purposes, Table 14-2 shows the physical data of a woven "Thornel" 300 fabric impregnated with 5208 (tetraglycidyl epoxy resin system). Note the lower, though more uniform, strengths in tension and compression for the woven graphite fiber/epoxy composite, as compared to the previous typical data for unidirectional carbon fiber tapes.

For aerospace fabrication activities, the use of prepregs for advanced composites is preferred. The prepregs are readily handled and are adaptable to automatic cutting to profiles and to the desired fiber orientation. There is a growing interest in handling and positioning the prepregs automatically in the tools for curing the composite assemblies.

Because of limited production, manual techniques have dominated the preparation of advanced composites. There are advantages to the manual assembly of prepregs prior to the molding and curing operation, in that special distributions and positioning of C/G fiber prepregs are required for complex shapes—particularly at abrupt changes in shape and in the positioning of elements that form parts of the final assembly. Final assembly is, in fact, a consolidation operation wherein the resin matrix flows and welds the entire assembly of reinforcing fibers, and sometimes other inserts, such as low density cores or

Table 14-1. T-300/5208 Batch Acceptance Test Results.[2]

PRELIMINARY DRAFT C-22-1379A/114 SPECIFICATION REQUIREMENTS		RESULTS OF TEST 1	2	3	4	5	AVERAGE
Area Weight (4) (3 × 3 in.)	139–149 g/m^2	146	148	147	147	–	147
Liquid chromatography (1)		–	–	–	–	–	–
Volatiles (2) (60 ±5 minutes	3.0 max. edge	0.30	–	–	–	–	–
at 350°F)	center	0.27	–	–	–	–	–
Dry resin content (4) (3 × 3 in.)	31-37%	36.3	35.6	36.3	36.4	–	[illegible]
Flow (2) at 350°F at 85 psi	9-18%	15.3	15.0	–	–	–	–
Gel time (2) at 350°F	Info. only, minutes	19.3	19.2	–	–	–	–
Cured fiber volume (3) 0.080 in. panel	60-68%	61.6	61.6	62.0	–	–	61.7
Cured fiber volume (3) 0.040 in panel	60-68%	62.4	62.5	62.0	–	–	62.3
Specific gravity (3) 0.080 in. panel	1.54-60	1.561	1.563	1.562	–	–	1.562
Specific gravity (3) 0.040 in. panel	1.54-1.60	1.563	1.562	1.561	–	–	1.562
Tensile strength, longitudinal (3) at 75°F	190 ksi, min., ind.	241	237	227	–	–	235
Tensile modulus, longitudinal (3) at 75°F	18.5 × 10^6 psi min., ind.	21.2	21.8	20.5	–	–	21.2
Flexural strength (3) at 75°F	210 ksi, min., ind.	260	271	280	–	–	270
Flexural modulus (3) at 75°F	18 × 10^6 psi, min., ind.	18.3	18.4	19.4	–	–	18.7
Flexural strength (3) at 180°F	200 ksi, min., ind.	211	216	243	–	–	224
Flexural modulus (3) at 180°F	16 × 10^6 psi, min., ind.	17.7	18.3	17.6	–	–	17.9
Short beam shear (3) at 75°F	13 ksi, min., ind.	18.8	19.6	19.2	–	–	19.2
Short beam shear (3) at 180°F	12 ksi, min., ind.	15.3	14.7	15.5	–	–	15.2
Thickness per ply (5) 0.080 in. panel	0.0046-0.0056 in.	0.0051	0.0051	0.0050	0.0050	0.0051	0.0051
Thickness per ply (5) 0.040 in. panel	0.0046-0.0056 in.	0.0054	0.0054	0.0054	0.0053	0.0053	0.0054

NOTES: Batch 1237
Date November 14, 1978
Lab report 351735

Table 14-2. T-300/5208 Graphite/Epoxy Fabric Characterization Test Data.[3]

TEMPERATURE CONDITION (°F)	75D	-67D	-67W	180D	180W
Warp tensile					
Strength (ksi)	73.3	66.5	55.9	77.7	76.3
Modulus (msi	9.1	8.7	7.7	9.0	9.8
Fill tensile					
Strength (ksi)	71.0	65.9	–	75.0	69.2
Modulus (msi)	9.7	8.9	–	9.1	8.5
Warp compressive					
Strength (ksi)	75.7	85.3	85.0	73.4	59.4
Modulus (msi)	8.8	8.9	8.1	8.5	8.6
Fill compressive					
Strength (ksi)	73.9	82.2	–	–	54.6
Modulus (msi)	8.8	8.4	–	–	8.2
In-plane shear					
Strength (ksi)	13.3	14.9	15.6	12.8	12.0
Modulus (msi)	0.76	0.87	–	–	0.61
Short beam shear					
Strength (ksi)	7.1	5.4	7.2	6.3	6.2
Poisson's ratio	0.055	0.070	–	–	0.072
Ply Thickness (in.)		0.0149			
Fiber Volume (%)		61.1			

D–Dry W–Wet

metallic bushings, into the final composite. Figure 14-2 illustrates a typical stack of prepregs which will be molded into a composite.

The techniques of advanced composite fabrication lend themselves to producing components of widely different geometric shapes and thicknesses in one unit—as contrasted to metal assemblies requiring the assembly and joining of many small units. For example, techniques have been developed at McDonnell Douglas Aircraft for manufacturing complex advanced C/G composites. Through the use of high expansion silicon rubber pads, large complicated cored components are shaped in closed metal molds. The entrapped rubber expands on heating, developing internal pressure forcing C/G prepregs against the mold profiles. The final product requires no subassemblies, as it is a one piece unit ready for assembly.

The methods for placing fibers may be classified into two categories.

1. *Random orientation.* Chopped fibers used in compression and transfer molding compounds fall into this category. Spray techniques, developed for glass fibers, are also adaptable to the carbon fibers, wherein a twin nozzle gun applies resin systems and fibers from separate units—a combi-

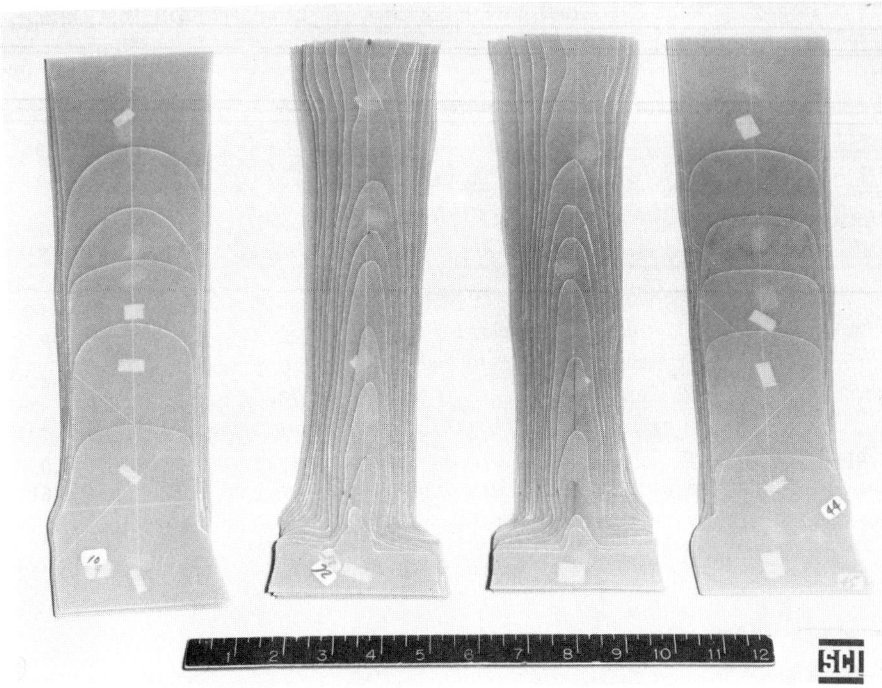

Fig. 14-2. Stacking plies to build up composite preform. (*Courtesy Structural Composite Industries, Inc.*)

nation being effected in an open mold. Mats would also be considered as randomly oriented fibers and are impregnated with resin before being placed in molds for reinforcement.

2. *Continuous fibers.* As already indicated from a description of impregnated tapes and woven products, continuous tapes and woven goods attain the best physical properties for structural applications. Techniques of pultrusion and filament winding, as described later in this chapter, achieve a good fiber orientation, and, depending upon the process, may require resin impregnation of the fibers at the time they are being shaped.

CURING AND POST-CURING

Curing processes for C/G fiber composites are numerous, and include the following.

1. Vacuum bag molding—either in an autoclave or in an oven.

2. Integrally heated tooling, involving the use of matched dies or molds. (Compression molding dies may receive their heat from hydraulic press platen.)
3. Integrally pressurized tooling with internal elastic members for developing molding pressure.

Vacuum Bag

The lay-up of C/G impregnated materials under a vacuum bag, either for cure in an oven or in an autoclave, requires a series of significant steps and the utilization of numerous ancillary materials. Figure 14-3, reproduced from a Fiberite Co. bulletin, illustrates typical manufacturing lay-up procedures, forming a good reference and checklist of essential steps to be followed. There are a few steps which should be highlighted, because considerable time may be wasted in preparing the composite if the measures are not carefully followed. For example, the outer envelope forming the vacuum bag encasement is subject to mechanical abuse as it is drawn down over abrupt changes in dimension or thickness, and

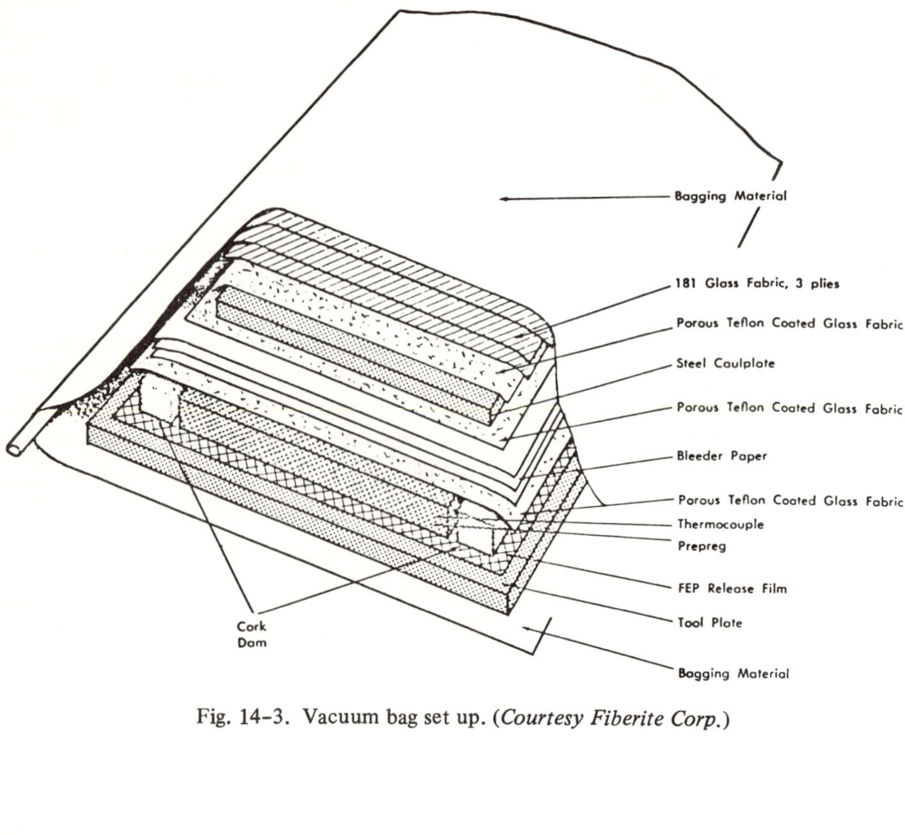

Fig. 14-3. Vacuum bag set up. (*Courtesy Fiberite Corp.*)

hence for the high temperatures required for processing, the strength of a flexible high temperature film, such as silicone, is preferred. If reinforcements are used in these vacuum bag envelopes, they should not interfere with the ability of the film to shape about the part being fabricated, when drawn down under vacuum. While a flat sheet specimen has been used in the illustration, this process is extrapolatable to larger component tools for aircraft parts. One can visualize the inclusion of several components on a large area composite structure which are, in effect, assembled manually and then welded together under the pressure of vacuum into an integral single body. The benefits of such consolidations have been apparent in aircraft and automotive applications, by reducing parts to be handled and minimizing the subsequent joining operations.

The seals which have been identified in the above illustrations are specific to vacuum bag technology, and to retain a resilient, tough, flexible consistency at the temperatures of cure is not easy to achieve. In practice, as the vacuum is applied and the flexible encasement forms about the composite being shaped, there may be wrinkles or air pockets which can be rubbed out at the same time as the seals are checked. This is particularly necessary when shaping large parts with multiple curves. After the resin impregnated fibers have been distributed according to plan, the entire assembly is placed into the pressure autoclave or oven to accomplish heat cure. A failure in the vacuum bag in the early stages of cure usually results in an unsatisfactory part and the loss of materials and assembly time—hence the concern with production and assembly details.

The removal of volatiles or by-products during the cure entails special techniques in the manufacture of advanced composites, and as noted in the illustration of vacuum bagging, there are porous release fabrics or papers between the impregnated fibers and the release film. The purpose of the release fabric is, of course, to provide paths for such volatiles and excess resin to be swept up into the vacuum system (into some suitable trap—depending upon the volatiles coming off). The fabrics have been pretreated with release agents which will prevent the matrix resin from adhering to it and, in the event of a large resin flow, a non-porous release film backing (such as "Teflon" will provide the separation. As the cost of these accessories and the lay-up time are added to the raw materials, it is apparent why advanced composites are expensive.

Cure Temperatures and Pressures

As observed in earlier chapters discussing resin matrices, there are numerous systems which have been developed for the manufacture of fiber reinforced composites. There are basic fabricating considerations which will influence the selection of the resin matrix (see Table 14-3).

Table 14-3. Considerations Influencing Selection of the Resin Matrix.

SYSTEM	ADVANTAGES	DISADVANTAGES
Low temperature cure resin systems (under 250°F)	Lower cure temperatures reduce thermal stresses in fabricating components. Less costly mold materials (such as epoxy base) may by used. Low pressure steam, as well as electric heating, is readily available for cure.	Low temperature cure resins are less stable at R.T. and require refrigeration before cure. R.T. cure systems are two component materials, requiring mixing immediately before use. Low temperature systems are not usually suitable for high temperature service and require post-cure.
Intermediate temperature cure resin systems (300° to 350°F)	One component resin systems with long working life permit long assembly times for large and complex parts Most prepregs fit into this category, and while some require refrigeration, they are adaptable to vacuum bag, autoclave, and press cure. Most chopped carbon fiber molding compounds cure in this temperature range.	Heating requirements and auxiliary equipment are more expensive. Will require higher post-cure temperatures for maximum performance at high temperatures. Increased thermal stresses as parts cool down from mold temperatures.
High temperature cure resin systems (400° to 600°F)	The new high temperature performing polyimide and poly(amide-imides) cure in this temperature.	Costly molds and considerable care are required to minimize built-in thermal stresses in composites. Still relatively new systems are involved and fabrication processes are in development.

Considerable attention has been given in recent years to the development of processes and tooling for operating at temperatures up to 600°F. While the activity in graphite fiber polyimide advanced composites is low as compared to other resin systems (such as epoxies and polyesters) some of the current papers on this subject are noted below to provide an up-to-date review on this expanding activity.

It was felt that bulk graphite tooling, by matching the thermal coefficients of the graphite fiber/epoxy resin composites, would be suitable for manufacturing advanced C/G fiber composites. Detailed descriptions were given of the tooling procedures by Curley and Penton.[4] Loyd[5] also described low expansion coefficient tooling for fabricating advanced composites. Cast ceramic tooling was reviewed, and the savings in cost by molding edges of graphite fiber/epoxy parts to net dimensions was claimed superior in quality to machined edges of composites.

Integrally heated graphite tooling and a hydroclave provided cost effective solutions for 550°F laminates. A 30 inch diameter graphite fabric/PMR-15 polyimide cylinder with multiple curvatures and sharp radii was fabricated in this tooling.[6] Mace *et al.*[7] described manufacturing techniques for fabricating, at 600°F, polyimide/graphite parts. The autoclave cure cycle used for these high temperature fabrication operations is reproduced in Fig. 14-4. Short beam shear strengths of 5.0 to 5.7 ksi and flexural strengths of 70 ksi were obtained on graphite cloth panels in temperatures up to 500°F. Autoclave processes for fabricating graphite fiber/PRM-15 polyimide prepregs were described by Sheppard *et al.*[8] Structural elements included hat sections, I-beams and honeycomb sandwich structures.

A good review of fabrication processes for advanced composites was prepared by Welhart *et al.*[9] Drawing upon aircraft experiences, he covered procedures primarily used for glass, boron, and graphite fiber reinforced epoxy systems. In

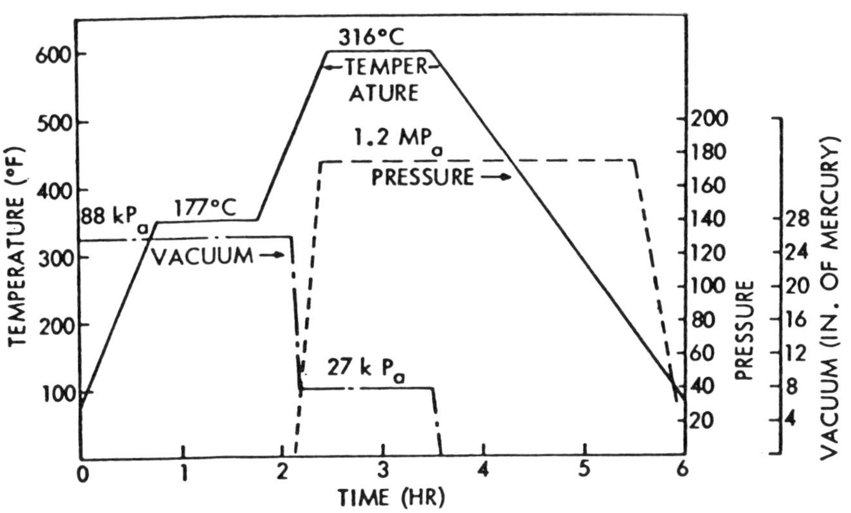

Fig. 14-4. Autoclave cure cycle used for typical polyimide composite.[7]

describing the fabrication of graphite fiber/epoxy composites by the vacuum bag/autoclave process, he summarized the following steps.

1. Application of mold release agent to surface of mold; bake at elevated temperature to eliminate volatiles and cure coated film; and buff surface to smooth, glassy surface. Repeat with each lay-up.
2. Templates are fabricated to facilitate cutting and sizing of the prepreg tape or broad-goods, following dimensions appearing on engineering drawings.
3. Cork or silicone rubber dams are placed around the composite prepreg lay-up. Apply each individual ply of prepreg to the tool, with the aid of a "Teflon"-lined squeegee and hot air gun to soften the ply as it is being positioned. A peel ply is required if the composite surface is to be ultimately bonded. Peel ply is removed prior to bonding. Add porous release film over the exterior laminate surfaces.
4. Check the bagged lay-up for leaks before placing in autoclave.
5. Cure schedule of 350° ± 10°F for two hours in autoclave, after raising to this temperature over a one to two hour period. Vacuum held at 27 to 29 inches of mercury, as the autoclave is pressurized to 125 psi. Cool to 150°F or lower.
6. Post-cure according to required schedule without vacuum restraint.

Other ancillary mold materials include gypsum plasters and ceramics, and metal powdered filled resins, though Table 14-4 more accurately represents current tools for advanced composites. It will be noted that the time to heat up and cool down tools and parts represents a large part of the fabrication cycle. With large and thick components, this should not be accelerated too drastically, as there are thermal stresses and cure considerations which are affected. The use of graphite reinforced molds has advantages, in that C/G components being produced will more closely match the expansion and contraction of the molds. This measure will alleviate some of the thermal stresses which arise, due to differences in coefficients of thermal expansion of metal molds and cured C/G components.

Reinhart[10] has outlined some of the goals in fabricating advanced composites for aircraft during the next decade:

1. Development of systems not requiring autoclaves and utilizing vacuum bags, where necessary, for cure, as well as the use of existing ovens;
2. Automatic monitoring of the cure cycle for the matrix component;
3. Increased automation in lay-up of impregnated fibers into forming tools; and
4. More emphasis upon fabrication of large component aircraft structures.

Table 14-4. Advanced Composite Tooling Materials.

TOOL MATERIAL	ADVANTAGES	DISADVANTAGES
Aluminum	Low material cost; good heat transfer; may be rough cast and machined.	Thermal expansion is much greater than that of C/G fiber composites.
Steel	Most durable—suitable for very large production runs; good for cure up to 600°F.	Most expensive tooling; long machining time; size limitations.
Epoxy/glass fiber molds	May be fabricated on master molds at low cost and rapidly; excellent for compound shapes and large areas up to 300°F.	Poor heat transmission as bonding fixture at high temperature; will change dimensions on repeated runs; limited production capabilities.
Reinforced graphite molds	Matches coefficient of thermal expansion of C/G reinforced composites; capable of high temperature cures and post-cures (to 700°F).	More costly, and may require machining for precision; limited production capabilities.

PULTRUSION PROCESS

Pultrusion is a processing technique for composite materials that has been practiced for over 25 years, though it has been relegated to a few specialty applications until recently. Pultruded products are composed of thermosetting liquid resins and fiber reinforcements, which are combined and pulled through a die rather than being pushed through or extruded. Advanced composite materials lend themselves to continuous processing by this method. Reinforced composites, such as I-beams, channels, wide flange beams, solid bars and rods, and round and square tubing, are among the types of pultruded parts being manufactured. References have also been made to these techniques in the manufacturing of advanced composite golf shafts and fishing rods (see the chapter on industrial applications). A large market for pultrusions is enjoyed by polyester/glass fiber composites, with the C/G composites beginning to enter this field by virtue of their high specific stiffness and strength, low expansion coefficients, and superior fatigue properties.

In reviewing processing trends in advanced composites, Goldsworthy[11] has indicated that the pultrusion process has brought advanced composites into competition with aluminum extrusion and steel structural shapes. The ability to produce structural shapes far in excess of any sizes manufactured by traditional extrusion or rolling methods may well represent the largest market segment for this process. Typical equipment manufactured by Goldsworthy Engineering for producing pultruded parts is illustrated in Fig. 14-5. It is significant that this group is engaged in a feasibility study for equipment to be installed in the cargo bay of one of the shuttles, which will be deployed in low earth orbit to continuously manufacture graphite fiber reinforced composites by the pultrusion process for a geodetic or triangular truss. Multiples of this truss would be joined together to produce a platform for space station experiments.

In the operation of the pultrusion machine, products are produced 5 to 36 inches and more per minute.[12] It is important that the proper impregnation of the fibers be maintained with the liquid resin matrix, and that the volume percent of the fibers remains constant. A typical process with a Goldsworthy machine uses a long, oil heated open steel die to cure the fiber reinforced matrix as it is being drawn through. With this equipment, molded urethane blocks

Fig. 14-5. Pultrusion equipment. (*Courtesy Goldsworthy Engineering Co.*)

shaped to fit the top and the bottom of the pultrusion grasp the product, pulling it through at a predetermined speed. The liquid thermosetting resin and curing agents are accurately metered into the resin trough, at a rate consistent with the usage, monitored to accommodate interruptions in the pultrusion cycle. The resin systems and fibrous reinforcements may be readily changed according to specification requirements, to utilize not only a variety of resin formulas, but also a broad range of fiber/resin volume ratios. The carbon and graphite monofilament fibers are fed continuously from multiple spools towards the resin impregnating bath (sometimes in angular orientation) and then the pultrusion die. The flexibility of the process lends itself to the preparation of hybrids by substituting other fibers for part of the C/G fibers being drawn through the machine. Glass fiber reinforced structural shapes are widely used in the chemical industry as frameworks, handrails, supports, walkways, etc., and millions of feet are reported used for their insulating qualities by the electrical industry. Further refinements have been possible through the introduction of decorative surface plies for the encasement of the product being pultruded.

Basic process modifications have extended the pultrusion technology, including novel molding equipment to shape the fiber bundle and attain permanently curved lengths as well as variable cross-sections.[13,14] This is accomplished by the addition of a molding station with mating segmented mold sections mounted upon caterpillar tread type conveyors to bring mold halves together to shape the cross-section of the pultruded part, as they move with it down the line through the final cure. As the profile moves through the heated mold set, the temperature is raised to complete cure of the issuing pultruded section. This process (and equipment) is available on a custom designed basis to suit specific applications, and will do much to accelerate the acceptance of C/G fiber technology in high volume industries. Dielectric (RF) cure may also be introduced to the composite before the pultrusion enters the heated metal mold.

Jones[15] has described the adaptation of pultrusion to several types of fibers, as well as tows, mats, braids, and cloths. He discussed a characteristic aerospace application using a stiffened hybrid composite. The core was essentially Owen-Corning HTS-904 glass fibers, impregnated with a bisphenol A epoxy/MPD curing agent, while the outer fibers were Hercules A-S graphite fibers distributed at a ±45° angle to form the skin. The same impregnant or matrix was issued for the inner and outer layers. At a resin temperature of 125°F, the resin viscocity was 500 centipoises for ease of impregnation. The die in which curing took place, as pultrusion was occurring, was at 350°F, and the rate of pulling was 12 inches/minute. Jones commented that internal mold releases were not effective in pultrusion and that the graphite fibers needed pretreatments to enhance bonds to polyester matrices.

FILAMENT WINDING

Filament winding is a manufacturing process adaptable to the manufacture of a multitude of containers, bottles, cylinders, tanks, and structural elements. It has been a process successfully developed for glass fiber reinforced plastics, and many thousands of units have been developed both here and abroad. The appearance of C/G fibers have made possible filament wound structures with improved fatigue strength and stiffness over glass fiber equivalents. In an earlier chapter, a number of examples of, and tests on, C/G fiber pressure vessels produced by filament winding were reported.

Procedures for filament winding and testing methods for specimens of glass fiber composites produced therefrom have been the subject of considerable study, and for those now venturing into the production of carbon fiber reinforced advanced composites, a 1962 ASTM Symposium[16] is recommended for review. Up-to-date reviews of filament winding machines appear in the *Modern Plastics Encyclopedia.*[17]

Typical of advanced filaments winding machines is the unit manufactured by Goldsworthy Engineering, shown in Fig. 14-6. The polar orbital winder with a gantry traverse achieves true longitudinal winding by rotating the mandrel through its longitudinal axis. This system assures precision in the helical and circumferential modes under heavy filament loads. An electronic-pneumatic

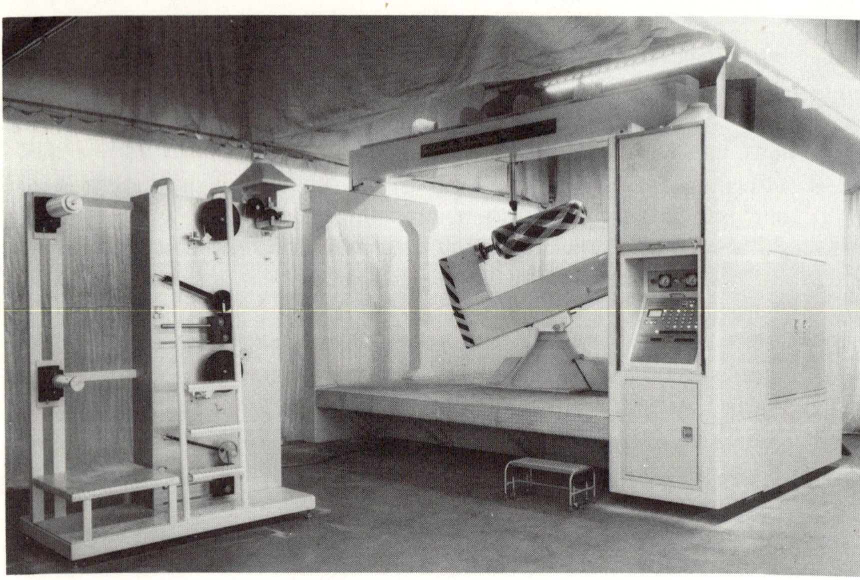

Fig. 14-6. Polar orbital winder with gantry traverse permits filament winding of complex non-uniform shapes. (*Courtesy Goldsworthy Engineering Co.*)

modular tension control system is maintained at the machine console, as shown in Fig. 14-6.

In the most common form of filament winding, the mandrel is rotated, drawing continuous fiber roving from a supply source through a resin bath onto the mandrel surface in a pattern designed to use most effectively the tensile strength of the filaments. Most filament wound parts use epoxy and polyester thermosetting resins. Another illustration of filament winding of cylinders (Fig. 14-7) shows the application of a narrow resin impregnated tape fed from several spools.

ADHESIVE BONDING

It has become quite evident that adhesive bonding of advanced composites has become an important adjunct to the assembly and application of C/G fiber advanced composites. While it is not the intent of this section to enter into an examination and comparison of the many adhesive systems available for the

Fig. 14-7. Filament wound vessel. (*Courtesy Structural Composites Industries, Inc.*)

bonding of advanced composites, a brief review of recent activity in this area will direct the attention of the reader to specific programs examining this subject.

The adaptation of adhesive bonding to space applications was updated by Mulroy and Mazenko[18] in a recent paper. This was done in an effort to select candiate adhesives for the bonding of specimens of advanced composites and metallic adherends. Four acceptable categories of adhesive were identified as follows.

	HIGH TENSILE	HIGH ELONGATION
1. Room temperature cure paste	Hysol EA 9321	Hysol EA 9313
2. High temperature cure paste	3 M's EC 2214	3 M's EC 2214
3. 250°F cure film	American Cyanamide FM-73	American Cyanamide FM-73
4. 350°F cure film	Narmco M-329	3 M's AF-143

Evaluations of the bonded adherends are in progress, though preliminary analyses were presented of the published data upon approximately 25 candidate adhesives. Considerable emphasis for space applications will be placed on outgassing in vacuum and high glass transition temperatures, as well as on tensile and elongation properties, and of course on lap shear and peel characteristics of bonded assemblies.

Guha and Epel[19] identified criteria for the selection of adhesives for the bonding of graphite fiber/glass fiber composites. An entirely different group of adhesive candidates appeared in their evaluation. In automotive transportation applications, the volume usage will be large, space and cost constraints will be great, and the work environment will be less suitable than the contaminant-free requirements of aerospace. Three candicate materials were selected for further evaluation:

1. Hysol's EA 934–R.T. set epoxy paste.
2. Goodyear's PG 6040/6045–two part urethane paste–R.T. set.
3. Loctite's Li 259/260–two part acrylic paste–R.T. set.

Among the selection criteria were minimum effect of bond line thickness on lap shear strength and ease of substrate preparation (such as scuff sanding of adherend surfaces). The ability to operate in a variety of hostile environments and to exhibit good fatigue strength remain to be evaluated in service. Most tests were conducted after a minimum cure of two days at room temperature. Two part urethane adhesive systems, in the author's experience, will demonstrate the highest lap shear properties and high peel strength characteristics.

Automotive applications require adhesive systems that will cure rapidly. The energy sources for rapid cure were evaluated by Mahoney.[20] Ultraviolet, infra-red, microwave (RF heating), and induction heating of metals were examined as potential energy sources for rapid curing. Most structural adhesives require the application of elevated temperatures to cure in short times–which may be accomplished by focused infra-red energy on the composite joints being bonded, or by RF heating in the absence of metal shielding. In some of the data presented, Mahoney demonstrated the relationship between fatigue life in cycles to failure (using bonded titanium tensile shear specimens), and the fiber modulus of reinforcing fibers used with two resin adhesive systems: a toughened, flexible adhesive (EA-9628) and a rigid, low elongation (2 to 3%) adhesive (EA-9653). With the latter rigid adhesive (see Fig. 14-8), improved fatigue life was not attained until the high modulus graphite fibers were present.

The feasibility of automation in speeding up the mixing and application of structural adhesives for the automotive industry has been emphasized.[21] Up to 600 linear inches/minute of a two part urethane adhesive were available from an automated system. Newly developing uses for adhesives include radiator supports, bumpers, frame cross members, doors, truck tail gates, truck cabs, and station wagon lift gates.

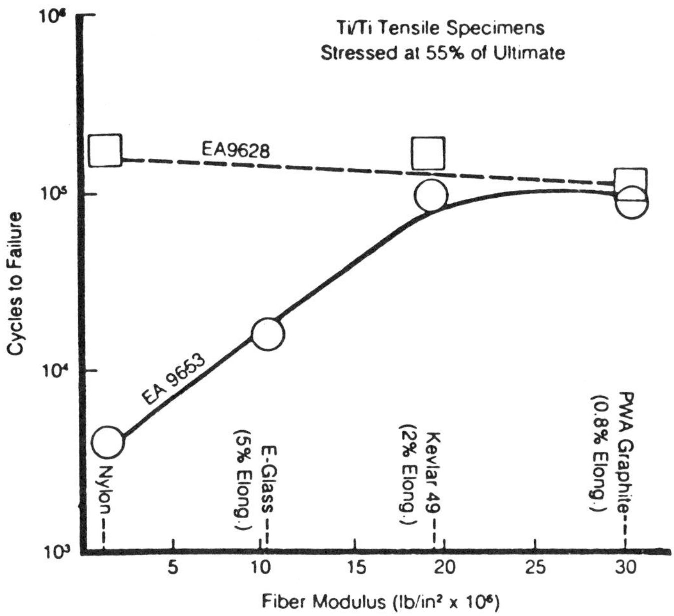

Fig. 14-8. Fatigue results with various fabrics and matrix resin systems.[20] *(Reprinted with Permission S.A.E.)*

REPAIR MEASURES FOR ADVANCED COMPOSITES

The utilization of an increasing number of advanced fiber composites by the aerospace and automotive industries has prompted studies on methods for effecting repairs on structural composites that have experienced impact damage. The various aircraft companies are in communication with one another under the sponsorship of United States agencies to compare experiences in repairing composites. This supportive effort will also provide useful information on automotive companies utilizing an increasing amount of C/G advanced composites.

A program was initiated to demonstrate that composite structures can be repaired, after having experienced large areas of damage, without significantly compromising basic structural integrity and reliability, including the ability of the repaired area to sustain structural loads after moisture infusion.[22] The tendency to delaminate around the fracture area, plus the effects of moisture, oil, paint stripper, and fuel must be taken into account during the repair. The relationship between glass transition temperature (T_g) and moisture content in the laminate was established, and the cure conditions and temperature for the repair patch of A-S graphite fiber/3501-5 were kept under the T_g of the laminate being repaired. This was necessary for large areas and thick panels (50 plies). A problem which arises is the long moisture conditioning time for thick patches, as many days may be required. Liquid epoxy resins were injected into blistered areas to fill voids in achieving a field repair to improve structural integrity.

The increased use of advanced composite materials in aerospace structures has encouraged the development of reliable repair measures. While there has been valuable technology available from the repair of glass fiber reinforced laminates, more attention has been given to the strength and stiffness properties of the repair of graphite fiber structures. Labor[23] reviews the various programs which have been initiated to repair graphite fiber/epoxy structures—including sandwich and hybrid construction. A typical flush scarf repair is illustrated in Fig. 14-9. Process specifications are being written to describe repair measures which may be made to damaged parts, including analytical methods for design and evaluation of repair procedures.

Available data on the repair and maintenance of composites will continue to accumulate as field experiences evolve. Developments in new specialty adhesives and coatings for carbon and graphite fiber composites will appear as greater acceptance of the products is attained. While it is easy to be quite definitive in discussing a specific resin system and fiber composite, broader terms are necessary to accommodate newer materials and processes which are appearing in the technology of carbon and graphite fibers. These materials have made possible new design concepts not anticipated with earlier materials. It is to the everlasting credit of the innovative scientists and engineers that they have recognized the new horizons and have accepted the challenges.

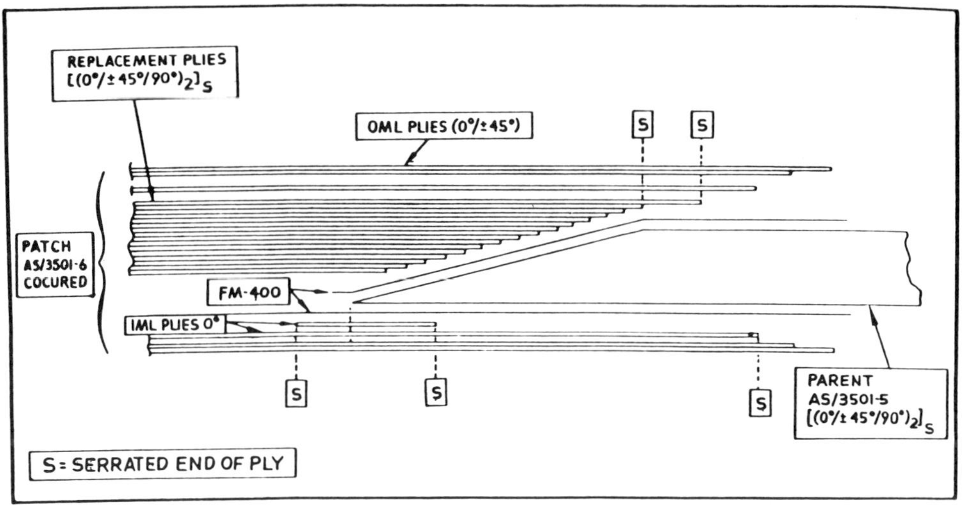

Fig. 14–9. Flush scarf joint repair. *(Courtesy Northrop.)*[23]

REFERENCES

1. Shepler, R., Towne, M., and Saylor, D.K., "Carbon Fiber Composites," *Machine Design,* May 1979.
2. Lockheed Corporation Report, LR28843, January 22, 1979 (Contract NASI 14,000), Table 2-8, page 2-28 (NASA-Langley Research Center), English, F.C. (Program Manager), Bohon, H.L. (NASA Technical Representative).
3. Lockheed Corporation Report LR28873, January 22, 1979 (Contract NASI 15,069), Table 3-2, page 3-7 (NASA-Langley Research Center), English, F.C. (Program Manager), Bohon, H.L. (NASA Technical Representative).
4. Curley, R.C. and Penton, A.P., "Manufacturing Methods for Re-entry Vehicles," *SAMPE Twenty-fourth National Symposium,* Vol. 24, page 190, San Francisco, California, May 1979.
5. Loyd, M., "Nonmetallic Castable Tooling for Advanced Composites," *Thirty-third Annual Technical Conference,* SPI, RP/C Institute, Section 21E, February 1978.
6. Mace, W.C., Sipes, C.L., and Bailie, J.A., "Fabrication of PMR-15 Graphite Cloth Cylinder," *SAMPE Twenty-third National Symposium,* Vol. 23, page 208, Anaheim, California, May 1978.
7. Mace, W.C., Wereta, A., and Sipes, C.L., "High Temp-high Complexity Graphite/Polyimide Part Manufacturing," *SAMPE Twenty-fourth National Symposium,* Vol. 24, page 217, San Francisco, California, May 1979.
8. Sheppard, C.H., Hoggatt, J.T., and Hunter, A.B., "Manufacturing Processes for Fabricating Graphite/PMR-15 Polyimide Structures," *SAMPE Technical Conference,* Vol. 11, page 40, Boston, Massachusetts, November 1979.
9. Welhart, *et al.* (McDonnell Douglas), "Advanced Composites–Fabrication Processes for Selected Resin Matrix Materials," N76-23360, March 1976.

10. Reinhart, T., Jr. (AFML), *SAMPE Conference on "Fibers Used in Composites,"* Newport Beach, California, October 18, 1979.
11. Goldsworthy, W.B., "Progress and Trends in Advanced Composite Processes," Page 101, *Conference on Advanced Composite Technology,* Technology Conference Associates, page 101, El Segundo, California, March 1978.
12. Hazenfield, R.C., "Pultrusion," *Modern Plastics Encyclopedia,* Vol. 53, page 360, October 1976.
13. Ewald, G.W., "Pultrusion," *Modern Plastics Encyclopedia,* Vol. 55, page 379, October 1979.
14. Anon., "New Technology for Continuous Reinforced Plastics Processing," *Modern Plastics,* Vol. 56, No. 9, page 82, September 1979.
15. Jones, B., "Design and Manufacturing Technology for Pultruded Composite Structural Elements," *Thirty-first Annual Technical Conference,* SPI, RP/C Institute, Section 17D, Washington, D.C., February 1976.
16. Symposium on "Standards for Filament Wound Reinforced Plastics," ASTM, STP 327, June 1962.
17. Ewald, G.W., "Filament Winding," *Modern Plastics Encyclopedia,* page 348, 1976/1977.
18. Mulroy, B.J. and Mazenko, D.M. (Lockheed), "Adhesive Bonding for Space Applications," *SAMPE Symposium,* Vol. 10, page 416, New York, October 1978.
19. Guha, P.K. and Epel, J.N., "Adhesives for Bonding of Graphite/Glass Composites," SAE Technical Paper 790149, Cobo Hall, Detroit, February 1979.
20. Mahoney, C.L. (Hysol), "Structural Adhesives for Rapid Cure Applications," SAE Technical Paper 790151, Cobo Hall, Detroit, February 1979.
21. Schmitz, W.R. and Rabito, T.G., "Automated Structural Bonding Systems for Graphite and Other High Strength Composites," SAE Technical Paper 790150, Cobo Hall, Detroit, February 1979.
22. Kiger, R.W. and Myhre, S.H. (Northrop), "Large Area Composite Structure Repair," AFFDL-TR-78-83, July 1978.
23. Labor, J.D. (Northrop), "Repair of Advanced Composite Structures," *Conference on Advanced Composites,* Technology Conference Associates, page 161, El Segundo, California, December 1979.
24. U.S. Army–DARCOM Pamphlet–No. 706-316 "Joining of Advanced Composites"–March 1979.

Index